The Analytical Approach

The Analytical Approach

Edited by

Jeanette G. Grasselli

AMERICAN CHEMICAL SOCIETY
WASHINGTON, D.C. 1983

Library of Congress Cataloging in Publication Data
The Analytical approach.

   Bibliography: p.
   Includes index.

   1. Chemistry, Analytic—Addresses, essays, lectures.
   I. Grasselli, Jeanette G. II. Series.

QD75.25.A5   1983   543   82-22618
ISBN 0-8412-0753-4
ISBN 0-8412-0755-0 (pbk.)   1-232   1983

Copyright © 1983

American Chemical Society

All Rights Reserved. The appearance of the code at the bottom of the first page of each article in this volume indicates the copyright owner's consent that reprographic copies of the article may be made for personal or internal use or for the personal or internal use of specific clients. This consent is given on the condition, however, that the copier pay the stated per copy fee through the Copyright Clearance Center, Inc. for copying beyond that permitted by Sections 107 or 108 of the U.S. Copyright Law. This consent does not extend to copying or transmission by any means—graphic or electronic—for any other purpose, such as for general distribution, for advertising or promotional purposes, for creating new collective work, for resale, or for information storage and retrieval systems. The copying fee for each chapter is indicated in the code at the bottom of the first page of the chapter.

The citation of trade names and/or names of manufacturers in this publication is not to be construed as an endorsement or as approval by ACS of the commercial products or services referenced herein; nor should the mere reference herein to any drawing, specification, chemical process, or other data be regarded as a license or as a conveyance of any right or permission, to the holder, reader, or any other person or corporation, to manufacture, reproduce, use, or sell any patented invention or copyrighted work that may in any way be related thereto.

PRINTED IN THE UNITED STATES OF AMERICA

Contents

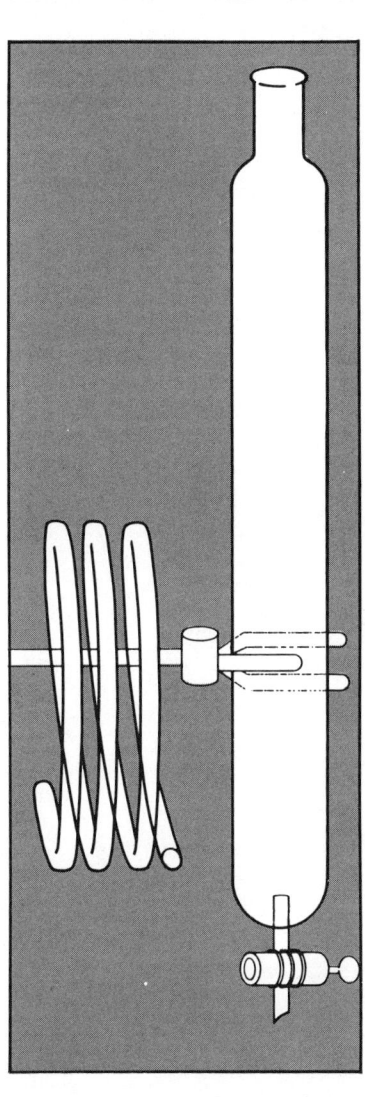
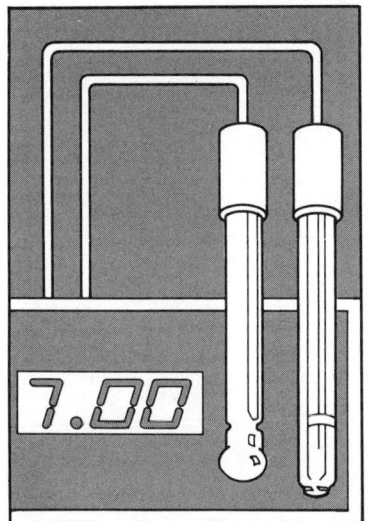

Preface ix

Production Processes

SIMS: Simplified Investigations of Messy Samples 3
W. H. Christie

Finding the Flavor Component: A Job for the Analytical Chemist 7
W. N. Einolf

Tracing the Origin of Off-Flavors in a Breakfast Cereal 11
M. G. Heydanek, Jr.

Characterization of Corrosion Scale from Nuclear Reactors: Approaching the Whole Problem 15
L. D. Hulett, J. M. Dale, H. W. Dunn, and P. S. Murty

Urgent Production Problem Solved by Unique Capabilities of Analytical Chemists 19
C. A. Lucchesi

Recognition and Solution of Production Problem in Vitamin Manufacture 23
E. MacMullan, R. Hagel, and R. Gomez

Identification of Impurity in Blood Preservative System 25
M. Marconi, M. Gara, and R. Dawe

Effects of Raw Material Change in Manufacturing Process Resolved 29
J. Mitchell, Jr.

Pure Steam Ahead! 31
D. F. Pensenstadler and M. A. Fulmer

Problem Solving in the Coatings Industry 37
T. K. Rehfeldt and D. R. Scheuing

Refinery Crisis Solved 41
S. A. Schmidt and V. F. Gaylor

Cross-Section of Analytical Problem Solving in the Home Appliance Industry 45
J. L. Wuepper

Products

Analytical Chemists at the Launch Pad 51
H. D. Bennett

Analysis of Orange Juice 55
S. A. Borman

Search for the Elusive Phosphorus Source 57
R. L. Campbell and R. Zuback

Over-the-Counter Drug Analyses with HPLC 61
M. A. Carroll, E. R. White, and J. E. Zarembo

Analysis of Products Used in the Electronics Industry 65
P. Cukor

Assuring the Quality of Honey: Is It Honey or Syrup? 69
L. W. Doner, I. Kushnir, and J. W. White, Jr.

Snap, Crack, Blister, and Peel: An Analytical Approach 75
G. L. Fix

Products — Continued

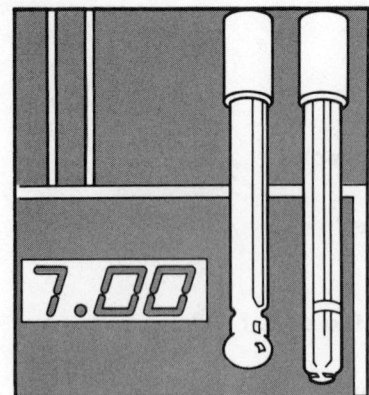

Multitechnique Approach Solves Construction Materials Failure Problems 79
W. G. Hime

Chromatopyrography for Polymer Characterization 83
J. C.-A. Hu

Analysis of Liquid Crystal Mixtures 87
T. I. Martin and W. E. Haas

Analysis of Rubber and Plastic Chemicals by LC/Spectroscopy 93
J. B. Pausch

Environmental

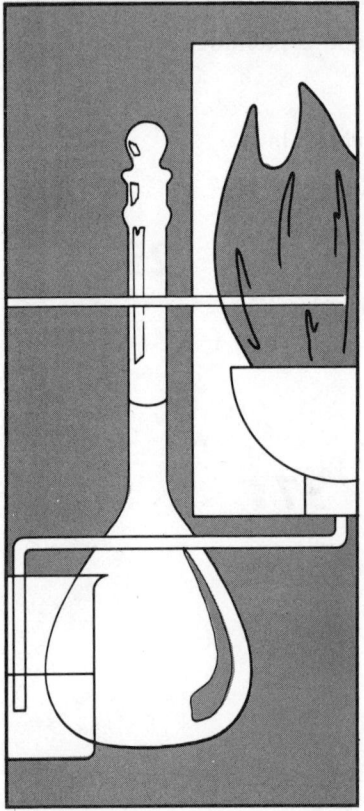

Who Spilled the Oil? 101
A. P. Bentz

Controversy Erupts over Mt. St. Helens Ash 105
S. A. Borman

Scientists Still Split over Silica in Ash 109
S. A. Borman

Ocean Pollution 111
S. A. Borman

Plutonium in Glaciers 113
S. A. Borman

Crystalline Silica in Mount St. Helens Ash 117
S. O. Farwell and D. R. Gage

Identification of Organic Compounds in an Industrial Wastewater 121
R. A. Hites and V. Lopez-Avila

Industrial Analytical Chemists and OSHA Regulations for Vinyl Chloride 127
S. P. Levine, K. G. Hebel, J. Bolton, Jr., and R. E. Kugel

Detecting and Defining Air Pollutants: One Laboratory's Experiences and Approaches 131
J. L. Lindgren, H. J. Krauss, and J. S. Mgebroff

Sampling and Analysis of Radioactive Solutions 135
D. H. Smith, R. L. Walker, and J. A. Carter

Toxicity

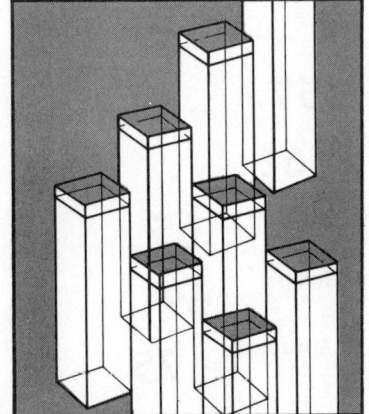

Analytical Chemists Vital in Commercialization of New Food Packaging Material 141
V. F. Gaylor

Industrial Hygiene: Safer Working Through Analytical Chemistry 145
R. E. Hemingway

Carcinogenesis Testing and Analytical Chemistry 151
E. A. Murrill, E. J. Woodhouse, S. S. Olin, and C. W. Jameson

In Search of the Cause of Legionnaires' Disease 157
I. H. Suffet and P. R. Cairo

An Unknown Salivary Morpholine Metabolite 163
J. S. Wishnok and S. R. Tannenbaum

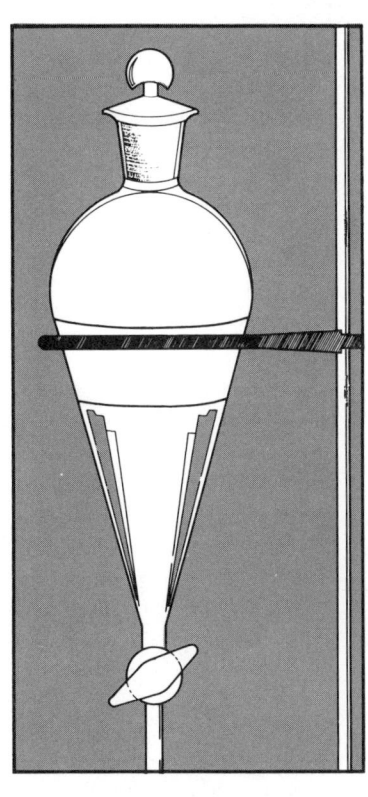

Forensic

Drugs of Abuse Detected in Hair by Radioimmunoassay   **169**
S. A. Borman

FBI Investigates Analytical Chemistry   **171**
S. A. Borman

The Case of the Great Yellow Cake Caper   **175**
P. A. Budinger, T. L. Drenski, A. W. Varnes, and J. R. Mooney

Analytical Techniques in Arson Investigation   **179**
M. J. Camp

Investigations of Clandestine Drug Manufacturing Laboratories   **183**
T. A. Dal Cason, R. Fox, and R. S. Frank

The Forensic Chemist: "An Analytical Detective"   **187**
L. S. Eichmeier and M. E. Caplis

JFK Assassination: Bullet Analyses   **191**
V. P. Guinn

Detection of Curare in the Jascalevich Murder Trial   **197**
L. H. Hall and R. F. Hirsch

Behind the Identification of China White   **201**
T. C. Kram, D. A. Cooper, and A. C. Allen

The Analytical Chemist as a Third Party Fact Finder   **207**
J. G. Montalvo, Jr.

Miscellaneous

Reliability in the Analysis of Rocks and Minerals   **215**
S. Abbey

Voyager Infrared Spectrometer   **219**
S. A. Borman

Authenticity of Medieval Document Tested by Small Particle Analysis   **223**
W. C. McCrone

The Bust of Nefertiti   **227**
H. G. Wiedemann and G. Bayer

Index   **233**

Preface

Analytical chemistry is fun and it is relevant! This collection of reprints represents a group of real-world problems that have been addressed through the use of The Analytical Approach. The papers provide unique insight into the scope, challenge, and excitement of analytical science as carried out daily in industrial and governmental laboratories. Examples cover the range of problems from product or process support and general industrial troubleshooting to forensic, toxicological, and environmental concerns.

The compilation highlights the excellent analytical methodology and expertise that is available today. By illustrating the relevance of what is being taught in academia to the demands of the job in industry and government, this collection can serve a very useful role as a teaching aid in higher education or as a guide in a corporate orientation or training program on analytical capabilities. Most of the chosen examples have a story to tell about the world in which we live with which the reader can identify. Topical subjects, such as the analysis of the JFK assassination bullets or Mt. St. Helens ash, are blended with less newsworthy but equally stimulating insights into areas such as the search for flavor changes in foods, odor in a package liner for cereal, or failure mechanisms in spacecraft parts. My description of the analytical chemist as a "super-sleuth" (1980, Vol. 52, p. 30A) is certainly apt.

Throughout these examples, elegance in using the analytical approach is featured. In 1974, Claude Lucchesi initiated The Analytical Approach column in ANALYTICAL CHEMISTRY. I have been editing the column since late 1980. In his introduction of the column in 1974 (Vol. 46), Lucchesi expressed the hope that the articles would be useful to teachers of analytical chemistry as pertinent and interesting examples of how the industrial scientist uses the teachings of analytical chemistry. He stressed that, in turn, the industrial scientist would be reinforced in his awareness of the complementary nature of many analytical methods—with their own strengths and weaknesses.

The analytical approach, so powerfully illustrated in these reprints, has been discussed from various viewpoints in the literature. D. Betteridge (1976, Vol. 46, p. 1034A) described the "analytical trinity" (the interplay between theory, technique, and problems) and declared that the successful analytical chemist must be adept at a number of techniques and must be at the forefront of chemical knowledge. The function of the analytical chemist is to provide information of sufficiently valid nature, that is, of the requisite statistical significance so that meaningful decisions can be made about materials or problems. This has been articulated by Elving, Grasselli, and Lucchesi.[1] In all cases, the essence of the analytical approach is:

- Defining the problem at hand.
- Ensuring that the samples available are representative of the problem.
- Interacting with the client to obtain his or her knowledge of the problem and to understand the requirements to be met by the solution in terms of needed accuracy and timing.
- Developing an analytical plan involving the sequence and number of methods and instrumental studies to be undertaken.
- Completing the work using the highest level of expertise and remembering the chemical theories relating to the work.
- Communicating answers, never data, including the precision and reliability on all numbers, and specifying any cautions or constraints on the best use of the answer.
- Interpreting the information and results in a clear, consistent, and meaningful report that can be used to help solve the problem.

This disciplined and logical attack should be the basis for the ongoing practice of analytical chemistry.

All of the papers included in this volume have come from The Analytical Approach feature except for one Instrumentation and seven Focus articles, all written by Stuart A. Borman of the ANALYTICAL CHEMISTRY staff, and added because the subject matter matched the general topic of this book. The papers are categorized under six headings: **Production Processes** outlines solutions to production problems in industries as diverse as food, home appliance manufacturing, and nuclear power generation. Under **Products** are discussed problems solved in evaluating the quality of products from orange juice to space vehicle components. The writings under **Environmental** describe recent findings in this important field and are spiced with measurements of plutonium in glaciers and silica in Mt. St. Helens ash. **Toxicity** brings home the challenge in isolating and identifying toxic materials based on case studies such as the search for the cause of Legionnaires disease. The **Forensic** section features the investigation of murder, assassination, and drug abuse by means of analytical chemistry and makes fascinating reading, even when the case cannot be solved from the evidence assembled. Finally under **Miscellaneous**, I have selected four intriguing general papers, covering the span from an analysis of a bust of Nefertiti to a discussion of the analytical instrumentation carried by the Voyager spacecraft. Indeed, there is something in this book for all tastes. A short descriptive guide is included before each section, pointing out significant aspects of the examples.

I must acknowledge the support of many people in assembling this book. First, Claude Lucchesi for initiating The Analytical Approach feature in ANALYTICAL CHEMISTRY and for capably serving as its editor for six years. Marcia K. Snavely and Jeane Beardsley of my own staff have contributed to the regular and expert editing of The Analytical Approach column and the organizing of this compilation. I must also acknowledge the ongoing support of the staff of ANALYTICAL CHEMISTRY, with a very special nod of recognition to Jo Petruzzi, Executive Editor, and Barbara Cassatt, Managing Editor. It has been a pleasure to work with them.

Jeanette G. Grasselli

[1] P. J. Elving, Journal of Chemical Education, 1977, Vol. 54, p. 269; J. G. Grasselli, Analytical Chemistry, 1977, Vol. 49, p. 182A; and C. Lucchesi, American Laboratory, 1980, October, p. 113.

Production Processes

Today's industrial processes are highly technologically oriented. Whether a production process is continuous or batch, it is important that it runs smoothly and produces products for a long period of time. Only in this way can the manufacturer achieve customer satisfaction and high levels of productivity and profitability. The analytical chemist plays a significant role in troubleshooting, establishing control procedures, evaluating raw materials, and effecting ongoing process optimization. An alert chemist can save his company many dollars by anticipating problems, identifying reasons for the changed conditions, and working with others to develop solutions. The timeliness of the answer can be a critical factor in solving problems originating in industrial processes.

"SIMS" (Secondary ion mass spectrometry) describes the use of this ion sputtering method in the handling and analysis of "messy," i.e., highly radioactive samples. The advantages of this method are the lack of sample preparation and the extremely small quantities of material required. A series of examples, including the analysis of a Three Mile Island sample, proved the merits of this versatile technique.

"Finding the Flavor Component" is a description of the combined efforts of spectroscopists, wet chemical analysts, and synthetic organic chemists to solve the problem of a valuable flavor component loss in tobacco manufacture. The loss of this volatile material was traced to a steam treating step. Analytical plans were developed to identify the compound and synthesis was used to verify its structure. The knowledge of tobacco chemistry was advanced by this study.

Flavor and odor problems in foods can be serious when only small amounts of impurities are involved. Despite multiple routine quality assurance tests, the nose or a flavor panel often detect problems not picked up in the testing. "**Tracing the Origin of Off-Flavors in a Breakfast Cereal**" describes the isolation and identification of a "piney" odor, which caused an adverse flavor in a large volume breakfast cereal. The odor was traced to a resin in the package liner by using sniff tests of chromatographic peaks. Specifications were set on the package liner to avoid future problems.

"**Characterization of Corrosion Scale from Nuclear Reactors**" illustrates the role of the analytical chemist as an active decision-making member of a group studying the source, cause and future prevention of scaling problems in nuclear reactors. A general approach to the analysis of complicated mixtures of solid corrosion products is presented based on a sequence of analytical tests, including X-ray fluorescence, elemental analysis, SEM, TEM, ESCA, and X-ray diffraction.

Pilot plants often encounter problems not seen at the laboratory bench scale. In "**Urgent Production Problem Solved by Unique Capabilities of Analytical Chemists**," the author describes the search for the solution of one such problem that arose in the start-up of a terephthalic pilot plant. A DMF-insoluble material was causing color and turbidity problems in the product. Seven analytical methods and a synthesis procedure were used to make tentative identification and "prove" its composition.

"Testing of vitamins is crucial before any release for sale can be made. In "**Recognition and Solution of Production Problems in Vitamin Manufacture**," an alert analyst in the plant recognized a problem in the solution clarity test of a particular material. Rather than simply rejecting the lot, he requested an analytical research study of the situation. A well thought out analytical plan resulted in the identification and synthesis proof of interfering material. Close cooperation among the analyst, the research staff, and production chemists led to process changes that resolved the problem. Phase analysis is highlighted in the analytical approach.

Liquid chromatography separation, followed by GC, GC/MS, IR, and NMR analysis was required in "**Identification of Impurities in Blood Preservative**" to prove the structure of an impurity in a fortifier used to prolong the shelf life of blood in blood storage banks. Synthesis was used as a final proof and manufacturing changes were made to eliminate the impurities.

Maintaining high levels of product quality in many industrial processes requires complete knowledge of compositions of chemical intermediates, including the type and concentration of trace impurities. Certain impurities may seriously affect product yield, as well as quality. In "**Effects of Raw Material Changes in Manufacturing Process Resolved**," an emergency task force was formed to determine the impact of a new feedstock on a complex process. Seven analytical methods were used to solve this problem in an efficient way.

"Pure Steam Ahead" describes the use of ion chromatography for the on-line control of the quality of the boiler water throughout a steam system. Condenser leaks and improper polisher operation were detected and cured using this technique. This valuable on-site analytical tool offers direct feedback on improper operating conditions in the power production industry.

Coating materials are complex mixtures of polymers, pigments, additives, and process residuals. In "**Problem Solving in the Coatings Industry**," the authors outline a general analytical sequence that can be used to evaluate nearly all coatings and to solve raw material product or environmental problems. Liquid chromatography is the pivotal method. Four examples are given to show the versatility of this analytical approach.

During the Arab oil embargo, corrosion in a refinery threatened to shut down production of badly needed gasoline. At the same time, high levels of chlorine were reported in light products. In "**Refinery Crisis Solved**," the authors describe how a nonstandard analytical method was given an inaccurately high chlorine number, while the real plant problem was a new corrosion inhibitor, which was blocking tower tubes, plus a high sulfur content in the crude oil. Returning to a proved inhibitor solved this problem.

In the appliance industry, the analytical approach involves dealing with the environment, product quality, crisis problem solving, materials characterization, product liability, and regulation. In "**Cross-Section of Analytical Problem Solving in the Home Appliance Industry**," the author describes four very different examples of where analytical chemistry solved important problems. The cases presented involve multiple technique applications in which an understanding of the basic chemistry is vital to the solution.

SIMS: Simplified Investigations of Messy Samples

W. H. Christie

Oak Ridge National Laboratory, Analytical Chemistry Division, Oak Ridge, TN 37830

Originally published in ANALYTICAL CHEMISTRY, 1981, Vol. 53, No. 11.

The technique of mass analysis by ion sputtering is called secondary ion mass spectrometry (SIMS). The method is characteristically used for the examination of inorganic surfaces. In recent years, the versatility of sputtering as a source of ions for mass spectrometry has been increasingly documented. The method is now applied to both organic and inorganic sample types. This article describes how ion sputtering combined with a sensitive mass spectrometer was safely used to analyze "messy" radioactive samples, including a Three Mile Island water sample. We apply the term "messy" because that is what these samples are when more conventional analytical methods are used.

The high sensitivity of the method allows analysis of sample aliquots sufficiently small that radiation is reduced to acceptable levels for safe handling in unshielded equipment. Ion sputtering as an ionization technique for mass spectrometry has the further advantage that isotopic analysis can be accomplished without sample chemistry. This is a distinct simplification if one is dealing with highly radioactive samples. With a minimum of sample chemistry (dissolution only), one can obtain quantitative information from these same radioactive samples.

The reader interested in the details of the SIMS method should consult excellent articles by McHugh (1), Andersen and Hinthorne (2), or Evans (3). In this technique, a beam of primary ions is focused onto the surface of the sample to be analyzed. These energetic bombarding particles dislodge (sputter) material from the surface as ions, neutrals, and various molecular species. One of the advantages of ion sputtering that allows circumvention of sample chemistry is the fact that the chemical form of a given element in a sample is usually of little consequence. During the sputtering process an appreciable fraction of an element will appear as singly charged elemental ions. When sputtering is used as the source of ions for a sensitive mass spectrometer, extremely low levels of many elements can be detected. Using ion implantation as a method for preparing calibration standards, Dobrott et al. (4) demonstrated that, for the case of B implanted in a silicon matrix, one B atom could be measured for each 1200 B atoms present in the sample. In this context only subpicogram amounts of B need be sputtered from a sample to give ~0.1% counting statistics in a modern pulse-counting spectrometer. This high sensitivity is easily realized for elements with low ionization potentials or appreciable electron affinities.

There are some instances in SIMS where one must monitor a molecular ion of the element of interest, but in most cases suitable elemental ion signals are observed. Monitoring an elemental ion avoids the necessity of making corrections for mass overlaps frequently observed in molecular ion clusters. The above observations can be summarized by saying that ion sputtering does the sample chemistry for you and that, by using this ionization method, the desired element is usually produced as an atomic ion suitable for direct mass analysis.

We became impressed with the simplicity of SIMS as an isotopic analysis technique when we used it to measure $^{10}B/^{11}B$ ratios in borosilicate glass samples (5). This work came about be-

Table I. SIMS Isotopic Analysis of Borosilicate Glass Powders

Sample no.	Atom ratio $^{10}B/^{11}B$ avgs.	SD[a]	%RSD[b]	n[c]	% absolute error $100(y-x)/y$[d]
1	0.2469	0.0005	0.20	10	0.00
1	0.2468	0.0005	0.20	10	−0.04
2	0.2495	0.0019	0.76	10	1.05
2	0.2472	0.0007	0.28	10	0.12
3	0.2469	0.0013	0.53	10	0.00
3	0.2478	0.0013	0.52	10	0.36
4	0.2467	0.0008	0.32	10	−0.08
4	0.2497	0.0006	0.24	10	0.41
5	0.2456	0.0008	0.33	10	−0.53
5	0.2474	0.0006	0.24	10	0.20
6	0.2457	0.0009	0.37	8	−0.49
6	0.2467	0.0009	0.36	10	−0.08
7	0.2455	0.0016	0.65	10	−0.57
7	0.2465	0.0009	0.25	10	−0.16
8	0.2470	0.0006	0.24	10	0.04
8	0.2463	0.0005	0.20	10	−0.24
9	0.2469	0.0010	0.41	10	0.00
9	0.2463	0.0006	0.24	10	−0.24
Avgs.	0.2469	0.0009	0.35		−0.01
SD	0.0009				0.34

[a] SD = standard deviation; [b] % RSD = percent relative SD; [c] n = number of replicates; [d] $x = {^{10}B}/{^{11}B}$ average, natural $^{10}B/^{11}B$ taken as $0.2469 \equiv y$

cause, in certain instances, solutions of fissionable uranium are stored in vessels that are of unsafe configuration from the standpoint of nuclear criticality. To ensure the nuclear safety of these containers, they are loaded with Raschig rings made of borosilicate glass. The ^{10}B in this glass has a large cross section for neutron adsorption and thereby reduces the nuclear reactivity of the stored solutions. These Raschig rings are sampled at regular intervals, and the ^{10}B/^{11}B ratio determined to verify that no significant neutron-producing event has occurred. The traditional solution chemistry, electron impact mass spectrometry technique for doing these samples was slow and somewhat undependable, and poor ion signals were frequently obtained. This resulted in inaccurate measurements. SIMS was the model of simplicity in solving this problem. The glass samples (0.1–0.2 g) were ground to micron-sized particles. Microgram amounts of each, dispersed in water, were distributed as thin layers on an electrically conducting sample mount and subjected to SIMS analysis. Twenty samples were easily determined in replicate in an 8-h day. Typical results for the isotopic determination of B in nine glasses containing natural B are reported in Table I.

We examined the applicability of SIMS for the quantitative measurement of B in these same glass samples. The borosilicate glass powders were weighed (0.1–0.2 g) and dissolved in dilute HF in plastic labware using the simple procedure of Burdo and Snyder (6). Ten µL of each sample, both as spiked and unspiked solutions, were loaded onto a sample mount and dried in air before being subjected to SIMS analysis. The quantitative results thus obtained from standard isotope dilution calculations are compared in Table II to results for the same samples using conventional atomic absorption methods. If one is interested only in a quantitative B analysis in these samples, SIMS offers no cost advantage over the atomic absorption technique. If, however, an isotopic analysis is desired, SIMS is clearly the method of choice as both isotopic and quantitative analyses can be performed on the same sample. Figure 1 is a schematic drawing illustrating how samples are handled using this method. Although we applied this technique only to the analysis of B and Li in the samples we studied, the method itself is very general in scope.

To summarize the advantages of SIMS for these analyses over more conventional mass spectrometries, we can state that, with SIMS: Quantitative and isotopic results can be obtained; no sample chemistry is re-

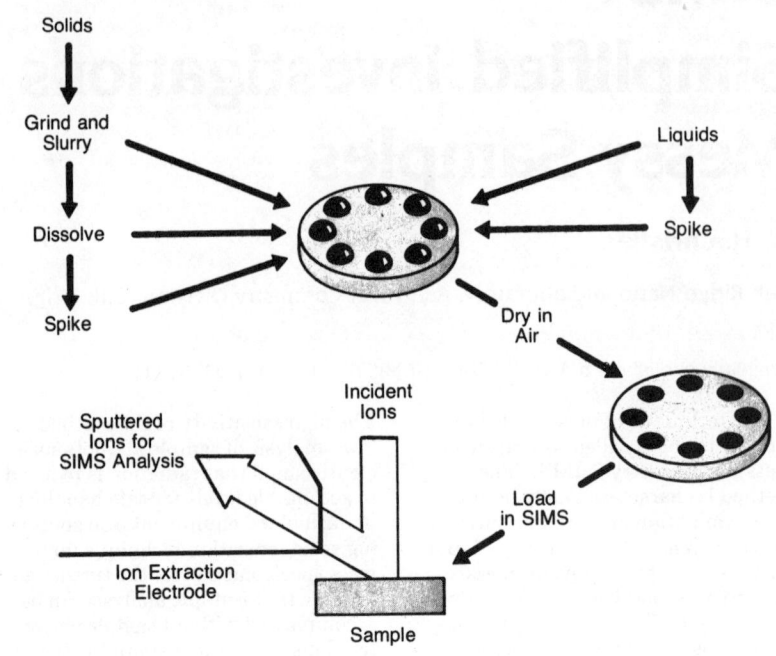

Figure 1. General scheme for SIMS analysis of difficult sample types

quired for isotopic analyses; dissolution of the sample is required only for quantitative determinations; turn-around time is rapid; and accuracies are typically ±½%.

After our success in analyzing borosilicate glasses, which previously had been difficult at best when using more conventional mass spectrometry, we tackled some Al_2O_3/B_4C composite pellets, which are used as neutron shims (control rods). They contain from 1.5 to 4.0 wt % B, which may or may not be enriched in ^{10}B. Lithium is of interest in these materials because ^{7}Li is formed by η,α reaction on ^{10}B during irradiation. The fact that these materials have been irradiated necessitates hot-cell handling. The associated radioactivity necessitates reducing operator radiation exposure to acceptable, safe limits and, hence, provides motivation for using high sensitivity, analytical techniques that require little or no sample chemistry. For conventional analysis the pellets are powdered in a ball mill and subjected to sealed tube chemical dissolution to dissolve B into solution. This step is followed by chemical separation to remove Al, which interferes with B ion emission during mass spectrometry. Using SIMS, we avoided the chemical steps described above and directly analyzed the powder produced in the ball mill in a manner sim-

ilar to that used for the borosilicate glasses. Only a few mR of radiation are associated with each sample loaded. Li and B isotopic data obtained in this fashion are reported in Table III for six irradiated Al_2O_3/B_4C composite pellets.

We also applied this method to the quantitative determination of B and Li in solutions of unknown composition that were generated in the Three Mile Island (7) nuclear reactor accident. The original sample consisted of a highly radioactive aqueous solution that indicated about $1Rh^{-1}mL^{-1}$ of β,γ activity at contact with the sample vial surface. Prior isotope dilution spark source mass spectrometry and optical emission spectrometry provided estimates of 2 µg cc^{-1} Li and 3400 µg cc^{-1} B in this solution. After acidification and dilution, 100 µL of this solution were obtained for SIMS analysis. Ten µL were spiked with a known ^{6}Li solution, and another aliquot was spiked with a known ^{10}B solution. These solutions were allowed to stand overnight to ensure isotopic equilibration. Approximately 1 µL of each was then loaded onto the sample mount; natural B and Li standards were loaded on the same mount. This resulted in a sample that gave a combined β,γ activity of approximately 1 mRh^{-1}, less than what one observes from a wristwatch with a luminous dial. The analytical results for both isotopic and quantitative analyses of B and Li obtained via SIMS are reported in Table IV and compared to results obtained from other techniques on aliquots of the same sample. Repetitive analyses of a normal Li standard demonstrated that the $^7Li/^6Li$ ratio could be determined with a precision of 0.5 to 1.0%.

A significant advantage of the SIMS method for this kind of analysis is the fact that only very low levels of radiation had to be handled because of the ultrahigh sensitivity of SIMS for Li and B. Furthermore, no instrument contamination of the sample was detected during the analysis. We estimate that less than one-ten-thousandth of the loaded sample was consumed by the sputtering process. The unconsumed portion remains on the sample mount where it can be disposed of safely. In conventional thermal or spark source mass spectrometry, a significantly higher fraction (from 1 to 100%) of the sample is spread about the instrument ion source region by the nature of the ionization process (evaporation or bulk sputtering).

Another study done in our laboratory suggests that SIMS may have applicability in the analysis of dissolver solutions produced in the reprocessing of spent reactor fuels. These samples are highly radioactive, which complicates their handling for conventional analysis. This possibility occurred to us when we used our SIMS instrument to investigate the mechanism of thermal ion emission from resin beads loaded with U (8). We were able to determine isotopic ratios on very small U samples. During this study the average of the $^{235}U/^{238}U$ ratios obtained was 1.012 ± 0.015 (1σ); the NBS certified value for the standard used was 0.9997. We estimate that only 20–60 fg of U was sputtered during this measurement. We feel that SIMS has a promising future in the analysis of nuclear materials that are frequently considered to be "messy." Of course, there are limits to the levels of activity that can be handled in an unshielded instrument, and this aspect of the problem must be given consideration.

References

(1) McHugh, J. A. In "Methods and Phenomena"; Wolsky, S. P; Czanderna, A. W., Eds.; North-Holland: Amsterdam, 1976.
(2) Andersen, C. A.; Hinthorne, J. R. *Anal. Chem.* **1973**, *45*, 1421–38.
(3) Evans, C. A. *Anal. Chem.* **1975**, *47*, 818 A–829 A.
(4) Dobrott, R. D.; Schwettmann, R. N.; Prince, J. L. Presented at Eighth National Conference on Electron Probe Analysis, New Orleans, La., 1973, paper No. 10.
(5) Christie, W. H.; Eby, R. E.; Warmack, R. J.; Landau, Larry. *Anal. Chem.* **1981**, *53*, 13–17.
(6) Burdo, R. A.; Snyder, M. L. *Anal. Chem.* **1979**, *51*, 1502–8.
(7) Blake, E. M. *Nucl. News* (La Grange Park, Ill.) **1979**, *22* (7), 32–6.
(8) Smith, David H.; Christie, W. H.; Eby, R. E. *Int. J. Mass Spectrom. Ion Phys.* **1980**, *36*, 301–16.

Research sponsored by the U.S. Department of Energy, Division of Basic Energy Sciences, under Contract W7404-eng-26 with the Union Carbide Corporation.

Table II. Comparative Quantitative Analysis of Boron

Sample no.	Isotope dilution SIMS wt % B	Atomic absorption wt % B
1	3.94	3.91
2	3.88	3.90
3	3.98	3.92
4	3.87	3.90
5	3.94	3.88

Table III. SIMS Isotopic Analysis of Li and B in Al_2O_3/B_4C Composite Pellets

Sample	ppm atomic ^6Li	Atomic % ^{11}B
1	45	83.4
2	35	83.4
3	38	82.9
4	49	82.9
5	46	83.8
6	43	84.3

Table IV. Comparison of SIMS to Other Techniques for the Analysis of a Three Mile Island Water Sample

Method	Total B wt, ppm	Isotopic B $^{11}B/^{10}B$ (atom)
Isotope dilution spark source mass spectrometry	3314 3565	~4
Microtitration mannitol procedure	3211 3235	
Isotope dilution SIMS	2978 3050	4.07

Method	Total Li wt, ppm	Isotopic Li atom fraction
Flame emission spectrometry	≤5	
Isotope dilution SIMS	4.64	^6Li 0.02% ^7Li 99.98%

Finding the Flavor Component: A Job for the Analytical Chemist

W. Noel Einolf

Phillip Morris Research Center, Richmond, VA 23261

Originally published in ANALYTICAL CHEMISTRY, 1975, Vol. 47, No. 6.

In the tobacco industry "total utilization" requires that all stages of manufacturing be reviewed to determine which material classified as waste actually has some value or contains compounds of value. Since millions of pounds of tobacco are processed annually at Philip Morris, the problem of recovery of valuable waste is particularly important. It is this continuing review of the manufacturing processes that prompted a team of analytical chemists to investigate the source of a potentially valuable compound. Isolation and identification of compounds are accomplished only through the combined efforts of chemists with expertise in spectral interpretation, wet chemical methods, and synthetic organic chemistry.

The problem presented here originated when the condensed water from steam-treated tobacco was observed to have a tobacco-like aroma. This water was collected in a process used to make the tobacco leaf more pliable, and the presumption was that the steam process was removing some of the volatile flavor components from the tobacco. Thus, compounds isolated from the condensed water should be present in the tobacco itself. The condensed water was passed over a carbon bed, and the carbon extracted with methylene chloride. The methylene chloride residue was dissolved in

Table I. Analytical Data for Unknown Flavor Component and Two Synthetic Compounds

UNKNOWN AND COMPOUND I		COMPOUND II	
IR			
Absorption band, cm^{-1}	Description	Absorption band, cm^{-1}	Description
1761	>C=O α,β-unsat. lactone	1802	>C=O lactone
1637	C=C	1701	C=C
UV			
Absorption max	Description	Absorption max	
208 nm	α,β-unsaturated γ-lactone	None	
NMR: 16 protons observed			
Chemical shift, ppm	Description	Chemical shift, ppm	Description
1.22	—CH$_3$	0.09	—CH$_3$ singlet
1.27	—CH$_3$	1.87	—CH$_3$ singlet
1.50	—CH$_3$	1.20	—CH$_3$ singlet
5.52	H—C—C	2.37	Singlet (2H)
2.0–1.5	Complex pattern	4.97	Singlet (1H)
		1.59	Multiplet (2H)
		1.69	Multiplet (2H)
MS (high resolution)			
m/e	Composition	m/e	Composition
180	C$_{11}$H$_{16}$O$_2$	180	C$_{11}$H$_{16}$O$_2$
165	C$_{10}$H$_{13}$O$_2$	165	C$_{10}$H$_{13}$O$_2$
152	C$_{10}$H$_{16}$O	123	C$_8$H$_{11}$O
137	C$_9$H$_{13}$O		
111	C$_6$H$_7$O$_2$		
Elemental analysis (calculated for C$_{11}$H$_{16}$O$_2$)			
	Calcd, %	Found, %	Found, %
C	73.30	72.90, 73.07	73.11
H	8.95	9.16, 9.31	8.88

8 The Analytical Approach

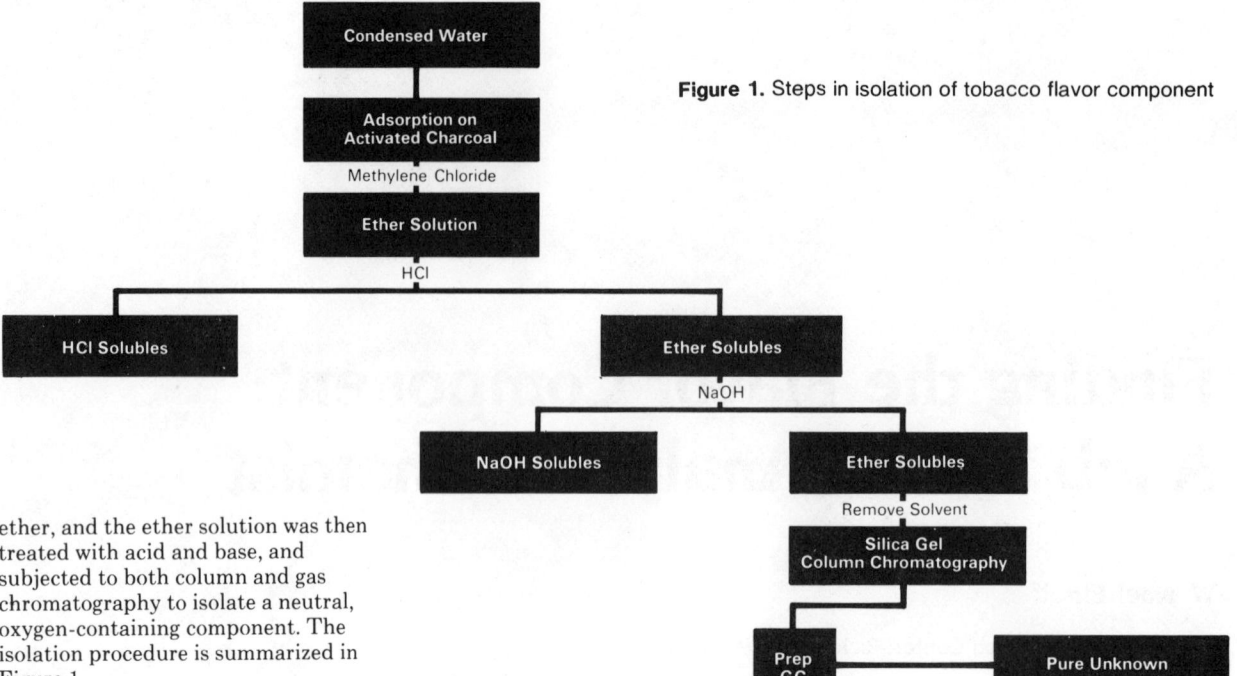

Figure 1. Steps in isolation of tobacco flavor component

ether, and the ether solution was then treated with acid and base, and subjected to both column and gas chromatography to isolate a neutral, oxygen-containing component. The isolation procedure is summarized in Figure 1.

Nondestructive Techniques Used First

Nondestructive analytical techniques were first used; i.e., IR, UV, and NMR spectroscopy, followed by the destructive techniques of MS and elemental combustion analyses. The low-resolution mass spectrum showed a loss of 43 mass units from the molecular ion, which could have been due to the loss of either an isopropyl or acetyl group. To determine which of the two moieties was involved, precise mass measurements were necessary to obtain the elemental composition of the major ions present in the mass spectrum. These and other results are given in Table I. Based on these results, the loss of 43 mass units was found to correspond to the acetyl moiety, and the compound was found to have a molecular weight of 180 with a molecular formula of $C_{11}H_{16}O_2$.

A carbonyl function was suggested by the absorption band in the IR spectrum at 1761 cm^{-1}, but resistance of the compound to silver oxide oxidation and the inability of the compound to form 2,4-dinitrophenylhydrazone, oxime, or semicarbazone derivatives ruled out the possibilities of an aldehyde or a ketone. The lack of a Cotton effect in a qualitative optical rotatory dispersion experiment (700–300 nm) also supported the latter conclusion. A double bond was suggested by the IR and NMR spectra and confirmed by catalytic hydrogenation, followed by the observation of an increase of two mass units in the mass spectrum of the hydrogenated product (MW 182). The UV spectrum of the original compound gave an absorption maximum at 208 nm, characteristic of an α,β-unsaturated γ-lactone. After hydrogenation the absorption maximum disappeared, and the IR carbonyl band shifted from 1761 to 1770 cm^{-1}. The original compound was resistant to hydrolysis in methanolic potassium hydroxide, ethylene glycol–potassium hydroxide, or sulfuric acid, even at elevated temperatures. In addition, no hydroxyl function was present by IR analysis. The techniques used and the

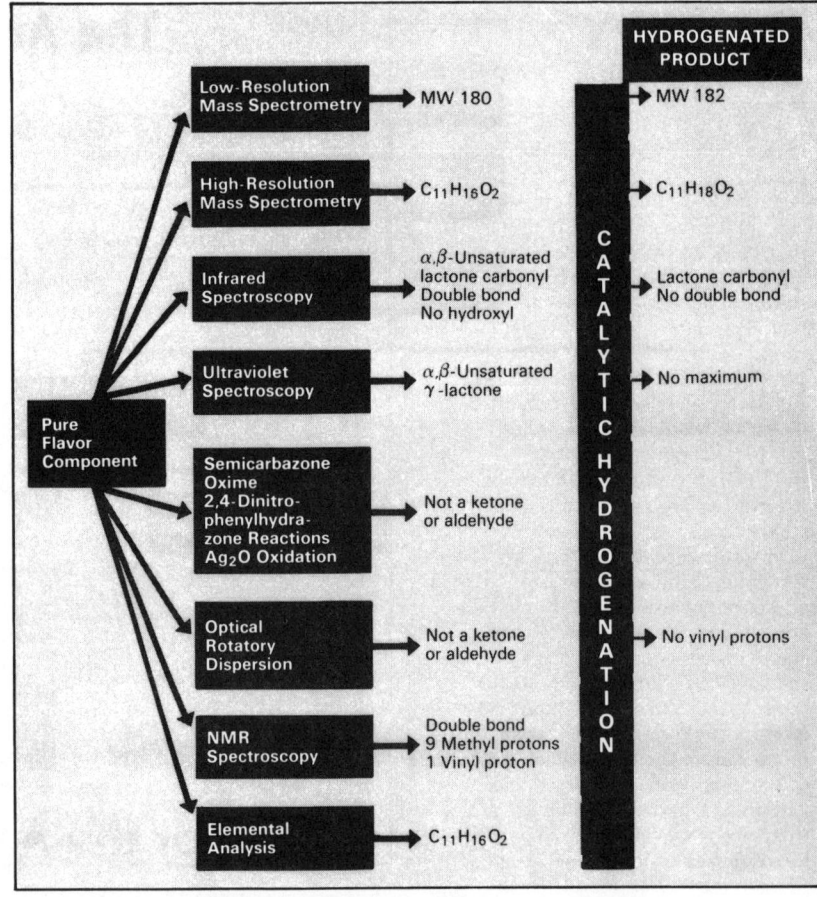

Figure 2. Techniques used in analysis of unknown flavor component

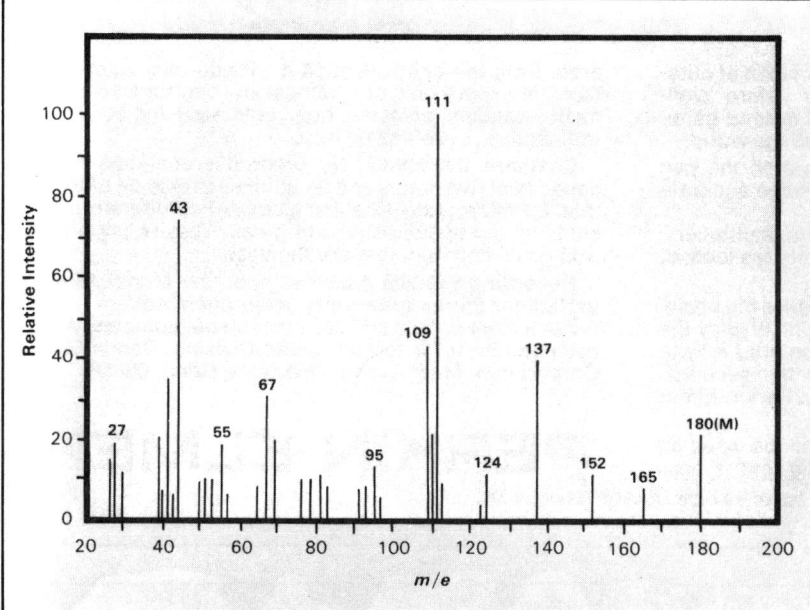

Figure 3. Possible structures of isolated flavor component

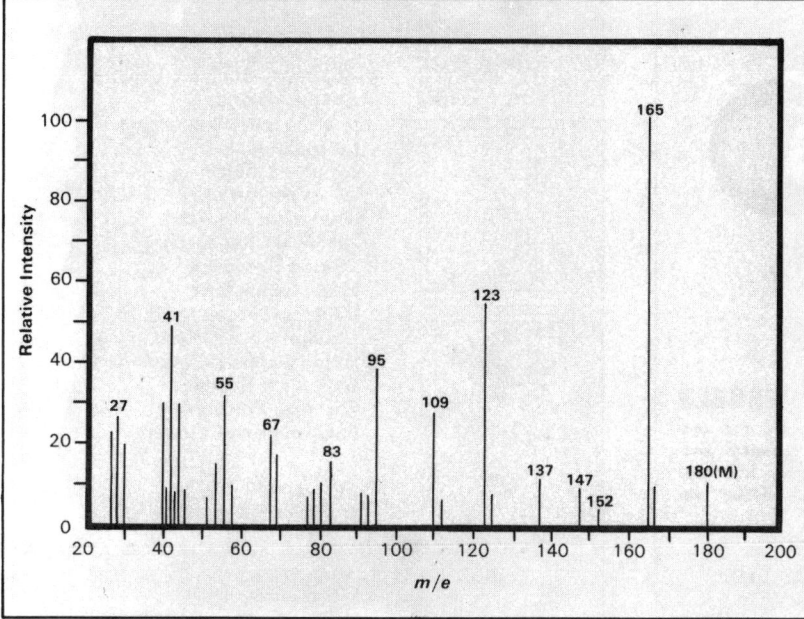

Figure 4a. Mass spectrum of Compound I, later identified as dihydroactinidiolide

Figure 4b. Mass spectrum of Compound II, 1,4,4-trimethylcyclohexan-2-one acetic acid enol lactone

conclusions drawn from each are shown in Figure 2.

Synthesis Clinches Identification

At this point, it was decided to synthesize two likely compounds, based on the physical data that had been obtained. These are Compounds I (dihydroactinidiolide) and II (1,4,4-trimethylcyclohexan-2-one acetic acid enol lactone), shown in Figure 3. A literature search complicated the problem somewhat because there were three reports covering the isolation and identification of Compound I, with each giving a different set of physical properties. Two of the three reports involved isolation of a compound from natural sources. Since large discrepancies existed between the chemical data obtained for the isolated compound and that reported in the literature for the same proposed structure, the best approach was to synthesize Compound I by at least two different routes to obtain an unambiguous structure. This was done, with the product from each synthetic route giving spectral characteristics identical to those of the unknown compound; the mass spectrum is shown in Figure 4a. The product was a viscous oil, which soon crystallized to give a mp 42–43°C. Hydrogenation of Compound I also gave a product identical in physical properties to the hydrogenated unknown.

Compound II was synthesized to conclude the investigation. As suspected, the physical properties of this compound, given in Table I, were considerably different from those of Compound I. The mass spectrum of Compound II is shown in Figure 4b for comparison with that of Compound I.

The discovery of dihydroactinidiolide in the condensed water as well as in bright and Oriental tobaccos has added to the knowledge of tobacco chemistry in addition to providing a potential source for a recoverable flavor component. Problems similar to this one and frequently of a proprietary nature occur on a continuing basis, providing the challenge which only a group of skilled and knowledgeable analytical chemists is able to meet via the analytical approach.

Acknowledgment

The solution of this problem is the result of the efforts of the following people. Details appear in these two papers: William C. Bailey, Jr., Ajay K. Bose, Robert M. Ikeda, Richard H. Newman, Henry V. Secor, and Charles Varsel, *J. Org. Chem.*, **33**, 2819 (1968); and Paul H. Chen, William F. Kuhn, Fritz Will, III, and Robert M. Ikeda, *Org. Mass Spectrom.*, **3**, 199 (1970).

Tracing the Origin of Off-Flavors in a Breakfast Cereal

Menard G. Heydanek, Jr.

Quaker Oats Co., John Stuart Research Laboratories, Barrington, IL 60010

Originally published in ANALYTICAL CHEMISTRY, 1977, Vol. 49, No. 11.

During the routine quality assurance examination of a packaged ready-to-eat breakfast cereal, a piney-spruce off-flavor was found. These quality assurance examinations routinely monitor all aspects of food products, such as package appearance, fill weight, product appearance, flavor, and odor, and usually total 25 or more attributes. These examinations or audits of product quality are designed to locate problems in production that may go unnoticed for some period of time if they are not monitored. They usually are most important in locating subtle, esthetic quality differences which, when absent or changed, will not present the food manufacturer's best tasting product to the consumer. The observation of this piney off-flavor in the cereal product was very important since subsequent taste panels showed that it contributed a highly undesirable flavor character. Location of the source of this adverse flavor and its quick removal were of paramount importance because of present day methods of manufacturing large amounts of food products at a single location.

Scheme A shows the steps taken to locate the piney off-flavor source. There were three possible sources:
• The ingredients of the cereal product
• The process, i.e., odor development during cooking
• The package or the external environment.

One advantage in the development of an analytical approach to this type of problem is that the off-flavor can be traced organoleptically with the nose as well as with analytical instruments. In fact, the nose is probably the most important analytical tool an analytical flavor chemist uses. Each of the ingredients used in the cereal manufacture was smelled, and no source of piney odor was found. The processing system, especially the water sources, was examined, and no piney odor source found. This was probably the least likely source of the off-flavor but could not be ruled out completely until the system was checked. When the packaging material used for the outer boxes was examined, it was not piney but the inner glassine liner contained a piney odor. A similar procedure was used to find the piney source in the components of the glassine liner. The liner is composed of two sheets of Kraft paper laminated with a resin in microcrystalline wax and overwaxed on both sides with paraffin wax. As seen in Scheme A, the resin used in bonding the paper layers is the obvious culprit.

Since the resin contains an odorous material, the method of choice for isolation of these components was a distillation technique. Vacuum distillation of the resin at 175 °C and 0.5-μ pressure was used to strip the volatile odor from the resin, and collection of the volatiles was accomplished in cold-finger traps cooled with dry ice–acetone baths.

The collected volatile material was rinsed from the traps with pentane, and the resulting solution produced the typical piney odor that was noted in the glassine liner and off-flavored product when examined organoleptically on perfume blotters (thick filter paper).

To determine precisely what chemical components were responsible for the off-flavor, a further fractionation of the distillate was undertaken (Scheme B). Again, with organoleptic evaluation as the criteria for following the piney off-flavor, gas chromatography was used to fractionate the components into discrete flavor entities. Gas chromatography was carried out

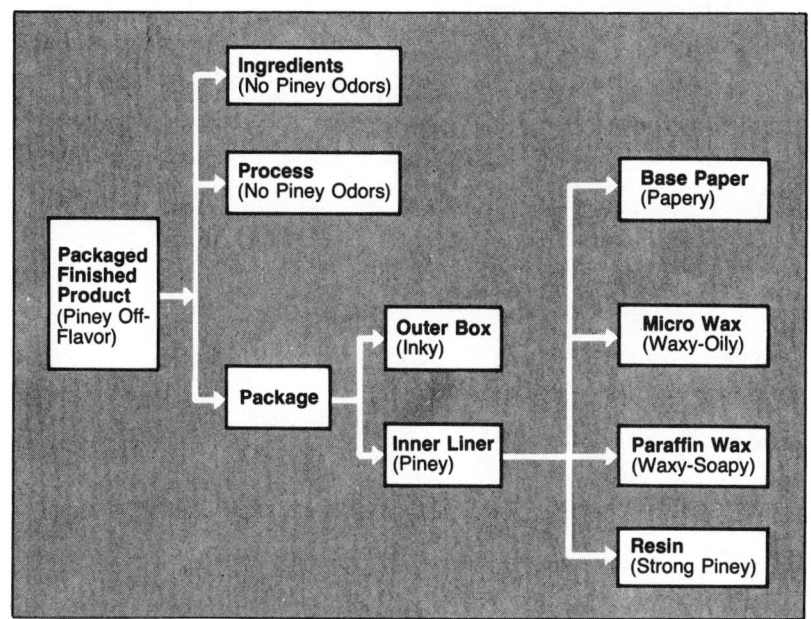

Scheme A. Odor source of piney off-flavor
Odor descriptions in parentheses

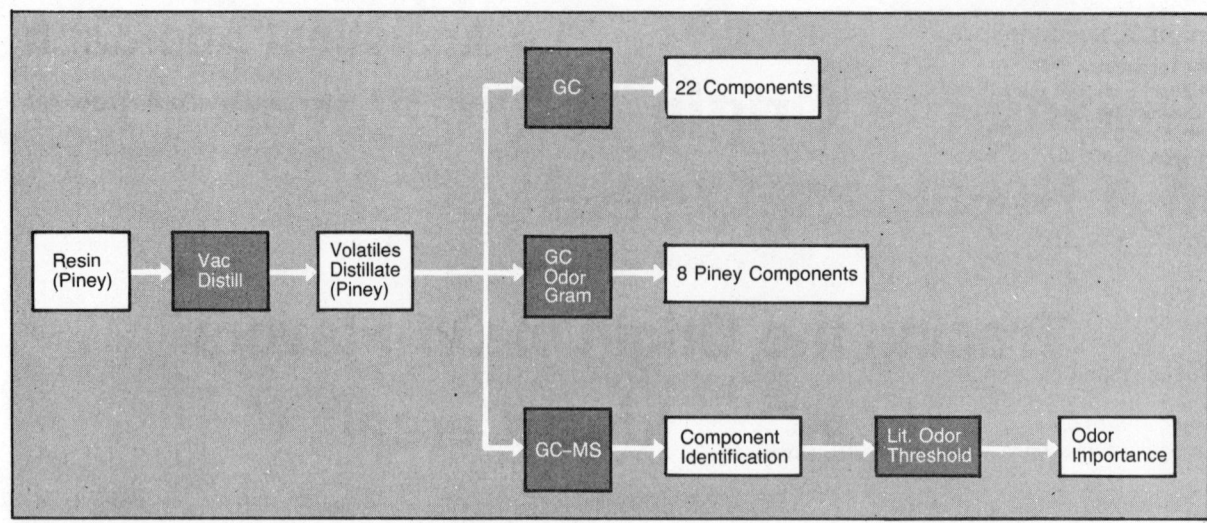

Scheme B. Evaluation of individual component/odor importance

on a Carbowax 20M column, and the separate components were organoleptically evaluated at the end of the column. The use of 3/1 exit splitter attached to a heated collection vent and exit sniffing port allowed each component to be organoleptically evaluated as it exited from the column.

From the chromatogram (Figure 1), at least nine of the separated peaks, including most of the major peaks, had odors associated with a piney character. Since more than one component had piney character, the identity of the piney odor components was needed to evaluate their individual thresholds. GC–MS analysis yielded the component identifications shown in Table I. All the components identified were common to pine oil or could be derived via hydrogenation of pine oil during resin processing and had published odor thresholds in the 1 to 8 ppb range (1).

At this point, even with the identities established, it was not known which of these components had a pronounced flavor effect in the finished product. Several avenues were now available to determine which of the chemicals or group of components were responsible for the off-flavors. Since there are ranges of flavor thresholds covering six orders of magnitude, it is not a safe assumption to regard the most abundant as the most odorous. One approach would be to obtain each of the chemicals and determine their odor thresholds in the product. Then one must determine the amount of transfer from the package to the product and limit the package to those levels or below for complete flavor compatibility. This would be a long and time-consuming task that would not necessarily yield any more practical results than the alternate approach we chose to use. The objective of our approach was to develop an appropriate quality control method and define specification limits for the purchase of the waxed glassine liner material. Since the ultimate desire was to prevent odor/flavor transfer to the product, we took advantage of this transfer property to help define the specification limits of the piney odor components.

To develop the appropriate flavor-instrumental correlations as outlined in Scheme C, samples of waxed glassine with varying levels of pine odors were subjected to organoleptic analysis for their ability to produce off-flavor in ready-to-eat cereals. They were ranked by a panel of expert judges into two classes: acceptable (no foreign flavor imparted) and unacceptable (definite foreign odor or flavor) resulting in substandard finished product. The latter category was subdivided into levels designated as weak,

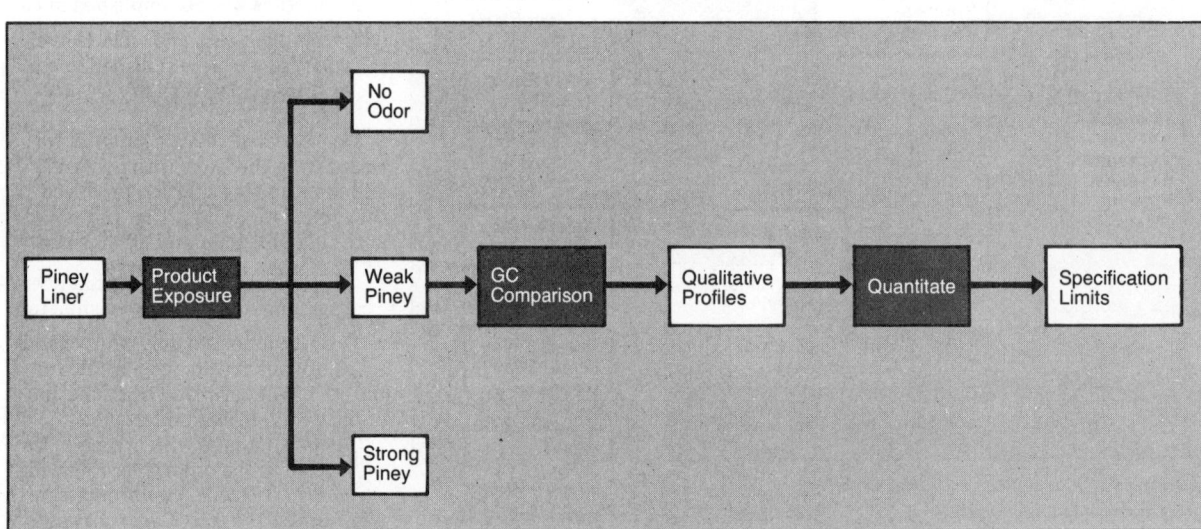

Scheme C. Determination of component limits in glassine liner for product compatibility

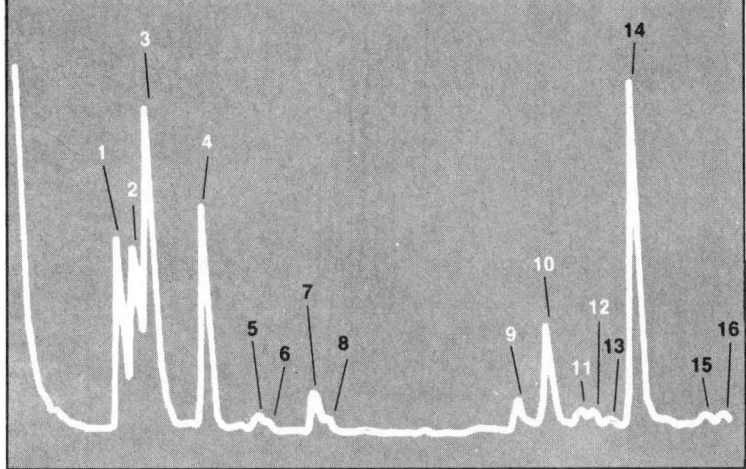

Table I. Component Identifications

Figure 1 pk. no.	Odor	Identification by GC–MS	Retention time of authentic compound
1	Weak pine	Menthane	
2	Weak pine	Menthane	
3	Slightly strong pine	Camphene	+[a]
4	Pine tree	8-p-Menthene	
5	Camphor	Limonene	+
6	Camphor	1,8-Cineol	+
7	Fragrant	P-cymene	
8	Sweet, pleasant	1,4-p-Menthadiene	
9	Sour, camphor	Camphor	+
10	Very strong pine	Fenchyl alcohol	
11	Licorice	p-Isopropyl anethole	
12	Fatty-citrus	...	
13	Flowery, green	...	
14	Pine, chestrub, strong	Borneol	+
15	Resinous, pine pitch	Borneol acetate	
16	Foul, acidic	...	

[a] + = authentic compound matches unknown retention time. Others were not compared or not available.

Figure 1. Gas chromatogram of resin volatiles

moderate, and strong. Also, 100 g of these same waxed glassine samples were subjected to hot jar-headspace analysis by the ASTM F-151 procedure. This method involves heating the sealed sample container, withdrawing a constant volume of headspace sample via syringe, and injecting it into a gas chromatograph. The gas chromatography procedure was calibrated by using α-pinene and borneol as standards for terpene hydrocarbons and oxygenated terpenes, respectively. The results showed a general increase in the level of pine components as the relative organoleptic strength of the product off-flavor increased.

From this type of data, the correlation between organoleptic analysis and objective gas chromatography can be established. In this case, there was good correlation between the detection of piney off-flavor and the presence of the major terpenes (peaks 1 to 4, Figure 1) and oxygenated species, borneol and fenchyl alcohol (peaks 10 and 14, Figure 1), at levels of 80 ppb or greater for each component. Therefore, routine quality control testing of incoming liner shipments with GC was initiated, and levels of these components were routinely determined with the understanding that at 80 ppb each or higher, the product might be subjected to off-flavor development during distribution.

The net result of instituting these tighter and more specific controls on the glassine liner was a marked reduction in the incidence of these piney odors in our standard packaging material. Several suppliers sought and found acceptable substitute resins that did not have this inherent odor and thus were more easily controlled. Others have met the specified levels, and very few problems have been seen since.

References

(1) Compilation of Odor and Taste Threshold Values Data, W. H. Stahl, Ed., DS48, 05-048000-36 ASTM Publications, 1973.

Characterization of Corrosion Scale from Nuclear Reactors: Approaching the Whole Problem

L. D. Hulett, John M. Dale, H. W. Dunn, and P. S. Murty

Oak Ridge National Laboratory, Analytical Chemistry Division, Oak Ridge TN 37830

Originally published in ANALYTICAL CHEMISTRY, 1976, Vol. 48, No. 14.

Corrosion is a serious problem that affects the cost, safety, lifetime, and overall practicality of nuclear reactors. Analytical chemists associated with nuclear energy research are heavily involved in the solution of corrosion problems. To properly do their jobs, they must assume more responsibility than just that of analysts. Since engineers or other scientists in charge of the corrosion studies often do not know how or when to apply the necessary analytical techniques, it is the responsibility of analytical chemists to help coordinate that part of the study. By doing this, they generate an extra dimension to their work by becoming active, decision-making parts of the overall research efforts. They become even more a part of the action if they interpret results and attempt to relate them to the cause of the problem. This is what we mean by the "whole problem approach". For the case history which follows, the whole problem approach assisted in determining the cause of scale that formed on the inside of a steam generator being tested for use with a nuclear reactor.

Corrosion scales must be characterized to determine their causes and possible prevention. This requires many different types of analytical techniques, and the information from each examination must be assimilated with that from the others so that a general description of the corrosion can be constructed.

The characterization questions posed are as follows:
(1) What elements are present?
(2) What are the chemical states (compound or ionic forms) of the elements?
(3) How are the elements and compounds distributed? Is the specimen homogeneous or heterogeneous?
(4) What are the sizes of the particles that make up the specimen?
(5) How do the particle surfaces differ from their interiors?
(6) What is the origin of the specimen (precipitation, corrosion, . . .)?

To collect this much information with the same solid sample and to preserve the integrity of the sample during analysis, nondestructive techniques must be used. The x-ray and electron physics methods fit these requirements quite well and were applied to this study of scale on an inconel (alloy 600) corrosion specimen from the inside of a steam generator. The methods used and their sequence of application are shown in Figure 1.

The first step in the examination (Figure 1) was to chip a small piece of the scale from the inconel substrate and analyze it by x-ray induced x-ray fluorescence (XR–XRF) to answer characterization question 1. This piece contained the elements of the substrate Ni, Cr, and Fe, but the proportion of iron was much higher. Also observed was bromine in a rather low concentration of no more than 0.1%. The next step was to mount a portion of the specimen in epoxy and to cut and polish it so that the scale and inconel substrate could be viewed in cross section by scanning electron microscopy to answer questions 3 and 4. Figure 2 shows a low-magnification scanning electron micrograph of the scale and substrate. Note that the outer portion of the scale is of considerably different texture from the inner regions. The overall thickness of the scale is 150 μm. The outer region of the scale, of different texture, is about 15 μm in thickness. X-ray fluorescence, induced by the beam of the microscope, was used to analyze various parts of the scale shown in Figure 2 and to answer characterization question 3. Phosphorus, silicon, and small amounts of calcium were found in addition to iron, chromium, and nickel. Isolated inclusions with high concentrations of manganese were also found in the scale. Another very significant observation was that the outer 15 μm

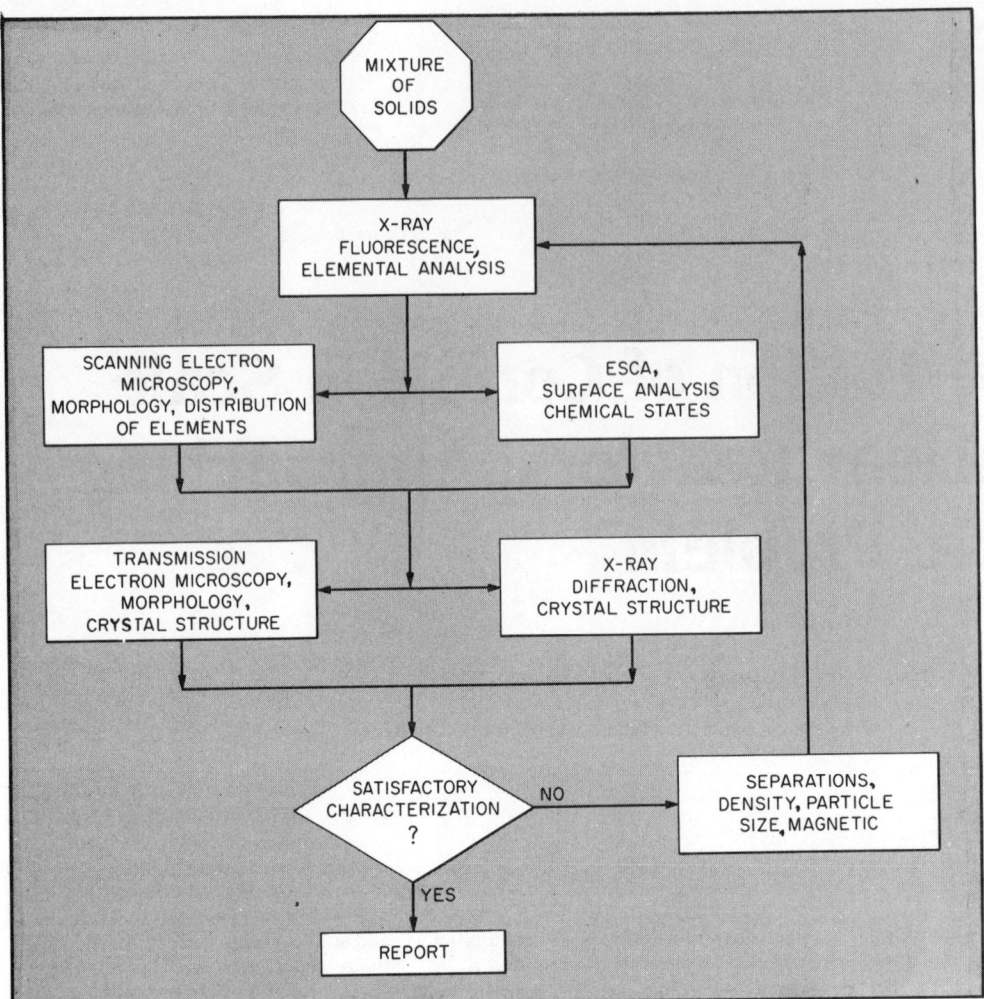

Figure 1. General approach to analysis of complicated mixtures of solids

of the scale, which had a different texture, contained predominantly Ni and Fe. Cr and P were the only other elements present in the scale, and their concentrations were very low. The summary of conclusions after XR–XRF and SEM–XRF examination is shown in Table I.

ESCA was then used for examining the scale. After the outer surface of the scale was analyzed, it was scraped away with a gold knife edge so that a thin layer of scale remained at the alloy scale interface. The ESCA spectrum of the scale at the alloy interface (Figure 3) was essentially the same as that on the outer surface; only one of the three transition metals, nickel, was detected. Iron and chromium, known from XRF data to be present in the scale, are not revealed in the photoelectron (ESCA) spectrum. The explanation for this anomaly is that the surfaces of the oxide crystallites that make up the scale are very different from their interiors. In addition to nickel, Figure 3 shows that sodium, phosphates, bromates, and silicates are present on the surfaces of the scale particles. The oxidation states of the phosphorus, bromine, and silicon were deduced from their chemical shifts. The ESCA study combined with XR–XRF applies to characterization question 5.

Another interesting observation from the ESCA studies was that carbon was present in two different oxidation states (Figure 4). The carbon spectrum of the scale is compared with that of $NiCO_3$. In both spectra of Figure 4 the peak having the higher kinetic energy is due to hydrocarbon vapors condensed from the air and the atmosphere of the spectrometer. The peak having lower kinetic energy is due to carbon in the form of carbonate. The carbonate form has slightly higher binding energy than that of hydrocarbon and releases its photoelectrons with lower kinetic energy. Some of the ESCA spectra of carbon on the scale showed a third carbon peak having even lower kinetic energy than the carbon of nickel carbonate. The chemical shift of this peak matched that of Na_2CO_3. Since sodium was found on the scale surface, we assumed that the third carbon was from Na_2CO_3. The conclusions from the ESCA study of the scale are indicated in the second entry of Table I.

After the ESCA examination of the scale, the x-ray diffraction technique was applied to answer characterization question 2. The scale was scraped

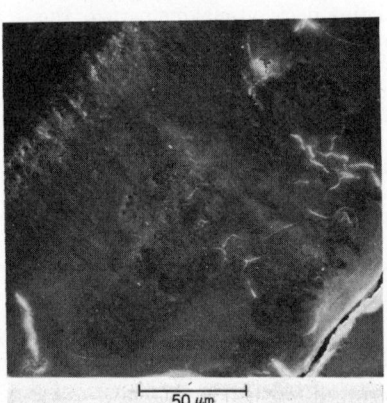

Figure 2. Scanning electron photomicrograph of scale

Table I. Summary of Conclusions

XR–XRF, SEM–XRF Examination
Fe, Cr, Ni, P, Si, Ca, and Br are present in scale. Ni and Fe predominate in outer 15 μm of scale. Mn inclusions in scale

ESCA Examination of Scale
There is a coating on particles that make up scale. Composition of coating is greatly different from that of scale particles

Main components of particle coating are phosphates, bromates, silicates, nickel carbonate, and sodium carbonate

X-ray Diffraction Examination of Scale
$Ni_xFe_{3-x}O_4$ is main component of particles that make up outer 15 μm of scale

$Ni_xCr_{3-x}O_4$ and $Fe_xCr_{3-x}O_4$ are main components that make up particles in interiors of scale

Ion Etch–ESCA Examination of Scale Alloy Interface
Corrosion process causes enrichment of Cr in scale

Origin of Scale
Whatever corrosion process that is taking place is causing a Cr enrichment at alloy-scale interface

Most of scale is due to precipitation rather than corrosion

Solid-state transformation is taking place at outer surface of scale, causing formation of $Ni_xFe_{3-x}O_4$

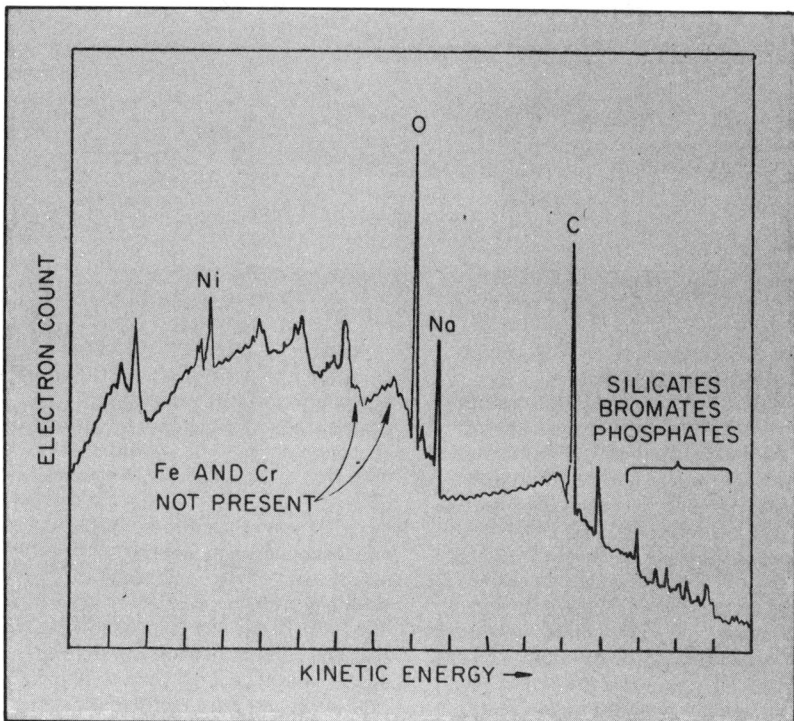

Figure 3. Photoelectron (ESCA) spectrum of surface of scale crystallites

from the inconel substrate for these studies. The only pattern that could be interpreted was that of a spinel. Spinels involving Ni, Fe, and Cr are well known. However, there are three different combinations of these elements that form spinels with essentially the same diffraction pattern: $Ni_xFe_{3-x}O_4$, $Ni_xCr_{3-x}O_4$, and $Fe_xCr_{3-x}O_4$. The Ni–Fe spinel is magnetic. When we examined the scale scrapings, we found a magnetic component which was easily separated from the rest of the scale. When an x-ray diffraction pattern was obtained for the magnetic and nonmagnetic components, the spinel structure was found for both. At this point, it was recalled from the SEM–XRF data that Ni and Fe were the predominant elements found in the outer 15 μm of the scale, and it was concluded that the outer 15 μm of the scale was the $Ni_xFe_{3-x}O_4$ spinel, while $Ni_xCr_{3-x}O_4$ and $Fe_xCr_{3-x}O_4$ compounds were present in the other parts of the scale. These conclusions are listed in the third entry of Table I.

The scraping of the scale with a gold knife edge left a thin film of scale next to the inconel substrate. It was felt that an examination of this layer might yield information about the corrosion process that occurred at the alloy interface, and the ESCA–ion etch technique was applied at this point. This involved the bombardment ("sputtering") of the surface with 1000-eV argon ions followed by photoelectron spectra measurement. The ion etch treatment removed material in a controlled fashion so that the composition of the scale as a function of depth could be measured. The results of this study are plotted in Figure 5. Note that the composition of the scale tends toward about 7% Ni, 5% Cr, and 1% Fe. This composition indicates that the scale at the interface is enriched in Cr due to the corrosion process. If the three metals had been in the scale in the same proportions as for the substrate, their percentages would have been something like 7.8% Ni, 1.4% Cr, and 0.8% Fe. This conclusion is also summarized in Table I. The quantitative interpretation of the ESCA data used for Figure 5 was done according to the scheme suggested by Carter et al. (1).

The first four conclusions of Table I regarding the character of the scale are summarized schematically in Figure 6. We shall now address characterization question 6, often the most important of all from the customer's point of view: *What is the origin of the scale?* A comparison of the thickness of the scale with the mass loss (decrease in thickness) of the substrate indicates that the mass of the scale is too great to be accounted for by corrosion. A large portion of the scale must therefore be due, not to corrosion, but to precipitation from the boiler environment. Our characterization studies support this conclusion. The manganese inclusions must necessarily have come from precipitation since manganese is not a component of inconel. The phosphate was probably from corrosion inhibitors put into the boiler. Calcium, sodium, silicate, and bromate must have come from precipitation also. The Ni_x-$Fe_{3-x}O_4$ forms a clearly distinct phase from the rest of the scale. This suggests that a solid-state transformation is taking place at the outer layers of the scale.

References

(1) W. J. Carter, G. K. Schweitzer, and T. A. Carlson, *J. Electron Spectrosc. Relat. Phenom.*, **5**, 827 (1974).

Oak Ridge National Laboratory is operated by the Union Carbide Corp. for the U.S. Energy Research and Development Administration.

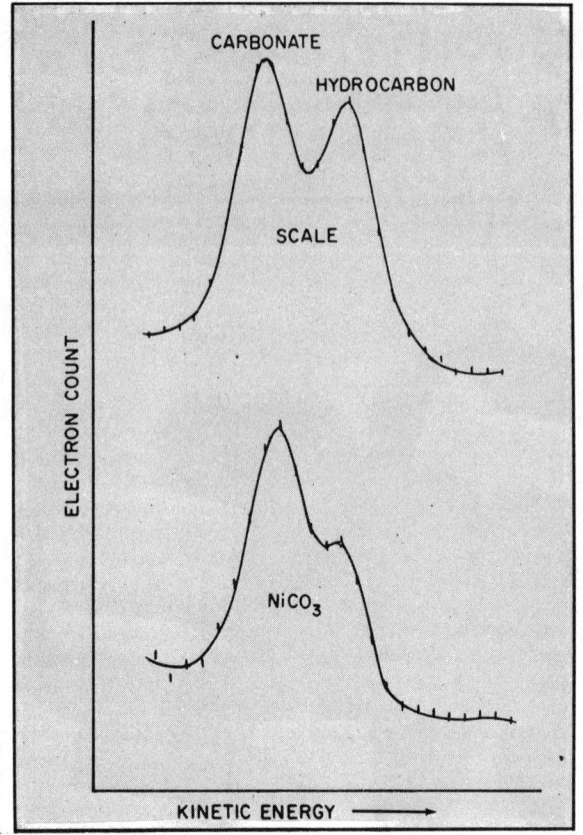

Figure 4. Comparison of carbon photoelectron peaks of corrosion scale with those of $NiCO_3$

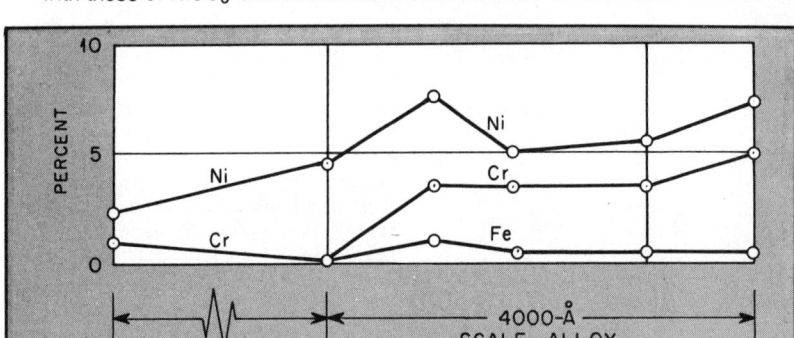

Figure 5. Concentration profiles of Ni, Cr, and Fe at scale-inconel interface

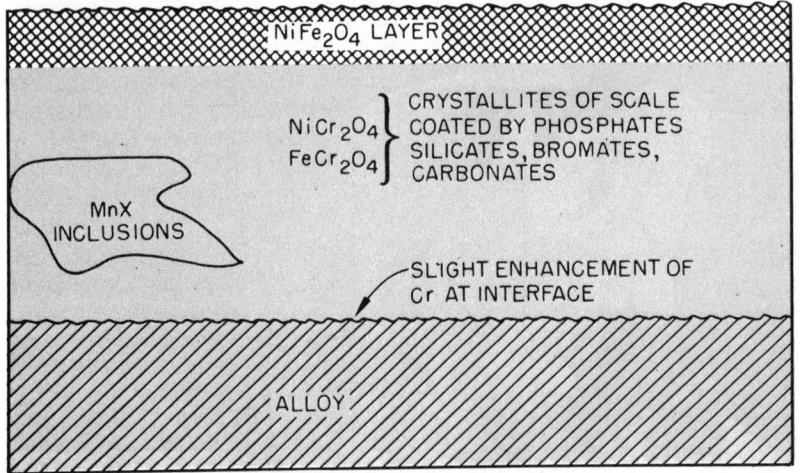

Figure 6. Overall character of scale

Urgent Production Problem Solved by Unique Capabilities of Analytical Chemists

Claude A. Lucchesi

Originally published in ANALYTICAL CHEMISTRY, 1974, Vol. 46, No. 4.

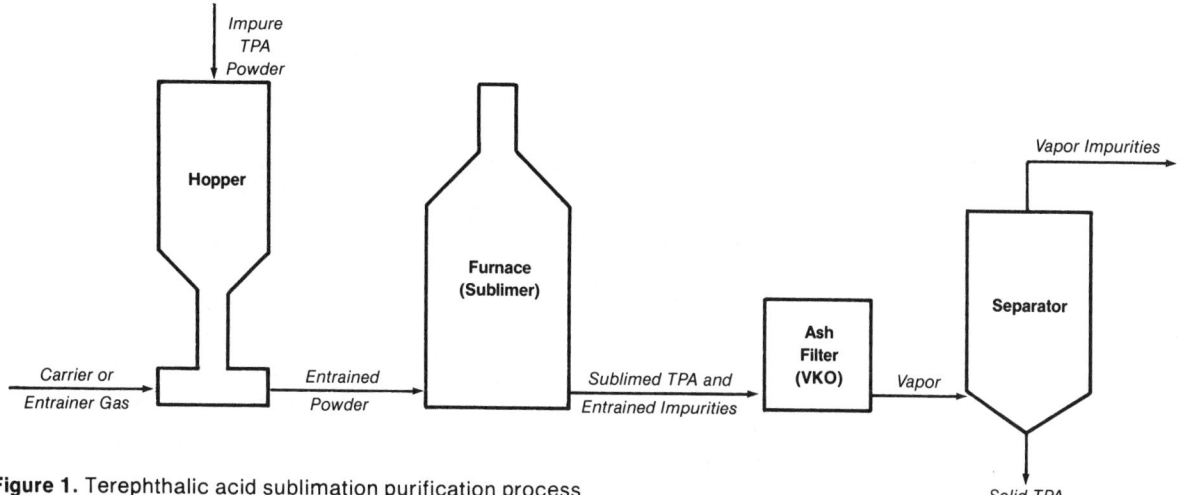

Figure 1. Terephthalic acid sublimation purification process

Analytical chemists at the Mobil Chemical Co. were confronted with a challenging problem during the start-up stages of the company's terephthalic acid (TPA) pilot plant. The Mobil TPA process consisted of two parts, an oxidation section and the purification section which is illustrated in Figure 1. Dry, impure TPA powder from the oxidation section was fed into a hopper where it became entrained in a high-velocity carrier gas and passed into a furnace where substantially all of the solids were vaporized. The effluent from the furnace was passed through an ash filter where entrained solids, including catalyst residues from the oxidation section, were removed from the vaporized material. The vapor then entered a condenser for fractional condensation of the solid TPA product.

The TPA was being produced as a potential replacement for dimethyl terephthalate (DMT), then the only source of the dicarboxylic acid in many polyester films and fibers. TPA had an expected cost advantage over DMT, provided it could be produced at an equivalent level of purity which was judged by a set of specification tests. One specification test involved the color of a 5% solution of TPA in dimethylformamide (DMF). And this is where the problem started. Its solution, in time to be of practical value to the company, required the integrated efforts of a team of specialists in IR, X-ray diffraction, X-ray spectrography, arc-spark spectrography, thermal methods, gas chromatography, and solution chemistry. In addition, a literature searcher was needed to find a synthesis procedure for the tentatively identified material to "cinch" an identification.

An early pilot plant run in Beaumont yielded TPA product which, not only failed the color test, but also did not completely dissolve in the DMF. The question was, "What is the DMF-insoluble material?" This was the question I was asked to answer when I was with Mobil Chemical as manager of the Analytical and Physical Chemistry Department. The then vice president of the Research and Development Division phoned and told me to go to Beaumont and find out what that DMF "turbidity" was. Although the immediate question was the identity of the turbidity, the critical question was how to prevent it, whatever it was, from forming. The problem had top priority, and for several weeks most of the specialists in the department did little else. Five pounds of TPA product which failed the DMF test was requested for the Research Laboratory in Metuchen, N.J., and I went to Beaumont, Tex., to become familiar with the pilot plant operation and to obtain test samples for study.

Nondestructive Tests Used to Survey Test Samples

Because only a few milligrams of the DMF turbidity could be isolated from several pounds of TPA product, initial tests on the DMF-insoluble material were limited to the nondestructive techniques readily available in our lab: X-ray diffraction, X-ray spectrography, and infrared spectroscopy. The same measurements were made with the Beaumont test samples, and the material removed from the ash filter and the DMF turbidity from the TPA product were almost identical (Figure 2). The two materials had identical infrared spectra and nearly identical X-ray diffraction patterns. Also, the X-ray fluorescence spectrographic measurements showed that the two materials contained the same metallic elements in roughly the same concentration ratios. These findings enabled us to do subsequent work with the pound or so of ash filter material rather than with the limited amount of DMF-insolubles.

Identification of Inorganic Part

The nature and concentration of the inorganic materials in the ash filter sample were established to the extent justified by the nature of the sample as illustrated in Figure 3. From all the data shown, it was estimated that about half of the inorganic material was $CaSO_4$. Thus, at most, only 3 or 4% of a metal was available to form a salt or chelate with TPA. Consequently, the data on the inorganic part of the ash filter sample supported the IR conclusion that the main constituent of the ash filter sample (and the DMF insolubles) was not primarily a salt or chelate of TPA. (For example, Ca(TPA) contains 11% Ca.)

Cobalt, iron, and calcium were found in the DMF-insolubles by X-ray spectrography. The ratios of the three metals in the DMF-insolubles and in the ash filter sample and the concentrations in the ash filter sample suggested that less than 0.1% cobalt was in the DMF-insolubles. Consequently, the DMF-insolubles could not have been a cobalt salt or chelate. The same can be said for iron.

Identification of Organic Part

Having convinced ourselves that only a minor part of the DMF-insolubles could have been of an inorganic nature, we concentrated on the or-

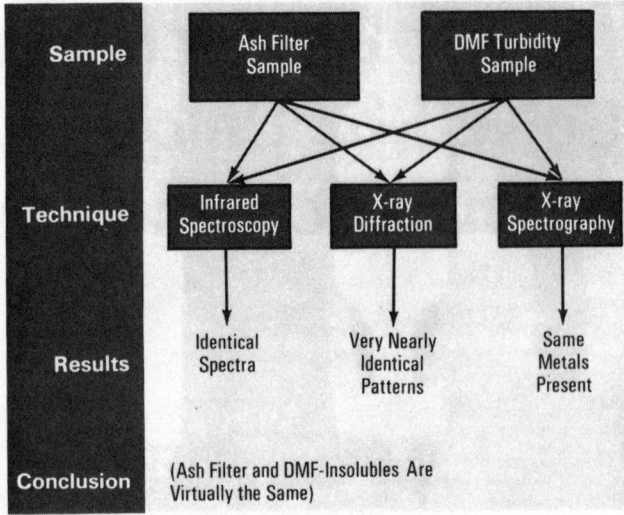

Figure 2. Nondestructive methods used to show DMF turbidity and ash filter material were virtually the same

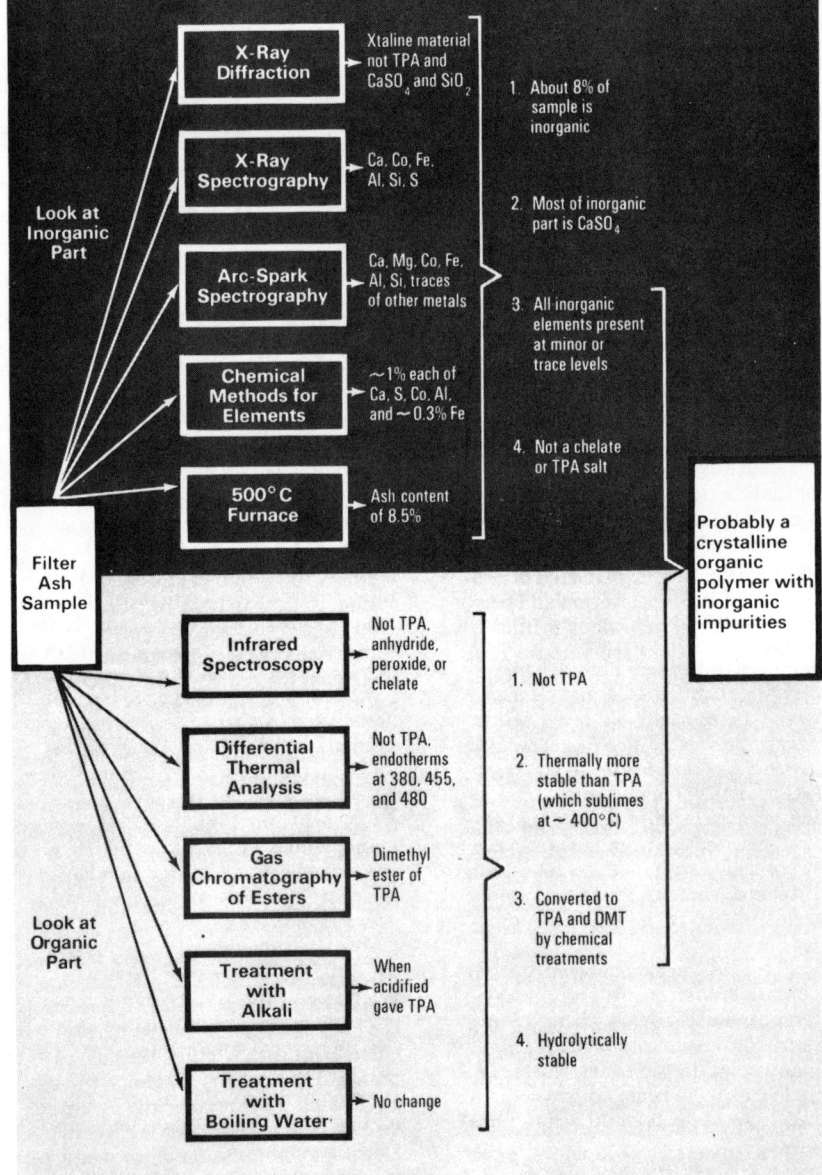

Figure 3. Techniques for inorganic and organic substances used in combination to show what material could and could not be

ganic part of the problem. The ash filter sample and several TPA samples were treated with methanol to produce methyl esters and other volatile substances which were extractable with chloroform and could be passed through a gas chromatograph. Although only about 20% of the sample was esterified, almost all of the material that got through the chromatograph was DMT.

Because of the gas chromatographic observation that the ash filter sample yielded only about 20% TPA ester when carried through the esterification procedure and because the IR spectrum of the residue was the same as the starting material, a hydrolysis study was made. The sample was treated with NaOH, and the insoluble residue was separated by filtration, dried, and weighed. When the filtrate was acidified, it gave a precipitate which was identified as TPA. About 62% TPA was recovered with the first treatment. After four treatments, 83% was recovered. This gave a residue of 17% which was in the ball park with the 8.5% ash figure obtained earlier.

Unfortunately, we did one more experiment that caused unnecessary confusion. We boiled the ash filter sample in water for at least 2 hr, and it remained insoluble. Nevertheless, we came to the conclusion that we had a highly crystalline polymer that was hydrolyzed with base or acid to give TPA and could be partially converted to DMT by a conventional esterification procedure.

At this point, I had a meeting with all of the specialists who worked on the problem, and we systematically decided what the DMF-insolubles could *not* be. The only possibilities we had left was an anhydride or a peroxide, and we concluded that the TPA more likely had formed an anhydride. But it was difficult to convince our organic chemists that TPA most likely had formed an anhydride. Consequently, one of our literature searchers went to the Chemists' Club Library in New York City and found a 1959 German article describing the synthesis of TPA anhydride. We translated the article and followed the recipe. The IR spectrum of the synthesized material matched the IR spectra of both the ash filter sample and of the DMF-insolubles. The X-ray diffraction patterns also matched.

When we were convinced that the DMF-insolubles indeed were poly-(TPA anhydride), we suggested to the engineers that steam be used as a carrier gas instead of nitrogen to prevent the formation of the anhydride. This solved the problem: no more DMF-insolubles and no more color.

Recognition and Solution of Production Problem in Vitamin Manufacture

E. MacMullan, R. Hagel, and R. Gomez

Hoffmann-La Roche Inc., Quality Control Department, Nutley, NJ 07110

Originally published in ANALYTICAL CHEMISTRY, 1975, Vol. 47, No. 4.

It is a truism that the first step in solving an analytical problem is recognizing that you have a problem. This was demonstrated recently in the Roche Control laboratories during the testing of a lot of thiamine mononitrate.

Thiamine (vitamin B_1) is an essential factor in human nutrition found naturally in rice hulls, cereal grains, yeast, liver, eggs, milk, and green leaves. It was originally isolated from rice bran; however, the thiamine used for dietary supplements is almost entirely derived from chemical synthesis. Thiamine mononitrate (I) synthesized at Roche must pass a rigid battery of chemical, physical tests to assure its purity and equivalence to the natural vitamin before being released for sale.

I

In the course of this testing, the analyst observed that one lot did not pass the solution clarity test. At a concentration of 2% in water, the solution was hazy whereas it should have been clear. In all other respects, the sample met specifications.

At this point, the analyst could have simply failed the lot and returned it to Production for reprocessing where an additional recrystallization would have undoubtedly brought the lot within specifications. But he recognized this as an unusual result worth a deeper investigation; therefore, he requested that the Analytical Research section look into the situation.

Defining the Problem

The general approach taken to define the problem was to:
- Determine the level of impurity in the sample
- Isolate some of the impurity
- Consult with the production chemist on likely impurities
- Subject the impurity to spectral and chromatographic analysis to determine its structure.

These steps are shown schematically in Figure 1.

The purity of the lot was measured by phase solubility analysis with procedures similar to those described by Webb (1) and MacMullan (2). Phase solubility analysis is a technique for determining the purity of materials based on a careful measurement of their solubility behavior. The method has its theoretical origin in the Gibbs Phase Rule in which absolute purity is defined as single component behavior. A solid material in equilibrium with a saturated solution which does not vary in composition as a function of the amount of excess solid in equilibrium with the solution is a pure solid. A solid whose saturated solution composition increases with increasing amounts of solid in equilibrium with the solution is impure to the extent of the increase in solution concentration.

The phase solubility analysis indicated an impurity level of 1.3% (Figure 2). This changed the picture significantly. What had been considered a trace amount of insoluble impurity was in fact a significant amount of a slightly soluble impurity. This was a serious problem and made the identification of the impurity even more urgent.

Identification of Impurity

To obtain more of the impurity, 11 grams of the thiamine mononitrate was shaken with 550 ml of water for 30 min. The resulting suspension was filtered, and 49 mg of dried impurity was recovered for study by UV, IR, NMR, and mass spectrometry. At the same time, discussions with production chemists led to the information that the impurity might be the immediate precursor in the synthesis (3), thiothiamine [3-(4-Amino-2-methyl-5-pyrimidinyl methyl)-5-(2-hydroxyethyl)-4-methyl-4-thiazoline-2-thione] (II), which is only slightly soluble in water.

II

This apparently was confirmed when thin-layer chromatography of the suspect lot showed an impurity spot at the

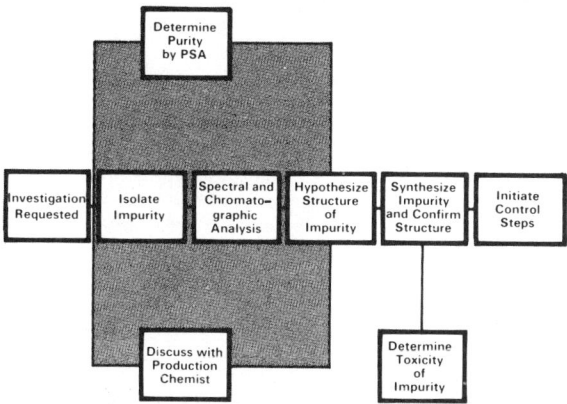

Figure 1. Analytical approach

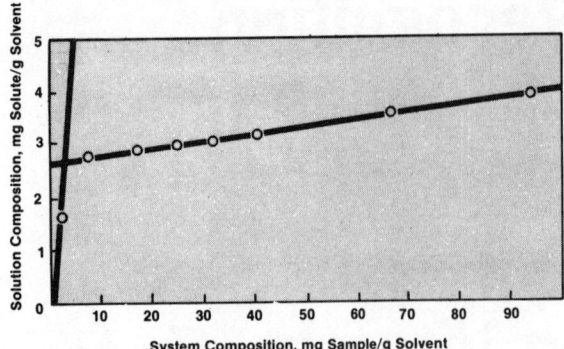

Figure 2. Phase solubility analysis of thiamine mononitrate
Sample: thiamine mononitrate lot A
System: methanol, 20 hr @ 25°C
Slope (computed by least squares with 95% confidence): 1.30 ± 0.08%
Extrapolated solubility: 2.63 ± 0.04 mg/g

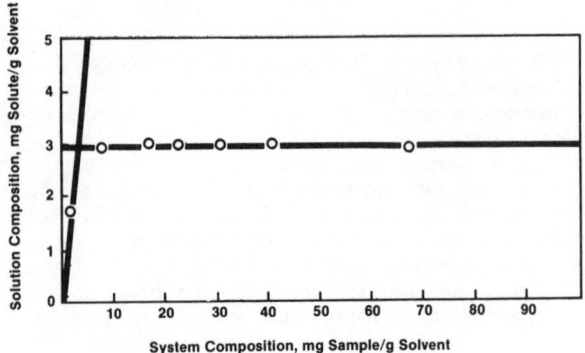

Figure 3. Phase solubility analysis of purified thiamine mononitrate
Sample: thiamine mononitrate lot A (phase purified)
System: methanol, 20 hr @ 25°C
Slope (computed by least squares with 95% confidence): −0.08 ± 0.3%
Extrapolated solubility: 2.92 ± 0.1 mg/g

same R_f as thiothiamine. However, when the plate was sprayed with thiochrome, the thiothiamine standard gave a fluorescent spot under longwave UV, but the impurity spot did not. Furthermore, the UV, IR, and NMR spectra were not consistent with the thiothiamine. The IR showed a carbonyl stretching band between two strong electronegative groups, and the NMR showed chemical shifts and splitting patterns consistent with a thiamine derivative containing a carbonyl in the 4-position of the thiazole ring. The identification data are summarized in Table I.

A hypothetical structure of the impurity was obtained by replacing the sulfur in thiothiamine with an oxygen. A search of the chemical literature showed that this compound had been made and was known by the trivial name thiamine thiazolone [3-(4-Amino-2-methyl-5-pyrimidinyl methyl)-5-(2-hydroxyethyl)-4-methyl-4-thiazoline-2-one] (III). Furthermore, it had been prepared from thiothiamine by oxidation in alkaline medium (4).

III

An authentic sample of thiamine thiazolone was prepared by the literature procedure, and the structure was confirmed by IR, NMR, MS, and elemental analysis. In all cases, the spectra of the insoluble impurity were identical to the authentic thiamine thiazolone.

Table I. Summary of Spectral and Chromatographic Results

Technique	Results	Conclusions
Thin-layer chromatography	Similar R_f to thiothiamine but different reaction to thiochrome	Impurity not thiothiamine
UV spectrum	Differs from thiothiamine	
IR spectrum	Carbonyl band between two strong electronegative groups	Hypothesized impurity as thiamine thiazolone
NMR spectrum	Consistent with oxygen analog of thiothiamine	
Mass spectrum	Molecular ion m/e 280 Fragmentation consistent with oxygen analog of thiothiamine	

Corrective Action

Once the impurity had been identified, the next step was to institute procedures to insure that it would not appear in subsequent lots. Further discussion with the production chemists indicated that the pH had probably gone up during the conversion of thiothiamine to thiamine mononitrate, allowing the formation of a small amount of thiamine thiazolone. The control steps instituted were:

- Better pH control of the reaction
- An in-process TLC done by the production chemist
- Purity determination of the finished lot by phase solubility analysis.

Another necessary step was to determine the toxicity of the thiamine thiazolone. The LD_{50} in mice was greater than 4000 mg/kg. This is even less toxic than thiamine which has a reported toxicity of 3000 mg/kg, and this represents a very low level of toxicity on an absolute basis (5). While we have instituted steps to eliminate this impurity, we have the added assurance of knowing it is a very nontoxic substance.

In retrospect, the critical steps in the solution of this problem were:

- Recognition of the problem by the control chemist
- Purity determination by phase analysis
- Close cooperation between analytical and production chemists
- Search of the chemical literature.

Given the success of these four steps, the successful laboratory solution of the problem was almost inevitable. These steps in some form or other are fundamental to the solution of any production problem.

Phase analysis had one more role to play in the process. This method can also be used as a separation technique since at solution equilibrium all the impurities end up in the solution phase, and the undissolved solids are essentially pure. This "phase purification" was run on a 50-gram sample of the impure lot and yielded 47 grams (94%) of material with a purity of 100.0% ± 0.1%. The analysis of the purified material is shown in Figure 3. This procedure could be scaled up for use in the production area.

References

(1) T. J. Webb, *Anal. Chem.*, **20**, 100 (1948).
(2) E. A. MacMullan, paper presented at Land-O-Lakes Conference on Pharmaceutical Analysis, August 1969.
(3) Maxion, U.S. Patent 2,844,579.
(4) T. Matsukawa and H. Hirano, *J. Pharm. Soc., Jap.*, **73**, 379 (1953).
(5) "The Merck Index," 8th ed., p 1037, 1968.

Identification of Impurity in Blood Preservative System

M. Marconi, M. Gara, and R. Dawe

Travenol Laboratories, Inc., Morton Grove, IL 60053

Originally published in ANALYTICAL CHEMISTRY, 1979, Vol. 51, No. 11.

The purine base adenine (Figure 1) is used as a fortifier to prolong the shelf life of blood in blood storage bags (1). Addition of adenine to blood storage solutions increases blood shelf life from 21 to 35 days. Since the active ingredients come into direct contact with the blood, this manufacturing of raw material must meet the rigid specification of containing no more than 0.1% purine impurities. Such impurities may interfere with the known key roles of adenine in biological systems. To ensure that adenine meets these requirements, a high performance liquid chromatography (HPLC) method with a strong cation-exchange column and a phosphate buffer mobile phase was developed in our laboratories. All incoming raw materials are routinely assayed by this method.

Several lots of adenine displayed a peak representing an unknown impurity that was estimated to be present at concentration levels up to 0.1%. The retention time of adenine under the specified assay conditions is approximately 5 minutes, while the unknown impurity had a retention time of 45 minutes. Several possible purine, pyrimidine, and nucleoside impurities were chromatographed in an effort to identify the impurity by matching retention times. This proved to be unsuccessful. Hence, our problem was one of isolation and identification of the unknown impurity. Specifically, our task was to determine whether the impurity was a purine.

Isolation of Impurity

Initial attempts focused on collecting the impurity as it eluted from the HPLC column and further study of the separated substance. However, it was a tedious task to acquire sufficient material for spectroscopic examination. Furthermore, removal of the impurity from the phosphate buffer mobile phase was problematic. Accordingly, a reverse phase HPLC separation with a C-18 octadecylsilane column and sodium heptane sulfonate as an ion pairing reagent was concurrently developed for the adenine raw material and the impurity. The impurity again had a longer retention time than adenine. Because of the chromatographic characteristics on both the ion-exchange and on the reverse phase columns, it was assumed that the impurity was more hydrophobic than adenine itself.

With the above idea in mind, a solvent partitioning was attempted to selectively concentrate the impurity. The adenine raw material was saturated in an alkaline aqueous solution that was then extracted with chloroform. The chloroform layer was backwashed with basified water to remove adenine, and the chloroform layer was examined by HPLC to ascertain whether the impurity peak was present. Indeed, the impurity was preferentially extracted into the chloroform with only a trace of adenine.

GC and GC/MS Analysis

The silyl derivatives of both the raw material and the chloroform extract described above were analyzed by gas chromatography (GC) with a 3-ft 3%

Figure 1.
Structure of adenine.
Molecular weight = 135

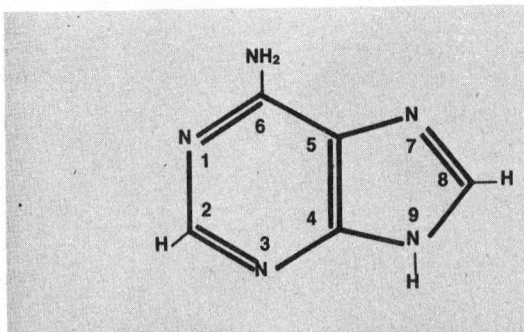

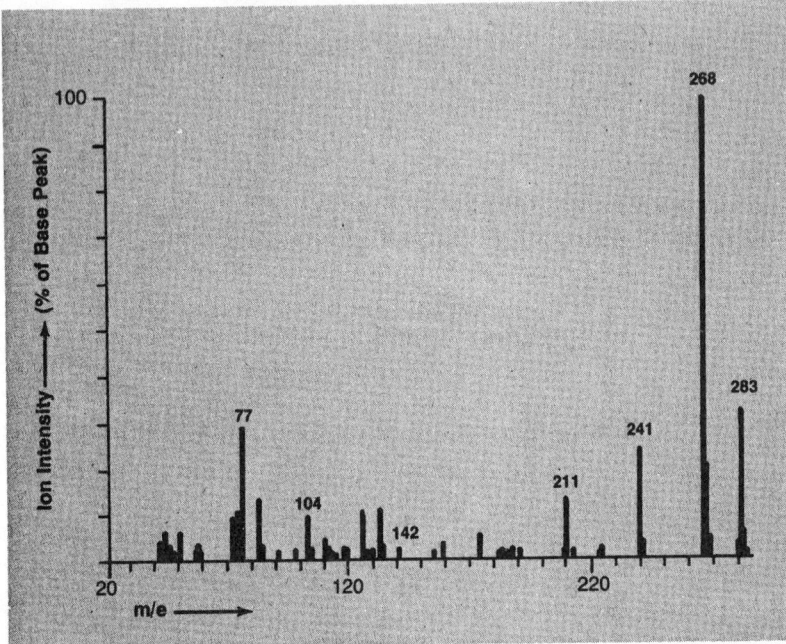

Figure 2. Mass spectrum of TMS derivative of isolated impurity

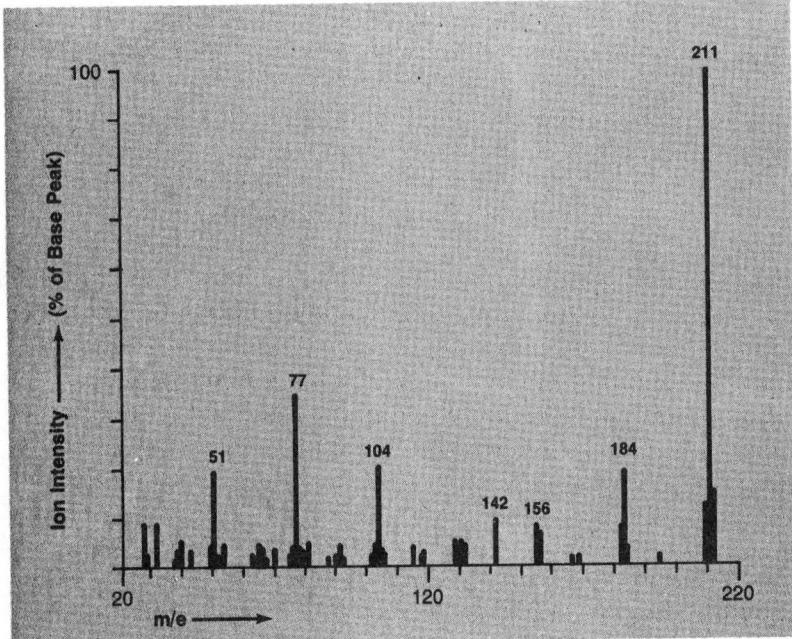

Figure 3. Mass spectrum of underivatized isolated impurity

OV-1 column. The trimethylsilyl (TMS) derivatives of adenine and associated purines were prepared by reacting the dry raw material with a mixture of N,O-bis(trimethylsilyl)-fluoroacetamide (BSTFA) in acetonitrile (2). Following derivatization, a gas chromatogram of the raw material was dominated by the bis-silyl derivative of adenine with several other minor components. Small amounts of mono- and tris-silylated adenine also were observed eluting immediately prior to and after the bis(silyl) derivative, respectively. The chloroform extract of the raw material after derivatization displayed only one prominent peak. Both samples were examined by combined gas chromatography/mass spectrometry (GC/MS). The mass spectrum of the TMS derivative of the impurity is shown in Figure 2. In addition, it was noted that the impurity extracted with $CHCl_3$ also could be chromatographed on the same GC column without derivatization. This peak also was examined by GC/MS and its spectrum is shown in Figure 3.

The mass spectrum of bis(trimethylsilyl)adenine is shown in Figure 4. A molecular ion is observed at m/e 279; however, the base peak is 264, which represents a loss of 15 mass units from the molecular ion. This loss is characteristic of silyl derivatives of purines (3). The mass spectrum of the TMS derivative of the impurity also shows a prominent M − 15 ion (assuming m/e 283 is the molecular ion). In addition, other characteristic ions of silylated purine include M − 15 − HCN at 241. Not characteristic of purine–TMS spectra is the small ion at m/e 77 that is usually present in monosubstituted phenyl derivatives.

The spectrum of the underivatized impurity shown in Figure 3 indicates the presence of a molecular ion at m/e 211 or 72 mass units less than the silyl derivative indicating one derivatizable hydrogen. Adenine itself predominantly forms a bis(silyl) derivative with silyl groups at 6-N and N-9. Steric hindrance retards derivatization of both 6-N hydrogens. The multiple losses of 27 (HCN), i.e., m/e 211 → 184, 184 → 157 are also indicative of a purine structure. Again, the distinct presence of the m/e 77 ion in this spectrum seemed to be indicating a phenyl group. The loss of 42 (NH_2CN) from 184 → 142 is indicative of a purine bearing an NH_2 substitution. The difference in molecular weight of 76 between the impurity and adenine (MW = 135) also strongly suggests phenyl substitution at either the 6-N or N-9 position. The use of ion-plotting techniques (mass chromatogram) permitted the verification of the presence of the silylated impurity at very low concentrations in the silylated raw material.

Infrared and NMR Spectra

More of the impurity was isolated from the raw material via the chloroform extraction technique and then subjected to infrared (IR) and Fourier transform proton nuclear magnetic resonance (NMR) spectroscopy. The IR spectrum of the impurity in a KBr disk was similar to, but distinct from, a file spectrum for 6-(N-phenyl)adenine (4). The use of Fourier transform NMR spectroscopy allowed the acquisition of a relatively clean spectrum of the isolated impurity in deuterated chloroform (Figure 5). The peaks downfield, at 8.37 and 8.23 ppm, are due to the purine ring hydrogens at C-2 and C-8. The peak at 7.6 ppm (integrating for 5 protons) represents a monosubstituted benzene ring. The broad signal at 4.4 ppm (integrating for about 2 protons) is probably due to a free amino group, because the peak disappeared after deuterium exchange with D_2O. All other peaks present in the spectrum are due to solvent impurities. Hence, the NMR data support the hypothesis of a phenylated adenine derivative and suggest substitution at the N-9 position.

In an effort to confirm the hypothesized structure, 9-phenyladenine was synthesized. Its mass and IR spectra were obtained and compared with spectra of the adenine impurity. The infrared spectrum of the synthesized N-(9-phenyl)adenine is shown in Figure 6. The fingerprint region of this spectrum is identical to that of the isolated impurity. The mass spectrum of the synthesized compound was the same as that obtained from the raw material contaminant. Once the impurity was identified, changes were made in the production process to ensure that 9-phenyladenine would not be present in the raw material. Identification of the impurity and elimination of it as a contaminant in the raw ma-

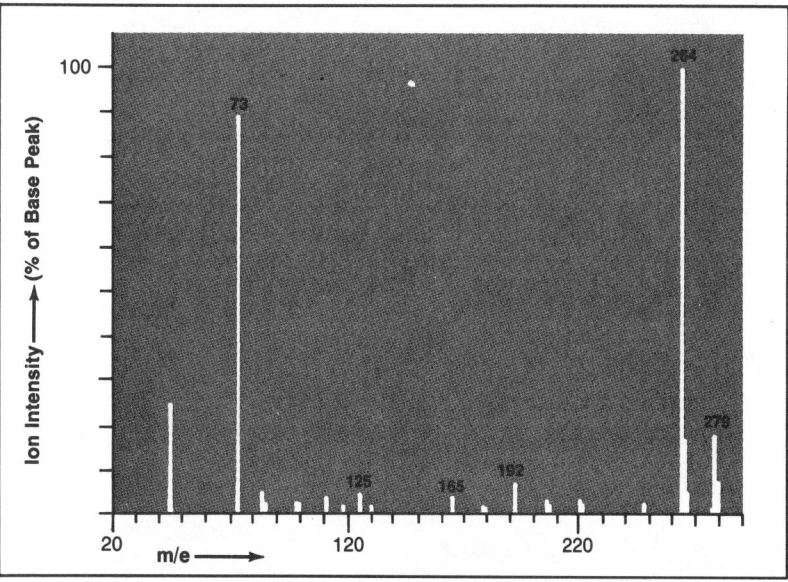

Figure 4. Mass spectrum of bis(trimethylsilyl)adenine

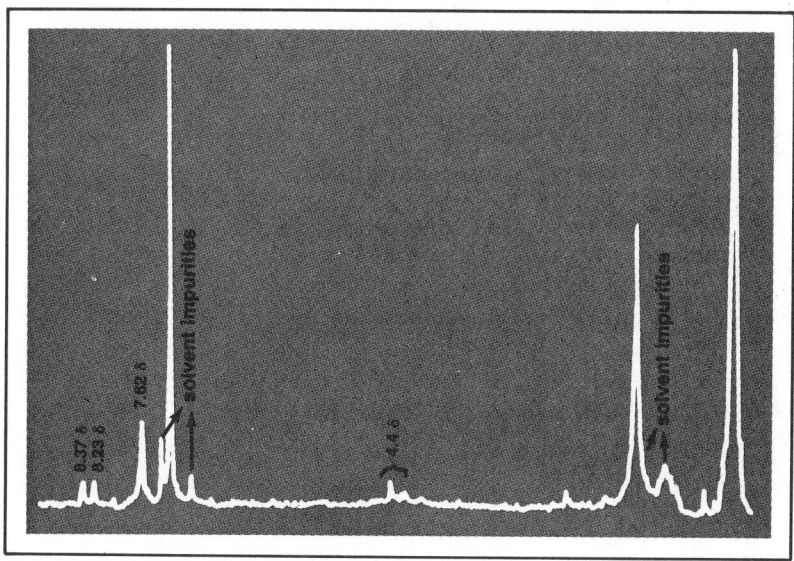

Figure 5. FT proton NMR spectrum of isolated impurity

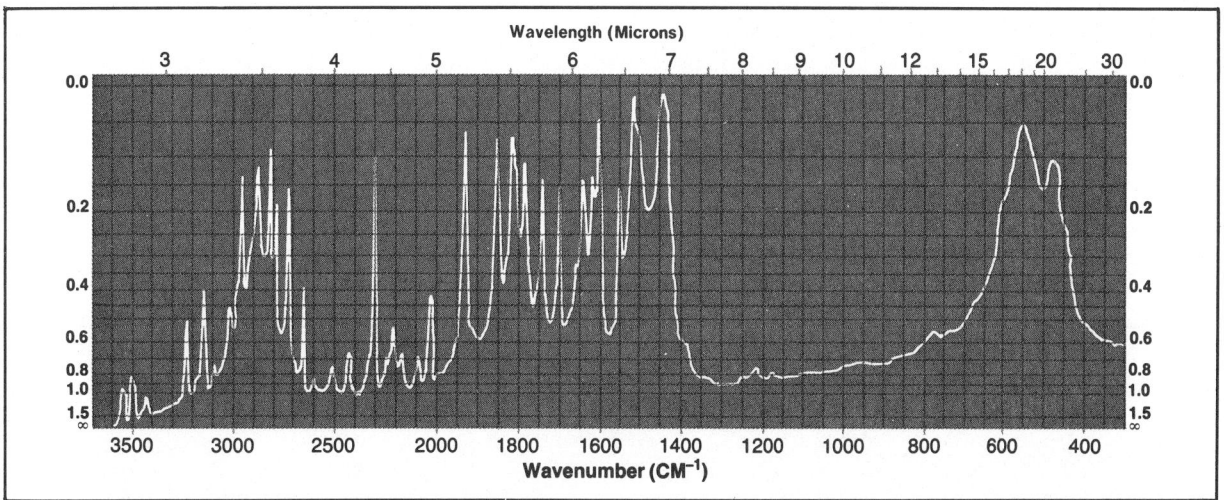

Figure 6. Infrared spectrum of synthesized 9-phenyladenine

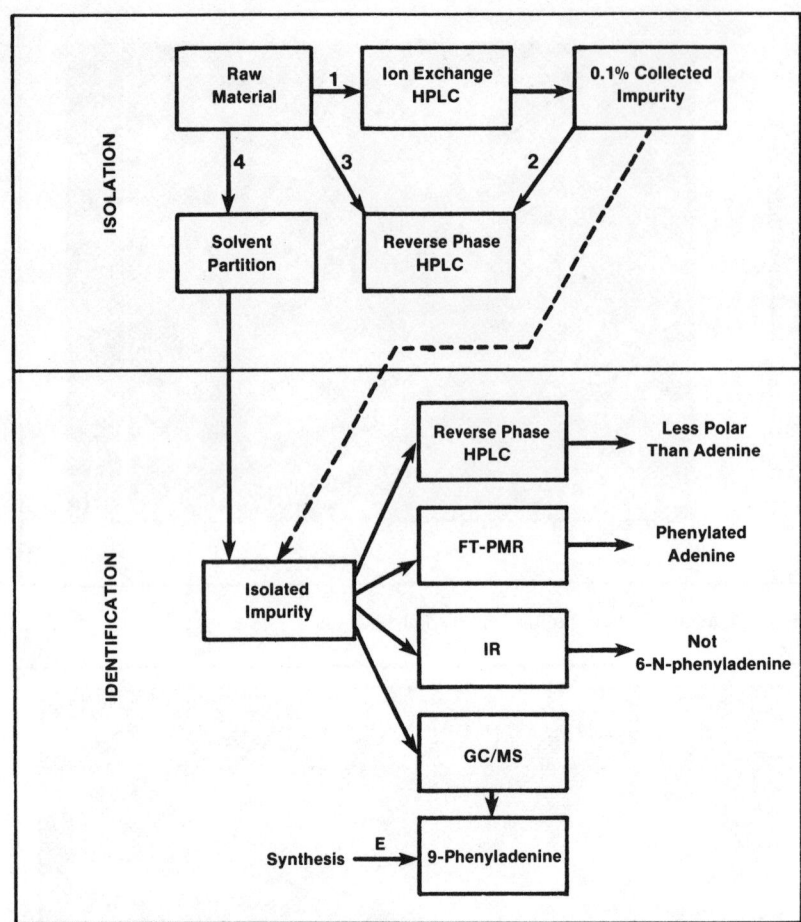

Figure 7. Flow diagram of problem solving

terial enabled production of adenine-fortified blood storage solution to continue.

The general scheme and approach to solving this particular problem is shown in the flow diagram in Figure 7. The most significant factor is that assumptions were made about the properties of the impurity based on its chromatographic behavior. This information was then used to make an educated guess about how best to isolate the impurity. Furthermore, the data obtained from several different instrumental techniques were combined, and a hypothesized structure of the impurity was fomulated based on the assembled data. The hypothesized structure was then confirmed by synthesis.

References

(1) G. B. Bartlett, "Changes of red cell phosphate compounds during storage of human blood in ACD," International Symposium on Erythrocytes, Thrombocytes, Leukocytes, Vienna, 1972, pp 139–143.
(2) C. W. Gehrke and D. B. Kaking, *J. Chromatogr.*, **61,** 45–63 (1971).
(3) J. A. McCloskey, Mass Spectrometry in Nucleic Acid Chemistry, in "Basic Principles in Nucleic Acid Chemistry," P. Tso, Ed., Academic Press, New York, 1974, pp 209–309.
(4) Sadtler Research Labs, "Infra-red Spectra, 1969," Spectrum No. 1559.

Effects of Raw Material Change in Manufacturing Process Resolved

John Mitchell, Jr.

E. I. du Pont de Nemours and Co., Plastics Department, Experimental Station, Wilmington, DE 19898

Originally published in ANALYTICAL CHEMISTRY, 1974, Vol. 46, No. 9.

Current shortages of many chemicals present a new challenge to the analytical chemist. Product quality in many industrial processes requires complete knowledge of compositions of chemical intermediates, including type and concentration of trace impurities. Active species may seriously affect product yield as well as product quality. This problem becomes acute when a plant finds it necessary to change the source of supply of a raw material, because maintenance of production requires rapid assessment of the nature of process stream impurities. A special emergency task force may be necessary to bring together the analytical expertise vital in resolving the effects of a raw material change on the manufacturing process. A plant faced such a problem in looking for another source of trimethylorthoformate used in a complex manufacturing process. Identification and determination of low levels of impurities were essential, requiring a multitechnique approach. Complete analysis required gas chromatography, infrared and ultraviolet spectrometry, nuclear magnetic resonance, mass spectrometry, distillation, and chemical methods.

Preliminary GC studies (Figure 1) indicated at least eight impurities. As indicated in Figure 2, GC/MS served to identify GC peaks 1, 2, and 3 as methyl formate, 2-methoxyethanol, and methanol, respectively. Compo-

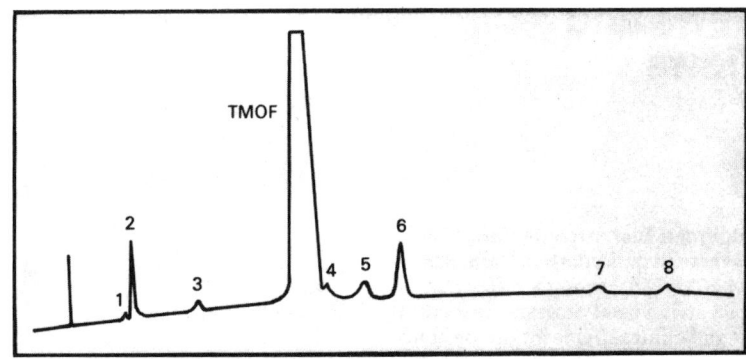

Figure 1. Gas chromatogram of trimethylorthoformate (TMOF)

Table I. Analytical Data on GC Peak No. 7

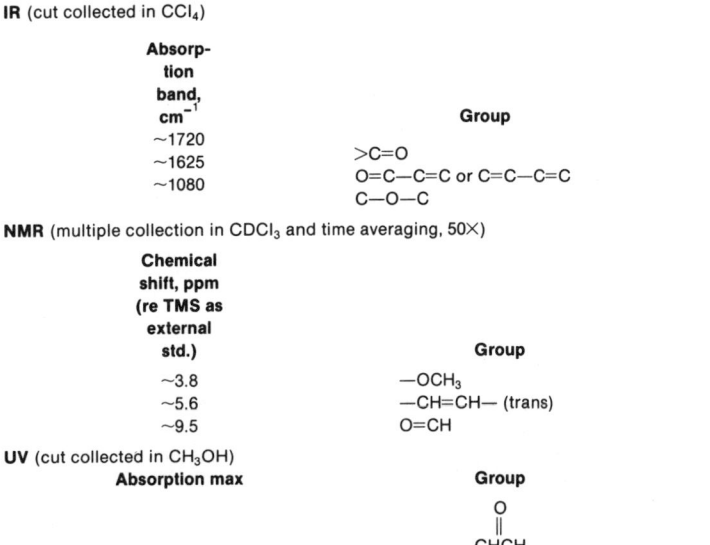

IR (cut collected in CCl_4)

Absorption band, cm^{-1}	Group
~1720	>C=O
~1625	O=C—C=C or C=C—C=C
~1080	C—O—C

NMR (multiple collection in $CDCl_3$ and time averaging, 50×)

Chemical shift, ppm (re TMS as external std.)	Group
~3.8	—OCH_3
~5.6	—CH=CH— (trans)
~9.5	O=CH

UV (cut collected in CH_3OH)

Absorption max	Group
	$\overset{O}{\underset{\|}{C}}$HCH

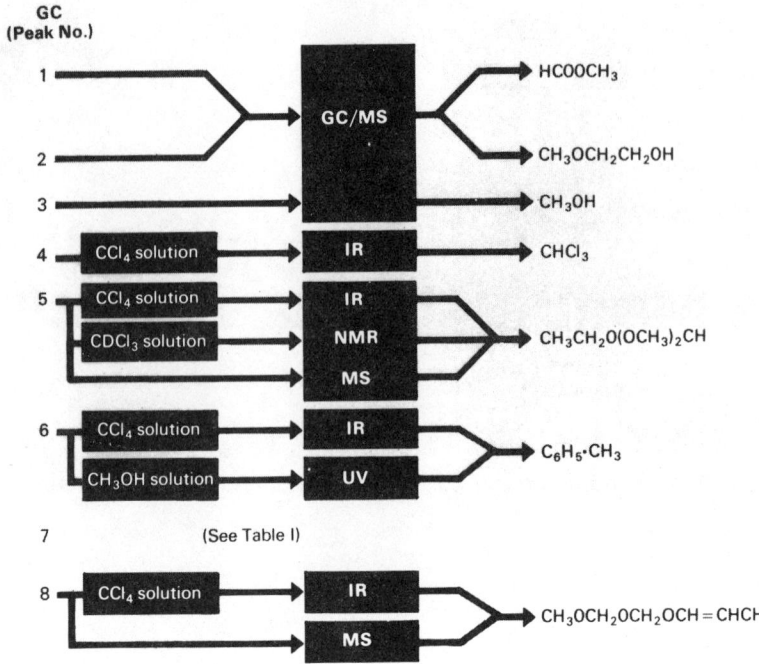

Figure 2. Identification of components separated by gas chromatography

nent 4 was identified as chloroform with the aid of IR. Fractions representing component 5 were collected in CCl_4 and in $CDCl_3$ for IR and NMR studies and MS was obtained directly. The spectra indicated ethyl dimethylorthoformate. Component 6 was identified as toluene. From GC, MS, and IR data, component 8 was identified as methoxymethoxymethyl-1-propenyl ether.

Component 7 proved to be the most challenging. Collection of analytical information is shown in Table I. Infrared spectra indicated C=O groups, conjugated unsaturation, and ether groups. NMR (with time averaging) showed methoxy, unsaturated methylene, and aldehyde group protons. Mass spectrometry indicated a molecular ion of 86; ultraviolet, an unsaturated aldehyde. The 2,4-dinitrophenylhydrazone (2,4-D) derivative, precipitated from ethyl alcohol solution, had a molecular weight of about 432 (Table II). It was orange in color and decomposed on heating, indicative of a dialdehyde. Elemental analysis of the 2,4-D supported the compound malondialdehyde. However, the data were inconsistent with direct results by IR, NMR, UV, and MS. Repeat of the direct analyses of the GC fractions verified the original results, including a methoxyl group. The suggested structure was $C_4H_6O_2$ (MW = 86), $CH_3OCH=CHCHO$ (3-methoxyacrolein). Formation of the di-2,4-D-derivative could be explained from cleavage of the methoxyl group in the strongly acid reagent to form the enol. The enol, in turn, would be expected to rearrange to the dialdehyde as the derivative was formed:

$$CH_3OCH=CHCHO \xrightarrow[H_2O]{H_2SO_4}$$

$$HOCH=CHCHO \xrightarrow[reagent]{2,4-D}$$

$$OCHCH_2CHO \longrightarrow 2,4-D$$

In summary, the eight compounds shown in Table III were identified. They are arranged in order of elution from the GC column.

In conclusion, the progressive analytical group associated with an industrial research organization combines expertise with versatility. Emphasis is on problem solving by the most efficient and effective means. Specificity is all important in detecting substances in concentrations varying from ppb to major amounts. The analytical chemist is presented with the challenge of contributing to special chemical structure needs in research, production, and marketing. This challenge is now all the more exciting as we must provide reliable results with respect to EPA, OSHA, and FDA requirements.

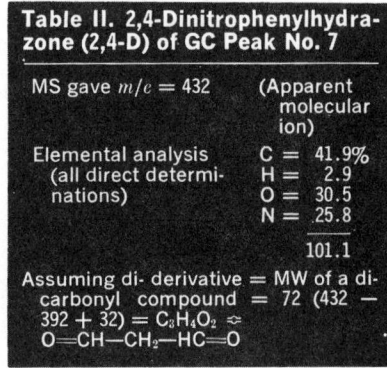

Table II. 2,4-Dinitrophenylhydrazone (2,4-D) of GC Peak No. 7

MS gave m/e = 432	(Apparent molecular ion)
Elemental analysis (all direct determinations)	C = 41.9% H = 2.9 O = 30.5 N = 25.8 ——— 101.1
Assuming di- derivative = MW of a dicarbonyl compound = 72 (432 − 392 + 32) = $C_3H_4O_2$ ⇌ O=CH—CH_2—HC=O	

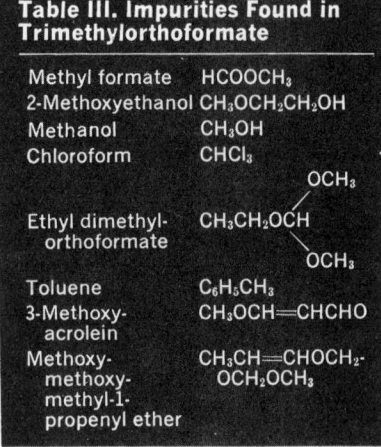

Table III. Impurities Found in Trimethylorthoformate

Methyl formate	$HCOOCH_3$
2-Methoxyethanol	$CH_3OCH_2CH_2OH$
Methanol	CH_3OH
Chloroform	$CHCl_3$
Ethyl dimethylorthoformate	$CH_3CH_2OCH(OCH_3)_2$
Toluene	$C_6H_5CH_3$
3-Methoxyacrolein	$CH_3OCH=CHCHO$
Methoxymethoxymethyl-1-propenyl ether	$CH_3CH=CHOCH_2OCH_2OCH_3$

Pure Steam Ahead!

D. F. Pensenstadler and M. A. Fulmer
Westinghouse Research and Development Center, Pittsburgh, PA 15235

Originally published in ANALYTICAL CHEMISTRY, 1981, Vol. 53, No. 7.

The purity of steam and water is very vital to the power production industry because the chemical environments within modern steam generators and turbines often cause turbine failures. Even trace (ppb) concentrations of corrosive impurities can cause excessive deposit formations, as shown in Figure 1, leading to corrosive attack or distress. Chemical analysis of these deposits has identified such inorganic species as chlorides, sulfates, and caustic.

In order to maintain product integrity and improve reliability for this industry, Westinghouse initiated a program in 1977 to monitor the chemical environment in operating systems and to determine the operating conditions in which corrosive ionic impurities enter the steam/water cycle of fossil fuel power plants. (See FOCUS, ANALYTICAL CHEMISTRY, 1980, 52, 1409–10 A).

The analytical approach taken involved the use of ion chromatography, a widely accepted method for cation and anion analyses. To make ion chromatography a useful quantitative analytical tool for the power production industry, it was necessary to adapt the technique for analysis at a low ppb level by preconcentrating samples. This adaptation will be described later.

The Westinghouse steam purity analysis program involves sampling various locations throughout the steam/water cycle of a power plant and using on-site ion chromatographic instrumentation for analysis of these samples. In typical field tests in operating power plants, a temporary ion chromatographic laboratory is set up for on-site sample analysis. Whereas condensate samples can be collected directly into sample containers, steam samples require heat exchangers to

Figure 1. Example of deposits on steam turbine blades

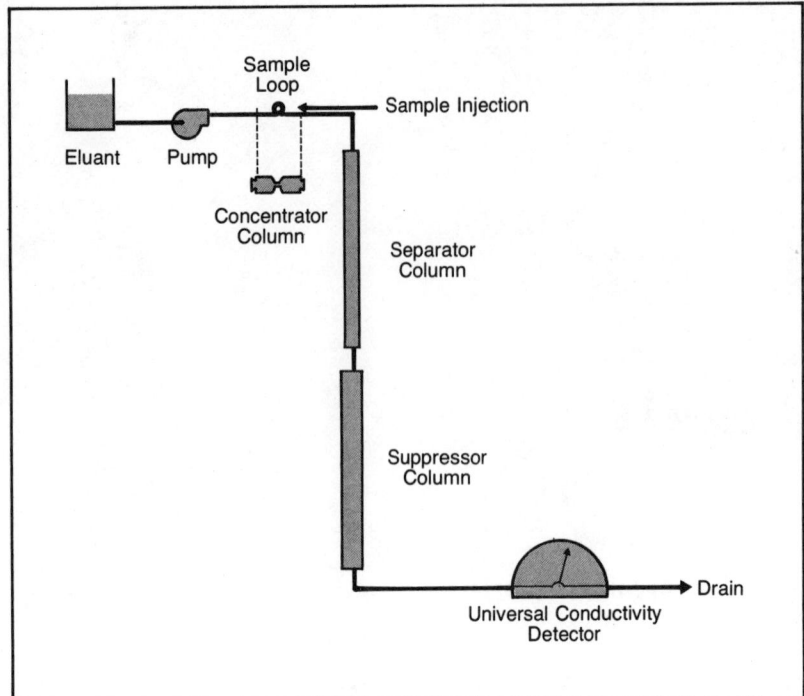

Figure 2. Schematic representation of ion chromatography

condense the steam prior to sample collection.

Since ion chromatography normally has a detection limit of approximately 100 ppb for common ions (with the standard 100-μL injection loop), it was necessary to improve the detection limit by a factor of 100 to get to the 1–10 ppb range required for steam purity measurements. The detection limit was improved to about 20 ppb simply by increasing the capacity of the 100-μL injection loop to 500 μL (0.5 mL). However, further improvement of the detection limit necessitated replacement of the injection loop by a concentrator column, which is now available commercially. A 10-mL sample of steam condensate is passed through the concentrator column, which contains an ion exchange resin. All of the cations or anions (depending on concentrator resin type) are retained on the concentrator column, but the volume of water retained, 0.5 mL, is 5% of the input volume. The concentrator column contents are thus concentrated by a factor of 20, decreasing the detection limit to 1 ppb.

For those unfamiliar with the ion chromatographic process, a schematic is shown in Figure 2. A sample is injected into an eluant stream, pumped through a separating column where the sample species are resolved by interaction with an ion exchange resin, and then pumped through a suppressor column where the ions of the eluant are suppressed. The species of interest are then detected by a conductivity cell. Either cation or anion analyses can be performed by utilizing separator and suppressor columns suitable for one or the other.

In an anion exchange process, the eluant is often a solution of $NaHCO_3$. The anions in the injected sample interact with the resin in the separator column: Resin-$N^+HCO_3^-$ + Na^+X^- $\rightleftharpoons$ Resin-N^+X^- + $Na^+HCO_3^-$, where X^- is an analyte anion. As the analyte anions flow through the separator column in a background of HCO_3^-, X^- and HCO_3^- compete for the resin-N^+ exchange sites. Separation of ions depends on their relative affinity for the exchange sites: The greater the affinity for the resin, the longer an anion is retained.

After the separation process takes place, the analyte anions and the eluant enter the suppressor column, where the following reactions take place: Resin-$SO_3^-H^+$ + $Na^+HCO_3^-$ $\rightleftharpoons$ Resin-$SO_3^-Na^+$ + H_2CO_3; and Resin-$SO_3^-H^+$ + Na^+X^- $\rightleftharpoons$ Resin-$SO_3^-Na^+$ + H^+X^-. The analyte ions, which could, for instance, be sulfate and nitrate, thus enter the conductivity detector as strong acids, sulfuric and nitric. The eluant is converted to a solution of carbonic acid, a weak electrolyte. Thus, the analyte anions can be easily detected against the carbonic acid background. Without the suppressor column, the analyte signal would be swamped by the strong conductivity signal from the concentrated bicarbonate eluant.

For anion analyses, the analytical or separator column thus contains a strong anion exchange resin, while the suppressor column contains a strong cation exchange resin in the hydrogen form. For cation analyses, a similar principle operates, except that the eluant is now a solution of a strong acid such as HCl, the analytical column contains a strong cation exchange resin, and the suppressor column contains a strong anion exchange resin in the hydroxide form.

With the concentrator column playing such an important role in sample analysis, it was desirable to qualify it for field use as completely as possible. Following any analysis, a concentrator column is normally still filled with the eluant solution. To test the effect of this prolonged storage, a concentrator column used for anion analysis was stored in eluant (.003 M $NaHCO_3$/.0024 M Na_2CO_3) for two weeks and then rerun. The chromatogram showed surprisingly high chloride and sulfate peaks, 25 ppb and 10 ppb, respectively. It appears that prolonged exposure to eluant leaches ions out of the glass or frees them from deep resin sites. Similar results were also found for cation concentrator columns, where the eluant used was .0075 N HCl.

It would therefore seem prudent to rinse all concentrator columns with fresh eluant prior to use if they have been stored in eluant for any appreciable length of time. Caution is also advised when using a concentrator column for a low-level sample after the column has previously contained a high concentration species (ppm level), e.g., a concentrated boiler blowdown sample. Unless it is thoroughly rinsed with fresh eluant, some contamination could result from ions of the previous sample that are being removed from deep resin sites.

The effect of prolonged storage of water samples on concentrator columns was also studied. The aqueous samples do not show the same leaching effect as eluant storage. The recovery of all species after prolonged storage was found to be statistically similar to normal reproducibility levels.

The efficiency of the concentrator column in removing and retaining all sample ions during loading was investigated. This test was conducted by pumping a 45-mL sample solution containing 15 ppb chloride and 180 ppb sulfate through two concentrator columns connected in series. The chromatogram of the second concentrator column showed no peaks, indicating total removal and retention of all ions on the first column.

The linearity of dilution from the ppm to ppb range of sodium, chloride, or sulfate was checked with progressively diluted standards. And to make certain that no unusual effects were being contributed by the concentrator columns, the linearity of peak height vs. concentration for chloride calibration in the 0–50 ppb range was determined.

The accuracy of sample analyses is based, in part, on the accuracy of standard solution preparation. Water used in the preparation of standards must contain less than 1 ppb of ionic species. Normally, distilled and deionized water contains ions at too high a level to be useful for standards, and it was necessary to fabricate a system that would produce ultrahigh-purity water. The system involved recirculating prefiltered water through an organic removal cartridge, two mixed bed deionizers, a UV sterilizer and a 0.2-μ filter. The system is capable of delivering 1–3 L/min of Type 1 reagent-grade water (<0.06 μS/cm at 25 °C). Water from this system is normally taken to field tests and used for standards preparation, sample equipment rinsing, etc.

During our field tests, the on-site ion chromatographs were always calibrated with standard solutions loaded on concentrator columns. Laboratory and field tests have shown that calibration does not change significantly over a several-day period unless the eluant is changed, which enabled us to minimize the frequency of recalibration.

One of the problems associated with obtaining an accurate chromatographic trace is interference from what is referred to as the "water dip." An in-depth discussion of this phenomenon is beyond the scope of this paper, but briefly, "water dip" refers to a decrease in background eluant conductivity that occurs when the eluted high-purity water sample passes through the conductometric detector. The elution time for this water dip, under our operating conditions, varies as a function of suppressor column resin exhaustion, moving toward earlier times as the suppressor column exhausts.

With a freshly regenerated suppressor column the water dip interferes with the chloride peak, resulting in an anomalously low apparent peak height. As the column exhausts and the water dip migrates away from the chloride peak, the peak appears to grow. At an exhaustion level of ~50% of the suppressor column, the water

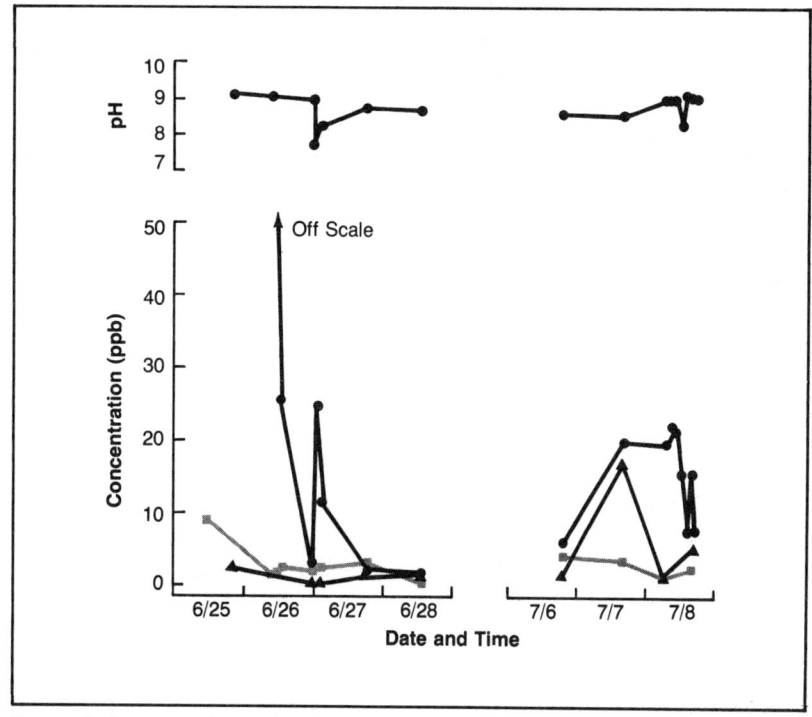

Figure 3. Concentrations of sodium (▲), chloride (●), and sulfate (■), and pH in boiler main steam

suppressor column exhaustion, but reproducibility is better. A comparison of peak heights for 10 runs on an unexhausted anion suppressor column showed a relative standard deviation dip is sufficiently resolved that it no longer has any effect on the chloride peak.

Not only is maximum peak height (and sensitivity) attained after ~50% of 22.0% for chloride, while the relative standard deviation was reduced to an acceptable 12.7% following 50% suppressor column exhaustion.

The effect of suppressor column exhaustion on sulfate analysis is insignificant, with data showing a relative standard deviation of 12.0% and 13.0% for 1% and 50% exhaustion, respectively. The relative standard deviation for seven cation samples showed a reduction to about 13.5% for sodium following an exhaustion of ~10% of the cation suppressor column. Slight exhaustion of the cation suppressor column is necessary, more so for stabilization of the baseline than for peak height changes.

The sample container is probably the greatest potential source of sample contamination, not only in the laboratory, but in the various operation–sampling–analysis procedures of a field test. Polystyrene was determined to be the optimum sample bottle material for this type of sampling.

Another material, polymethylpentene (PMP), is currently undergoing evaluation and shows promise as a viable sample bottle material. If this can be borne out, PMP would be a substantial improvement over the relatively brittle polystyrene material, which easily crazes (temperature effect) and cracks upon shipment to and from sampling sites, resulting in the occasional loss of a valuable sample.

Figure 3 shows the results of graphic analyses of samples collected at one of the sampling locations for a 580-MW plant with a Babcock and Wilcox 2400 psia once-through boiler operating with 1000 °F reheat temperature. The extreme range of concentrations indicates poor control over steam chemistry. The usefulness of on-site ion chromatography was demonstrated here, as our testing confirmed a suspected condenser leak. In addition, the detection of ionic impurities at another sampling location, the polisher header, made plant personnel aware of improper polisher operation.

Elimination of the condenser leak and a change in polisher operation were directly reflected in the analysis of polisher effluent samples following a shutdown and subsequent restart. The concentration of all species was below 5 ppb following these changes.

In contrast, Figure 4 presents four

sample sets of analytical data collected from various locations throughout a plant that had over eight years of corrosion-free operation with good control of the steam chemistry. This plant was a 326-MW Combustion Engineering 2000 psi natural circulation drum boiler operating with 1000 °F reheat temperature. With the exception of a limited number of suspect points, the data from 72 ion chromatographic analyses show concentrations of all species were consistently maintained below 5 ppb. The accuracy and reproducibility of the data were checked by monitoring impurity concentrations at three selected sample locations throughout an entire test. In over 120 separate anion and cation analyses, only one high SO_4^{2-} analysis could be attributed to sample contamination or technique.

Figure 5 is a plot of impurity cleanup for sodium and chloride following a unit startup where load was increased to normal operation over a 4-h period. The first three points at each load level represent analysis of water phase samples and show relatively high ini-

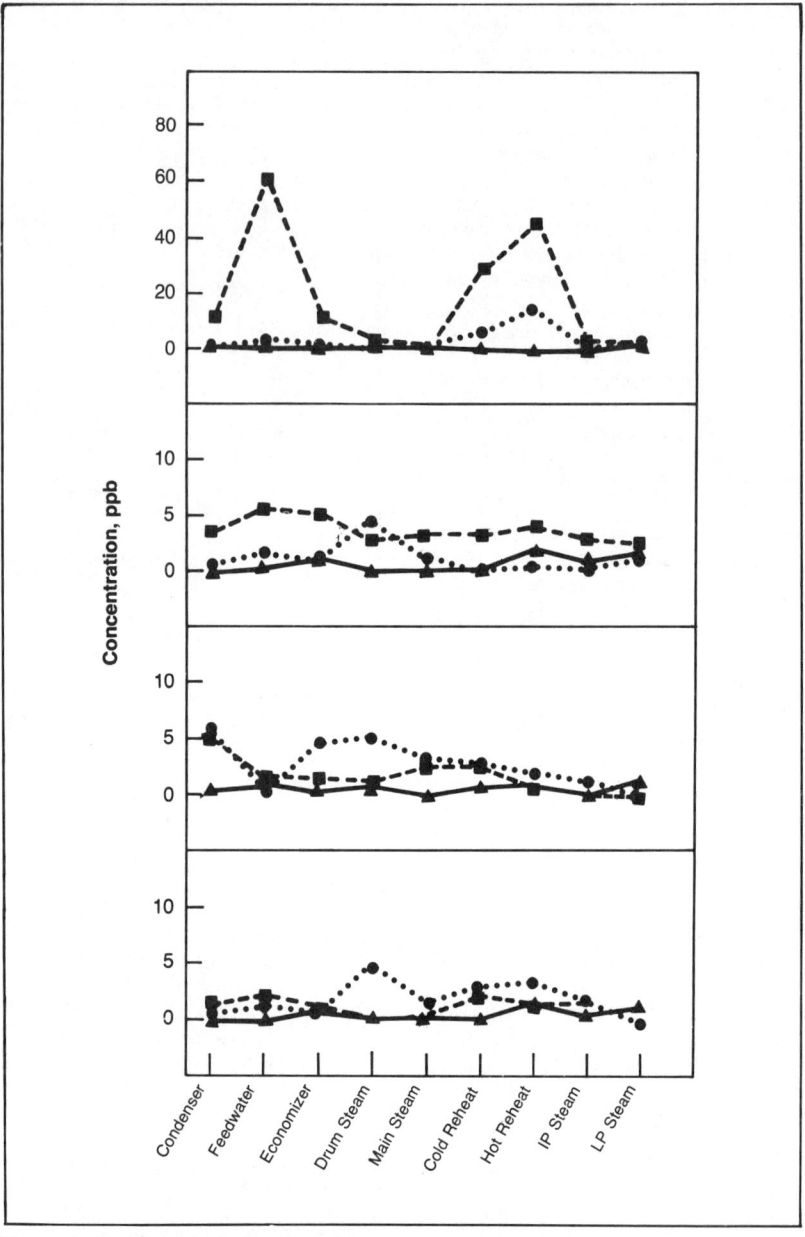

Figure 4. Base load impurity levels throughout water/steam cycle during drum boiler plant test. Na^+ (▲), Cl^- (●), SO_4^{2-} (■)

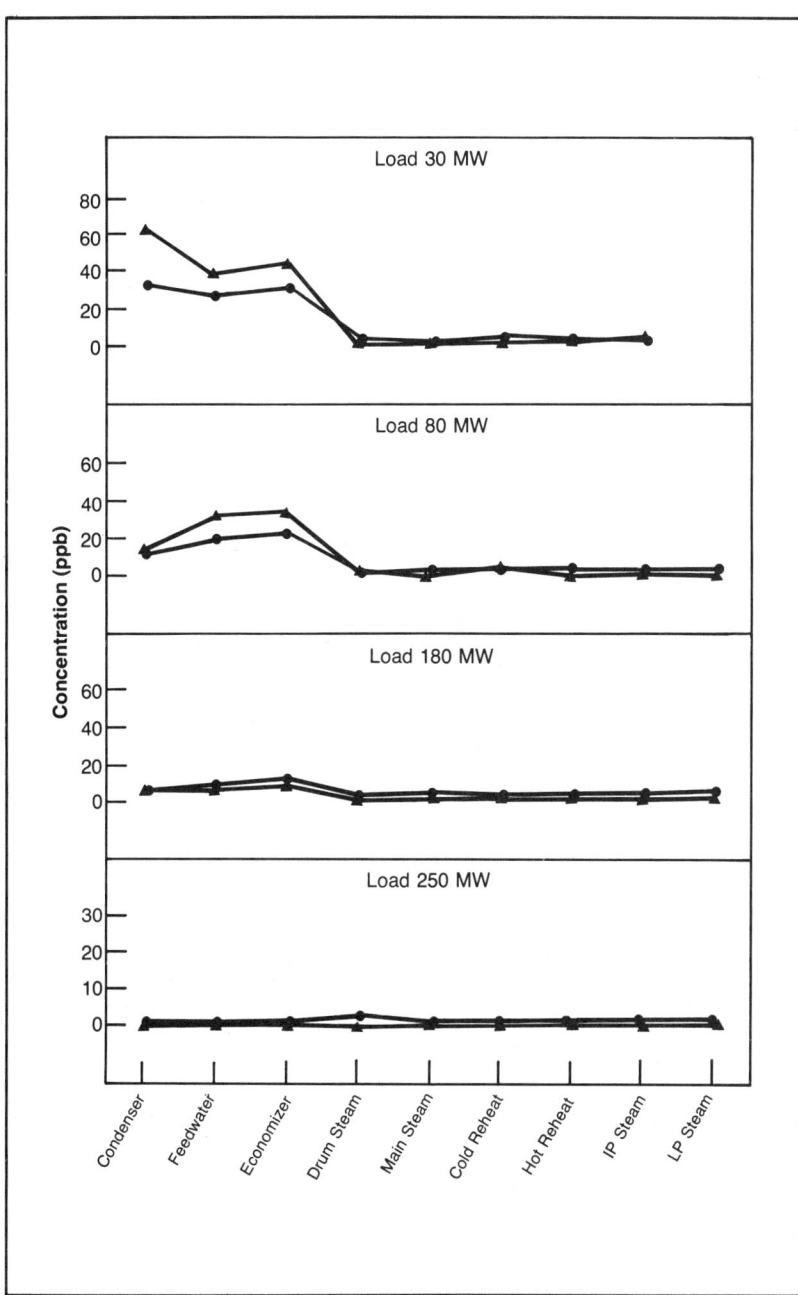

Figure 5. Impurity level cleanup for sodium and chloride during drum boiler plant startup. Na⁺(▲), Cl⁻(●)

tial impurity levels. The remaining points indicate analysis of steam samples and show the effectiveness of the drum in maintaining steam purity levels.

All of these analyses have shown that ion chromatography is a very viable analytical approach for the determination and quantification of ionic impurities in the voluminous steam flow typical of most power plants (2×10^6 to 1×10^7 lb/h). Ion chromatography is a valuable on-site analytical tool, offering plant personnel direct feedback for detecting problems or improper operating conditions in the power production industry.

Acknowledgment

The authors wish to acknowledge the assistance of Steven H. Peterson and James C. Bellows in the steam purity program; the efforts of S. L. Anderson and W. E. Snider in collecting and analyzing samples; and the many helpful discussions with Gerald L. Carlson. Also, the cooperation of Decker Power Plant (Austin, Tex.) personnel is greatly appreciated.

Problem Solving in the Coatings Industry

T. K. Rehfeldt and D. R. Scheuing

Sherwin-Williams Research Center, Chicago, IL 60628

Originally published in ANALYTICAL CHEMISTRY, 1978, Vol. 50, No. 11.

Materials encountered in the coatings industry are often complex mixtures presenting myriad problems for the analyst. The analytical chemist may be asked for a total analysis of one of many types of paint or to characterize a raw material for such a paint. He may be asked to develop a quality control procedure or to conduct one-time only pioneering research. The typical sample is heterogeneous and contains a broad range of chemical species of varying molecular weights. The questions may involve the entire sample or some fraction of the total. Matrix effects are always present and often large. Consequently, separations are almost always necessary, and liquid chromatography is used extensively. Presented here is a general classification of coatings industry problems and an analytical approach to the problems with liquid chromatography as the pivotal method.

The problems are classified according to the fraction of the sample that is of interest. In a total paint, for instance, we may wish to examine the polymer, the pigment, the additives, or process residuals. Other types of samples, such as a raw material or intermediate or environmental material, also may be of interest. In all of these cases, the same classification and analysis sequence shown in the figure is used to approach the problem. This

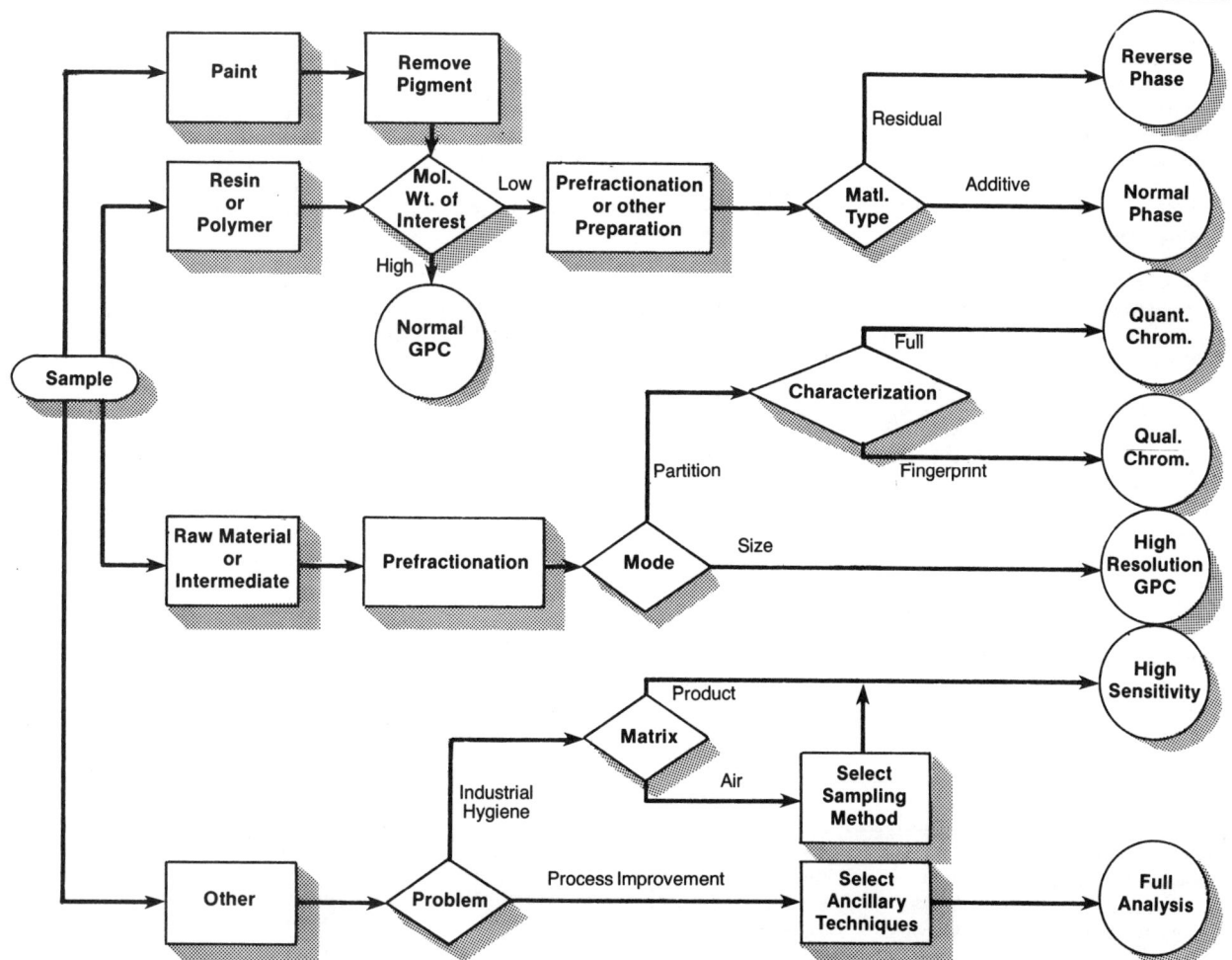

approach is illustrated by the examples which follow.

Process Residuals Problem

An alkyd resin made with linseed fatty acids produced a soft, low-gloss film that required a long drying time. It was suspected that these poor properties were due to the extent of reaction of the conjugated unsaturation of the linseed fatty acids. Accordingly, it was desired to know the quantity of residual fatty acids, i.e., the amount left in the alkyd at the end of the reaction process. Also, the composition of the residual as compared with the input composition of the linseed fatty acids had to be determined.

The quantity of unreacted fatty acids was obtained by a size separation via gel permeation chromatography. Two columns packed with microparticle porous polystyrene material with 100 Å nominal pore size were used. The pore size was small enough so that most of the polymer eluted at the exclusion volume. The length and pore size of the columns provided enough resolution to separate the fatty acids from the resin solvent, but not from each other. Thus, a single chromatographic peak was observed for the fatty acid fraction. This peak material was collected, and the identity confirmed by infrared spectroscopy. The concentration of the fatty acids was determined from a calibration curve produced with known concentrations of the fatty acids.

To determine concentrations of the individual acids in the fatty acid fraction, more fatty acid peak material was collected and further separated on a specialized phase of moderate polarity designed specifically for fatty acid analysis. This was an isocratic separation with a mobile phase containing tetrahydrofuran, acetonitrile, and water. The mobile phase composition was adjusted to optimize the resolution of the individual fatty acids. An isocratic method was chosen because it was necessary to use refractive index detection to measure both the saturated and unsaturated fatty acids. The original raw material also was analyzed in the same way, and its composition compared with that of the collected fraction. The composition of the residual acids was different from the original material. The residuals contained more saturated acids which indicated a preferential esterification of the unsaturated acids into the resin.

As a result of this work, the reaction processing conditions in the plant were changed to increase the extent of incorporation of the saturated fatty acids into the alkyd resin to give a product with the desired properties. In addition, the method developed to separate the fatty acid fraction was used as a quality control test for the alkyd.

Raw Material Color Problem

A coating developed an undesirable yellow color when an aqueous concentrate of a particular raw material, 2-amino-2-hydroxymethyl-1,3-propanediol, was substituted for the crystalline form of the same chemical. There was a cost advantage in using the aqueous concentrate; therefore, it was necessary to analyze the concentrate to discover how to eliminate or minimize the yellowing in the finished product.

The general approach to the analysis involved the comparison of the liquid chromatograms of the liquid concentrate and of the crystalline form of the raw material. A nonpolar column consisting of an 18-carbon aliphatic chain bonded to a silica substrate was used as the immobile phase, and a constant composition mixture of methanol and water was used as the mobile phase. Both refractive index and UV detection were used. A number of components were found in both raw materials, but the aqueous concentrate contained several components not present in the crystalline material. The refractive index detector indicated that the additional components in the concentrate were present at very low levels, and the UV indicated that they were high UV absorbers. Since the only problem in the finished product was the yellow color, these high-UV absorbing components were almost certainly the culprits.

The solution to the problem consisted of using the more expensive crystalline material for the more demanding applications where the yellow color was objectionable, and using the concentrate in dark-color finished products where the yellow color was of no consequence. In addition, a quality assurance test in which the liquid chromatography "fingerprint" of the high-UV absorbing components was used to distinguish an unacceptable from an acceptable raw material was devised. The beauty of the test is that it provides an index of acceptability without the necessity of knowing the nature of the deleterious components sometimes found in the raw materials.

Process Improvement Study

A new resin intermediate product with amine functionality had excellent properties and promised to be a useful component of a variety of new coatings systems. However, the processing time was so long that the economic feasibility of producing the intermediate was in doubt. What was needed were an analysis and characterization of the resin intermediate and a study of its composition as a function of processing time. In this way, the optimum processing time for the desired composition could be found.

Several factors affected the analysis. Molecular weights were expected to be small and of narrow range; therefore, size separation was unlikely to be suitable, but no prefractionation or cleanup step was necessary to remove interference from polymeric material. Since the material was water sensitive, aqueous mobile phases could not be used. This effectively eliminated reverse phase separation schemes. A wide polarity range was expected, but not all suspected components had ultraviolet absorbance. Thus, it was necessary to use both UV and refractive index detection, which prevented the use of a gradient elution method because of the limitation of the refractive index detection. Consideration of these factors led to the selection of a normal phase, isocratic separation with methylene chloride and heptane as the mobile phase. The functionality of the product suggested a column with bonded —NH_2 groups.

The intermediate contained at least 10 components when the mobile phase was adjusted to maximize the resolution of the components. This finding created some new problems. The original material was thought to be a relatively pure product. But did all 10 components behave alike in subsequent use, or was the active agent only one of the 10 components? To answer these questions the component peaks were collected, and the peak materials were examined by the IR microattenuated total reflectance technique. Several of the components were residual starting materials, and a few were by-products. Two of the major components had an infrared spectrum which was consistent with the expected structure of the desired product. In the case of the residual raw materials, identification was confirmed by measurement of the chromatographic retention volume of the pure components. In the case of the presumed by-products and desired product, multiple chromatographic collection yielded enough material to be examined by NMR. The structure information obtained confirmed the identity of the desired product. Two of the major peaks were isomers of the major product, and both isomers were acceptable for subsequent applications of the product. It was also determined that the major by-product would not interfere with end-product use.

After the above characterization work, a set of experiments was conducted to optimize the reaction and scale-up to factory production. The liquid chromatographic method was used to monitor laboratory and pilot plant syntheses. Samples of the reaction mixture were taken at intervals during the synthesis, and it was shown that production of the desired product was sufficiently complete after only 6 h instead of the 18 h previously thought to be necessary. This savings in time was enough to make the product economically viable.

Industrial Hygiene Concern

Of great and increasing concern are the health and safety of products and the production environment. The case in point involves isocyanate monomers. The problem was to adequately control these monomers to protect both the customer and the production personnel. Since the molecular size of isocyanate monomers is significantly smaller than the polymeric intermediates made from these monomers, the size separation mode of liquid chromatography was favored to measure the residual isocyanate in finished products and to monitor the industrial environments. However, the analysis was complicated by the high reactivity of the isocyanate moiety. This problem was overcome by deactivating the functional group with diethylamine. Isocyanates react with amines to form ureas by the following reaction:

$$R'-N=C=O + NHR_2 \rightarrow R'-NH-\overset{\overset{O}{\|}}{C}-NR_2$$

During development of the method, infrared spectroscopy was used to monitor the —NCO moiety of the sample after addition of diethylamine. The —NCO absorbance disappeared within 10 min. The final procedure includes addition of diethylamine, a 10-min wait, and chromatographic analysis on a single size column packed with 100 Å nominal pore size polystyrene gel. Isocyanate prepolymer intermediates and finished products are routinely monitored by this method, and the isocyanate monomer levels are strictly controlled.

The safety of our production personnel during processing of isocyanate materials naturally also was of concern. To safeguard these employees, the airborne isocyanate monomer concentrations had to be controlled. Because gas chromatographic methods were not general enough for our purposes and thin-layer chromatography was too subjective, we turned to liquid chromatography. In this instance, advantage was taken of the high reactivity to facilitate collection of the airborne material. Air was drawn through midget impingers containing a nitroaromatic derivatizing reagent [as described by Dunlap et al. (1)] that rapidly forms urea derivatives of the isocyanate groups. A general method for several possible isocyanate monomers was needed; therefore, it was necessary to use gradient elution to adequately resolve the components of interest. Since the derivatives were strong UV absorbers at 254 nm, UV detection was used. The column was silica, and the solvent was a methylene chloride to 10% isopropanol gradient in 20 min. This procedure completely resolved the four isocyanates of interest with a detection limit of 5 $\mu g/m^3$ based upon a 20-L air sample.

This method was subsequently used to measure airborne isocyanate concentrations around resin processing kettles, paint mixing tanks, storage tanks, and during spray gun application of finished coatings.

Literature Cited

(1) K. L. Dunlap, R. L. Sandridge, and Jürgen Keller, *Anal. Chem.*, 48 (3), 497 (1976).

Refinery Crisis Solved

S. A. Schmidt and V. F. Gaylor

Standard Oil Co., Cleveland, OH 44128

Originally published in ANALYTICAL CHEMISTRY, 1976, Vol. 48, No. 12.

It began with an innocent-sounding telephone call. "Please attend a Corporate Engineering meeting to discuss a refinery corrosion problem." That meeting was the beginning of a crash analytical effort that involved nearly all the R&D analytical groups.

The crisis occurred during the beginning of crude oil and energy shortages when our refineries had to operate with whatever crude oils the Purchasing Department could supply. And they had to *keep* operating to supply the gasoline-hungry public. But one of our refineries was in trouble. Abnormally high corrosion rates were being measured in the naphtha handling system just downstream from the crude oil distillation tower. High corrosion rates were accompanied by solids buildup in a crucial part of the system. Both problems were so alarming that shutdown of the crude oil distillation tower was a very real threat. As an additional complication, the light crude naphtha was reported to be high in total chlorine content (10–100 ppm), and some inorganic chlorine-containing compounds can easily damage postdistillation naphtha refining facilities. The refinery engineers were convinced that all of their troubles were chlorine derived!

Diagnostic Plan

Figure 1 is a simplified schematic of the part of the refinery directly involved in this crisis. Crude oils delivered to refineries contain significant quantities of water and hydrolyzable chloride salts. Most of the water and salts are electrostatically removed in the desalter. Trace quantities of water and chloride salts are, however, present in the feed to the crude oil distillation tower and can generate HCl during distillation. An amine corrosion inhibitor is therefore injected into the light hydrocarbon overhead to neutralize the corrosive acid. The amine–HCl salt is subsequently removed in the water knockout tank in the water phase. The acid-free hydrocarbon stream is then fed to the naphtha splitter, where the crude naphtha is separated into light and heavy naphtha fractions.

The crisis centered around the naphtha splitter reboiler furnace. Tubes in the furnace were plugging (high-pressure readings), corrosion meters indicated a corrosion rate of $\frac{1}{16}$ in. per year, and crude naphtha from the splitter was reported to be high in total chlorine.

In cooperation with Corporate Engineering, we outlined possible reasons

for the three refinery problems (Figure 2) and planned the analytical approaches accordingly. Both chlorine- and sulfur-containing compounds can corrode refinery equipment. Therefore, our analyses included both types of compounds.

Was It Chlorine Corrosion?

The initial work listed in Table I seemed to eliminate chloride salts as the cause of corrosion. The desalter was working properly; we found no unusually high levels of Cl^- and no unusual metals in the desalted crude oil. Furthermore, no unusual levels of HCl were generated on distilling the crude oil in the lab. Acidity level of the undistilled crude oil was also normal.

Chlorinated hydrocarbons are sometimes used for secondary crude oil recovery from wells, and the presence of these materials could account for total chlorine found in the naphtha. But we could find no chlorinated hydrocarbons by polarography.

Was the total chlorine determination valid? That was a question asked early, and the answer turned out to be "No". The refinery lab was using a nonstandard, combustion/microcoulometric method (Dohrmann) for total chlorine. With the refinery's procedure, we confirmed apparent high concentrations of chlorine in the naphtha. But further experimentation showed that high levels of sulfur interfere with this chlorine method. After reducing total sulfur concentration by precipitating H_2S and RSH with Cd^{2+}, the apparent total chlorine concentration dropped to insignificant levels.

There was no "high chlorine" problem. It was analytical error—a false trail for diagnosing the corrosion problem and due to sulfur interference with the Dohrmann chlorine measurement.

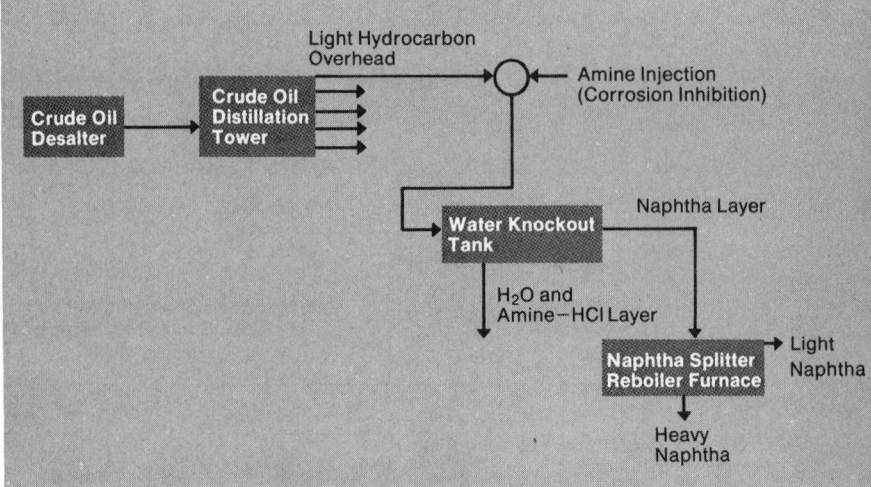

Figure 1. Block diagram of portion of refinery involved in crisis

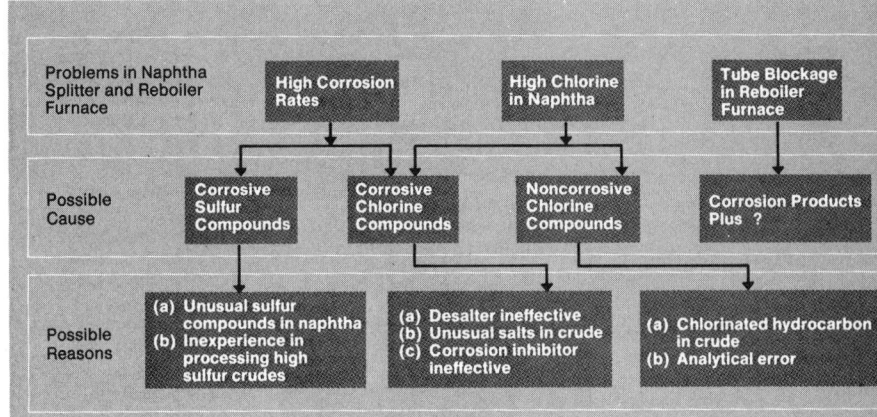

Figure 2. Possible causes of refinery crisis

Table I. Analysis of Desalted Crude and Crude Naphtha Streams for Chlorine- and Sulfur-Containing Compounds

Method	Results	
	Desalted crude	Crude naphtha
CHLORINE		
Titrimetry and selective ion electrode	Normal Cl^- levels. Desalter effective	
Emission spectrography	No unusual metal salts present	
Titrimetry and polarography	Normal acidity levels	No acidity detected
HCl evolution test	Normal HCl levels generated	
Polarography		No chlorinated hydrocarbon detected
Dohrmann microcoulometry on untreated samples		Apparent high total chlorine
Dohrmann microcoulometry after RSH and S^{2-} removal		Total chlorine insignificant
SULFUR		
X-ray fluorescence spectroscopy	High total sulfur	High total sulfur
Titrimetry	High H_2S and RSH	High H_2S and RSH
Corrosive sulfur evolution test	High levels of corrosive sulfur generated on heating	No corrosive sulfur generated on heating. (No unusual sulfur compounds present)

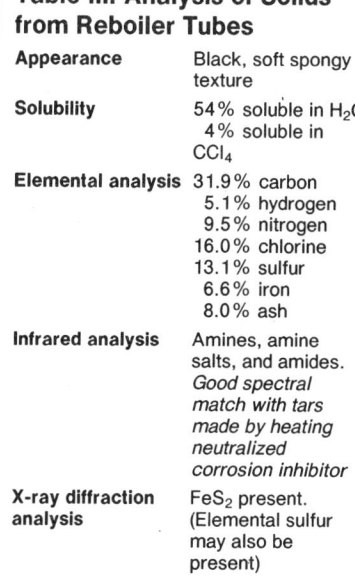

Table II. Analysis of Corrosion Inhibitor

IR analysis	GC–MS analysis	Neutralized with HCl	recovered salts
Complex mixture of cyclic amines in aromatic solvent	Approx. equal amounts of five cyclic amines, C_8–C_{12} benzenes, and C_{10}–C_{14} naphthalenes	Partitioned between H_2O–HC	Heated 6 h at 250 °F
		>20% into HC phase	Brownish black tar, contained chlorine, water soluble

Was It Sulfur Corrosion?

There were high levels of corrosive sulfur compounds (H_2S and RSH) in the crude oil, and more were generated on heating (Table I). Therefore, the naphtha also contained high levels of corrosive sulfur compounds. We concluded that a major part of the corrosion problem was due to processing high sulfur crude oil.

What Was Blocking the Reboiler Tubes?

Buildup of corrosion products contributes to tube blockage in the reboiler furnace. However, the engineers were convinced that the pressure buildup rate far exceeded that expected from corrosion rate measurements. The corrosion inhibitor was the major remaining unknown in the whole system. Corporate Engineering had no prior experience with the inhibitor in use; it was not used in any other company refinery.

Table III. Analysis of Solids from Reboiler Tubes

Appearance	Black, soft spongy texture
Solubility	54% soluble in H_2O 4% soluble in CCl_4
Elemental analysis	31.9% carbon 5.1% hydrogen 9.5% nitrogen 16.0% chlorine 13.1% sulfur 6.6% iron 8.0% ash
Infrared analysis	Amines, amine salts, and amides. *Good spectral match with tars made by heating neutralized corrosion inhibitor*
X-ray diffraction analysis	FeS_2 present. (Elemental sulfur may also be present)

We characterized the inhibitor by the tests summarized in Table II. The inhibitor was a very complex mixture of cyclic amines and aromatic hydrocarbons. Partitioning of the inhibitor–HCl salts between water and hydrocarbon was unfavorable; this partitioning experiment suggested that all the amine salts were not removed in the water knockout tank (Figure 1). Some amine salts were carried into the furnace reboiler. And the thermal stability test (Table II) showed that the amine salts would indeed produce tars on heating in the tubes.

We concluded that tube blockage was caused by the corrosion inhibitor and recommended the refinery change to a company-proven inhibitor.

Epilogue

A brief shutdown of the naphtha splitter provided an ultimate test of our diagnoses. Refinery management concluded the partially plugged reboiler furnace had to be cleaned before switching to the recommended corrosion inhibitor. Engineers responsible for the cleaning job found voluminous quantities of a black, spongy solid in the furnace tubes and retrieved a representative sample for analysis. It was a most unusual "plant crud" sample (Table III) with an ash content of only 8% and a high water solubility of 54%! FeS_2 and maybe elemental sulfur were present, confirming our diagnosis of sulfur-caused corrosion. Large quantities of nitrogen and chlorine, as well as carbon and hydrogen, pointed to HCl salts of the corrosion inhibitor. This was nicely confirmed by infrared analysis which identified amines, amine salts, and amides. And the spectral match with laboratory prepared tars from the neutralized corrosion inhibitor was excellent (Figure 3).

Three Problems—Three Solutions

High chlorine problem: Analytical error

Corrosion problem: Corrosive sulfur compounds from crude oil

Furnace tube blockage: Caused by corrosion inhibitor

Acknowledgment

The authors are grateful to T. E. Andrews, N. R. Anthony, P. A. Budinger, L. M. Klimas, J. E. Sucher, M. S. Vigler, and R. Zuback whose analytical work helped us solve this problem, and to Gary Greeves, the engineer member of the troubleshooting team and an active participant in solving the problem.

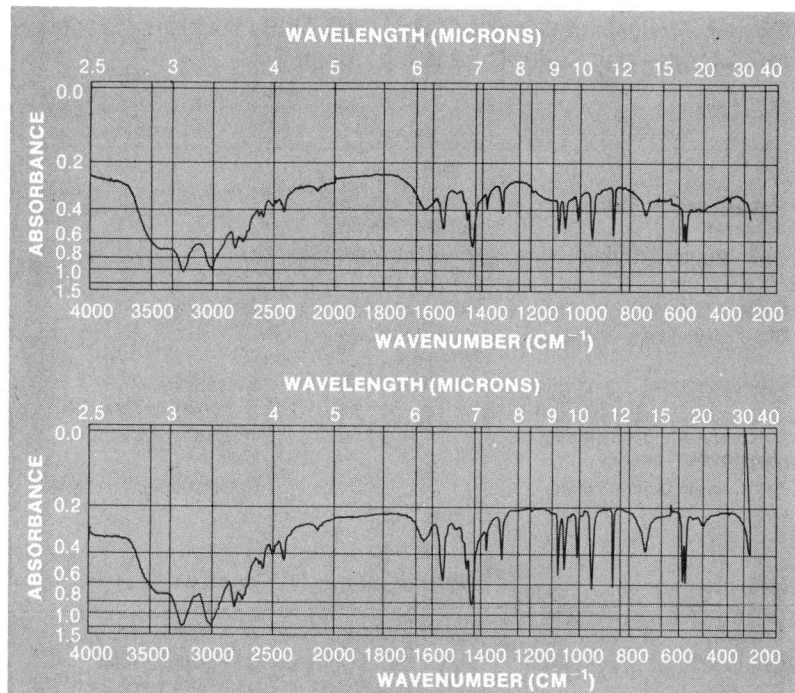

Figure 3. Infrared spectra of: upper: solids from naphtha reboiler furnace tubes; lower: tars formed on heating HCl salts of corrosion inhibitor

Cross-Section of Analytical Problem Solving in the Home Appliance Industry

J. L. Wuepper

Whirlpool Corp., Research and Engineering Center, Benton Harbor, MI 49022

Originally published in ANALYTICAL CHEMISTRY, 1977, Vol. 49, No. 11.

Like other consumer-oriented manufacturing industries, the appliance industry presents a bewildering array of situations where the analytical approach is actively involved. Typically, the analytical chemist can be involved in areas dealing with the environment, product quality, crisis problem solving, materials characterization, product liability, and regulatory considerations. In most cases, the problem rather than the analytical methodology is paramount. In these instances, analytical expertise is called upon because it can provide reliable information on which decisions can be based and because it can transform speculation into justified opinion.

I have chosen to discuss the solutions to several problems which contained that certain element of analytical elucidation that seemed to clear up all the mystery: leaf necrosis, resistor corrosion, polymer volatile fingerprinting, and a metal plating problem.

Leaf Necrosis, An Environmental Problem

In the late summer of 1975, we received various kinds of leaf samples for analysis. These samples were submitted by an environmental engineer from one of our operating divisions. Several plant species were represented, and all had been obtained near a metal finishing solid waste disposal site about 700 × 300 × 4 ft. The solid waste consisted of insoluble metal hydroxides, silicates, sulfates, carbonates, etc., that constitute the solid product from a modern industrial wastewater treatment facility. The leaves showed evidence of marginal necrosis as shown in the cottonwood leaf comparison in Figure 1. Concern was expressed over the fate of the trees and other vegetation in the area. In the interest of preserving the environment and preventing a similar problem in the future, our analytical group was requested to investigate the phenomenon and determine the vegetation distress cause if possible. Pre-

Figure 1. Comparison of healthy and unhealthy cottonwood leaves
Left: Healthy leaf. Right: Unhealthy leaf showing marginal necrosis

liminary investigation showed that air pollution or defoliants were not involved.

As a first approach, the leaves were dried at 110 °C, pulverized in plastic vials with a Wig-L-Bug, and the resulting powdered samples were analyzed by DC arc emission spectroscopy. Several elements were observed in the spectrum, but only the presence of boron seemed unusual. At this point no definite conclusions were drawn because several species of leaves had been compared, and we could not be sure that the differences in observed boron concentrations in the leaves were due to comparison of dissimilar species or if the presence of boron was actually implicated in the phytotoxicity. Work was halted for lack of good samples and the approaching winter.

In the spring and summer of 1976, we resumed our work with fresh leaf growth from which we chose samples.

Emission spectroscopy and atomic absorption were used to study a variety of leaves, ground water, soil, and solid waste samples. This time we were careful to limit our comparisons to healthy leaves and unhealthy leaves of the same plant species. Emission spectroscopic analysis of these leaves showed striking and repeatable differences in boron concentrations as shown in Table I. A literature search (1, 2) indicated that while boron is an essential element for plants, toxic effects—including the telltale marginal necrosis on old leaves—are observed even in tolerant species when the boron concentration in ground water is above 5 ppm. Ground water in the area contained as much as 180 ppm boron, which is very unusual for a geographic region that is not rich in borax deposits. Analysis of the solid waste from the metal finishing waste lagoon showed that it contained boron in sufficient quantity (0.1–1%) to cause high

Table I. Boron Concentration in Healthy and Unhealthy Plant Leaves

	Boron concn, ppm	
Plant	Healthy leaves	Unhealthy leaves
Cottonwood	<50	1200
Horsetail (snakeweed)	<50	8000
Staghorn sumac	<50	200
Raspberry	<50	500
Milkweed	NA[a]	~2000[b] ~500[c]

[a] Not available. [b] Outer leaf. [c] Inner leaf.

ground water levels due to rainwater or surface water leaching. Eventually, we were able to trace the source of the soluble boron material all the way back to the manufacturing process from which it originated. After presenting our data to the Environmental Protection Agency, we came to an agreement on how best to cure the problem by closing off the disposal site. In addition, future disposal and wastewater treatment plans will profit from this work.

Resistor Corrosion Failure Explained

A recent problem involved resistor failure in an appliance control. The resistor was of the wire-wound type in which nichrome wire of appropriate diameter and length is wound onto a ceramic core and then covered with insulation. Resistance measurements on failed resistors showed infinite resistance, suggesting discontinuity in the wire. The approximate point of discontinuity was located with a volt-ohm-milliammeter, and the area was examined by scanning electron microscopy (SEM) and x-ray energy spectrometry (XES). A photomicrograph of the point of discontinuity is shown in Figure 2. The remarkable irregular character of the break suggested a corrosion or metal dissolution problem. X-ray multielemental analysis of the corroded and uncorroded areas produced the spectra shown in Figures 3A and 3B. The elements found included aluminum silicon, chlorine, sulfur, chromium, and nickel. Only the presence of sulfur and chlorine could not be explained as originating from either the nichrome wire or residual insulation. Also, the presence of these two elements in the failed areas and their absence in other areas implicated them as the active corrosive agents.

These findings led to the elimination of chlorine and sulfur from the manufacturing materials and prompted improved resistor design.

Techniques that compete with the SEM–XES approach include the laser microprobe and the electron microprobe. The resistor problem would not have been solved if the laser microprobe had been used since chlorine and sulfur would not have been detected. The electron microprobe would have detected the chloride and sulfur, but the photomicrograph quality would not have been as good. Thus, the adequacy and limitations of competing analytical tools must be realized in selecting a particular approach and in interpreting the data.

Fingerprinting Polymer Volatiles for Source Identification

In any end-use manufacturing operation, a given polymer type is often obtained from several different suppliers. If it then becomes necessary to identify finished parts due to quality differences, it may not be enough to know, for example, that the part is made of polypropylene. It may also be important to identify the polymer source. If any time has elapsed since manufacture of the part, chances are excellent that the supplier identity has been lost. Since materials specifications can be the same regardless of supplier, the polymers may be remarkably similar with respect to infrared spectra, melt flow, molecular weight distribution, hexane solubles, stabilizer package, filler, rheology, and processing characteristics. Nevertheless, performance characteristics may still differ.

In attempting to identify the source of a polypropylene part whose history had been lost, the similarity of polypropylenes from various suppliers was rapidly discovered. The fact that polymer manufacturing processes differ with respect to solvents and catalysts suggested that evidence of those differences might still be contained in the various polypropylenes in the form of trace solvents or differences in propylene oligomers. One-gram samples of identifiable unprocessed polypropylenes were placed in 100-mL bottles and covered with crimped foil-lined septa. Samples from five different suppliers of the polypropylene part were heated to 170 °C, and 1-mL samples of the headspace gases were chromatographed isothermally at 100 °C on 10% Apiezon L. The differences in the chromatograms shown in Figure 4 were obvious and sufficient to serve as an identification technique. A chromatogram of headspace vapors from the unknown provided a good match with that of one of the standards, thus identifying the unknown polymer supplier.

Metal Plating Problem "Solved"

Our analytical group was recently requested to help solve a problem that was occurring in a nickel plating bath. The steel parts to be plated are immersed via conveyer into a series of baths. They first enter a cleaner bath comprised mainly of a metasilicate solution. They then pass through a series of neutralizing and rinsing baths and finally into the nickel plating bath. The problem was that incompletely plated, splotchy irregularities showed in the finished parts. Close examination of the parts, as well as the

Figure 2. Photomicrograph of failure point on wire-wound resistor. Magnification 1000×

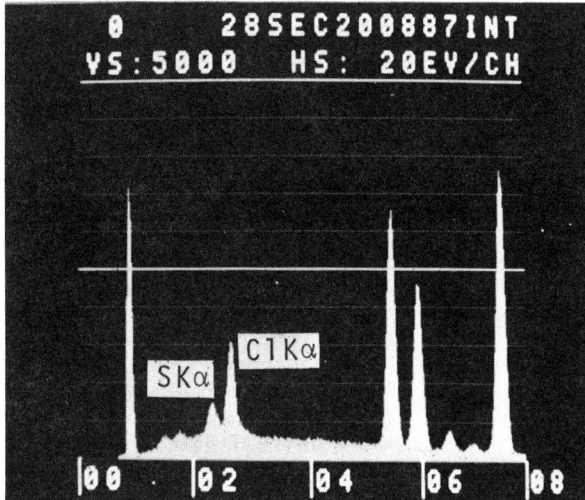

Figure 3A. X-ray spectrum of failure point on wire-wound resistor

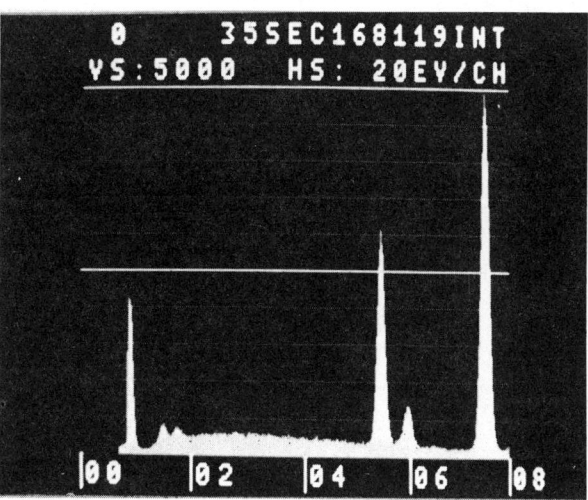

Figure 3B. X-ray spectrum of uncorroded area on wire

nickel plating bath surface, showed the presence of a foreign oily sludge that attached itself to the parts and prevented a good plate from forming.

Chloroform extraction of the foreign solids followed by infrared analysis showed about 15% oil which was traced to carry-over from the initial cleaner stage. The solid portion of the material was more difficult to trace. An infrared spectrum of the washed and dried sludge solids showed the absence of organic absorption bands and the possibility of a fluosilicate-containing material. The emission spectrum showed strong silicon lines and several metals. Emission and infrared spectra of all the materials used in every phase of the nickel plating operation were obtained, but none matched that of the foreign substance found in the plating bath which appeared to be a mixture of fluosilicate salts.

At this point two water solutions were prepared. One contained the salts used in the cleaning tank, and the other all the salts used in the nickel plating bath. When the two solutions were mixed, an insoluble precipitate formed. The infrared and emission spectra of the precipitate were very close to that of the substance which had been obtained from the production line plating bath. It was immediately obvious that the sludge solids in the plating bath were the result of reaction between metasilicate cleaner carried over with the parts from the cleaning bath, metal cations, and a fluoride brightener in the nickel plating bath. Further wet chemical and infrared analysis confirmed the identity of the mixed metal fluosilicate salt sludge.

A proposed reaction for the sludge formation was described as follows:

$$SiO_3^{2-} + 6F^- + 3H_2O$$

Carry-over from cleaning bath Brightener from plating bath

$$\rightarrow SiF_6^{2-} + 6OH^-$$

$$Oil + SiF_6^{2-} + M^{2+} \rightarrow \frac{M(SiF_6)\text{--oil}}{\text{sludge}}$$

from cleaning bath Various metal cations present in system

Once the sludge formation chemistry was understood, steps were taken to prevent its formation. A surfactant was added to the plating bath to keep oil in solution and allow trapped solids to drop to the bottom. Good rinsing of parts was ensured to minimize carry-over of cleaner. Production of quality parts was maintained, and the problem never returned.

Acknowledgment

Contributions from my colleagues at the Whirlpool manufacturing divisions and the Whirlpool Research and Engineering Center, especially Dick Matthias and Bill Powe, are gratefully acknowledged.

References

(1) "Water Quality Criteria", California State Water Resources Control Board, Publication 3-A, 1963.
(2) "Determination of Maximum Permissible Levels of Selected Chemicals That Exert Toxic Effects on Plants of Economic Importance in Illinois", Illinois Institute for Environmental Quality, PB-237-654.

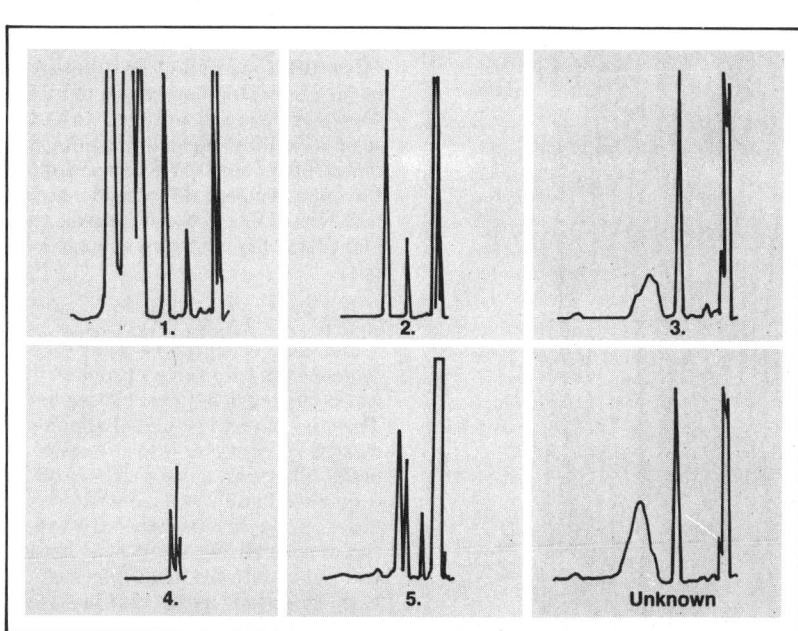

Figure 4. Headspace analysis of hot polypropylenes (170 °C) from different suppliers and unknown

Products

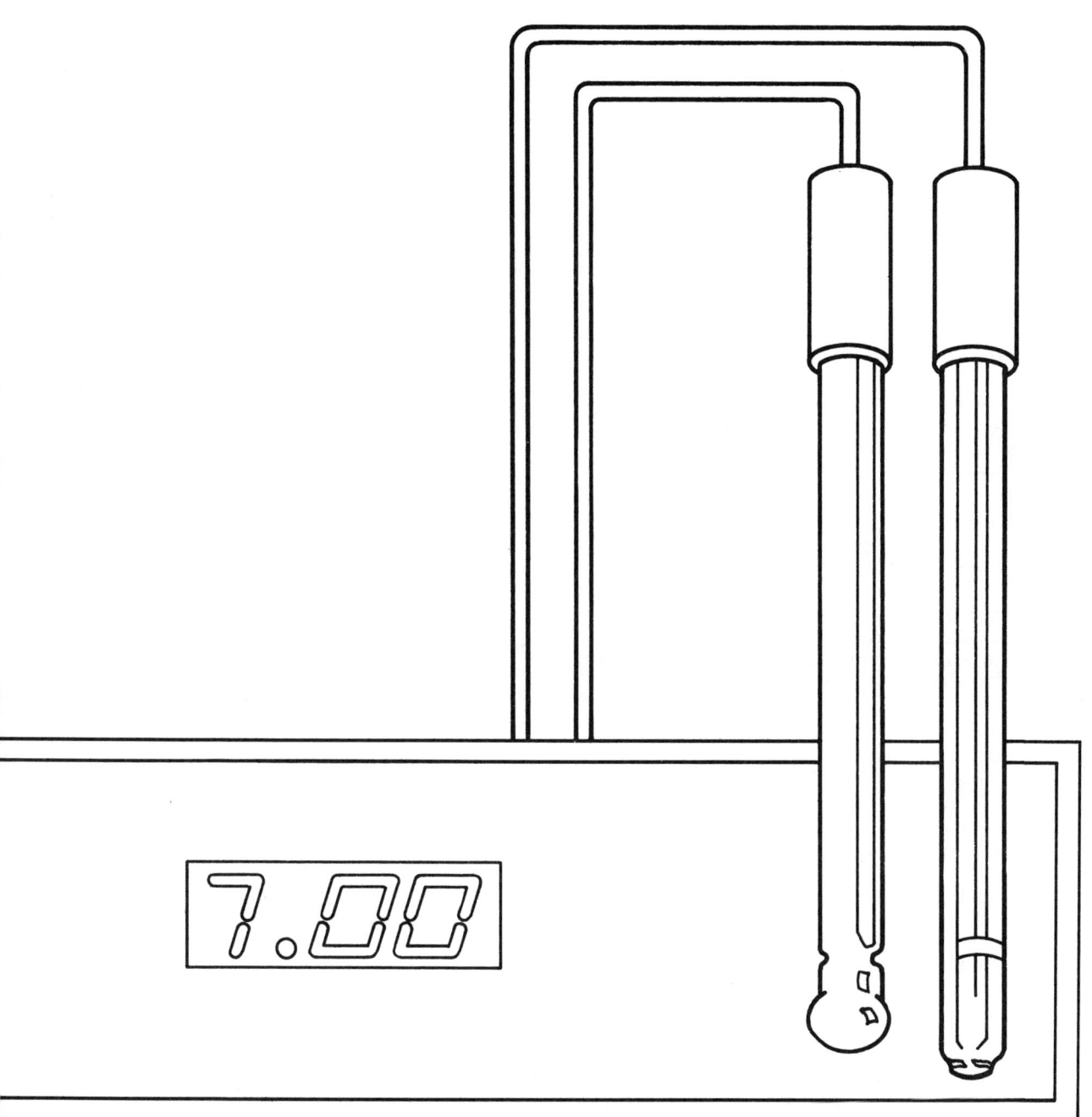

The role of the analytical chemist in product quality assurance and control begins with the initial research and pilot plant stages, and continues throughout the life of a product. The introduction of new raw materials and/or new operating conditions in processes or plants can have a major impact on product quality. The analytical chemist can assist in determining this and in interpreting the results of the analysis. Besides confirming the structure of compounds or the identity of materials, the analyst is constantly asked to answer questions like: "Why is this sample good and this one bad?" "What is the color due to?" "How much additive is present?" "Can you correlate this structural feature to this desirable physical property?" and "Identify my competitor's product!" The versatility of the analytical scientist is evident in the approaches to answering such questions.

The Microchemical Analysis Section at the Kennedy Space Center conducts multidisciplinary failure analysis in a highly instrumented laboratory. In "**Analytical Chemists at the Launch Pad**," a number of examples where analysis was important to the space program are reviewed. An important service is interpreting analytical chemistry to engineers who have little knowledge of this area of science.

As in the case of honey described below, orange juice can be adulterated by the addition of sugar. In "**Analysis of Orange Juice**," the work of a government chemist, often on a fundamental basis, is described as he develops methods to uncover this deception. The use of $^{13}C:^{12}C$ ion ratio mass spectrometry is now being tested to detect this form of fraud.

A mundane problem becomes a detective story when unusually high levels of phosphorus are noted in unleaded gasoline in "**Search for the Elusive Phosphorus Source**." Because phosphorus can damage catalytic mufflers, the source of contamination had to be found. Refinery problems were ruled out. High phosphorus levels were found in pumps with low throughput. Extraction of phosphorus from the rubber hoses was shown to be the problem, and further work was done to set specifications for hose composition to avoid future problems.

Over-the-counter drugs are complex mixtures of multiple ingredients. In "**Over-the-Counter Drugs Analyses with HPLC**," the authors describe the power of high performance liquid chromatography (HPLC) separation and analysis of fractions by mass, IR, UV, and NMR spectroscopy in a series of examples. This methodology is versatile and readily applied in this difficult area of analysis.

Many analytical problems are involved in the analysis of inorganic materials in integrated circuit manufacture. The article "**Analysis of Products Used in the Electronic Industry**" emphasizes the analysis of organic materials, which are finding expanding usage in this important industry. The scope of such work is outlined and several analytical approaches to specific problems are described in detail. The importance of methods that are rapid and effective over a wide range is emphasized.

Adulteration of honey through the addition of low-cost sugar cane and corn-derived syrups is a violation of law, unless the mixture is appropriately labeled. In "**Assuring the Quality of Honey—Is it Honey or Syrup?**" government scientists described two possible approaches to detecting violation. In the first, the adulterant is identified directly, while in the second one the dilution of a component or property known to be always present in honey is detected. The latter method is not useful in practice because of the variability in a range of honeys. A routine method involving isotope ratio mass spectroscopy, looking at the $^{13}C:^{12}C$ ratio has proved to be the best method and it can be applied routinely.

In "**Snap, Crack, Blister and Peel**," the author describes the analytical difficulties in identifying failure mechanisms in electronic and military hardware. In most instances, the type of failure mechanism determines the selection of the analytical technique applied to a problem. Examples of cohesive, electrical, and crazing failure are discussed. The analyst must look for and find the root cause of the problem, and then develop an analytical plan that outlines the sequence and type of separation and analysis technique to be used.

"**Multi-Technique Approach Solves Construction Material Failure Problem**" describes how petrographic microscopy to discover the mechanism of concrete failure can be combined with analytical techniques to ascertain the causative agents. Experienced microscopists–chemist teams can solve 90% of construction material failure problems using the approach suggested. As the examples show, both experience and science are vital elements.

A new one-step, two-shot application of pyrolysis gas chromatography is shown to be an important industrial quality control method in "**Chromatopyrography in Polymer Characterization**." The method received its test by fire when the source and composition of an unlabeled rubber part was needed urgently to maintain delivery schedule on an important contract. Management made an unprecedented visit to the laboratory to emphasize the need for rapid response. Fortunately, the scientific staff had done the basic work to allow a rapid identification. Further examples of use of the method are presented.

In "**Analysis of Liquid Crystal Mixtures**," LC/MS methods are shown to be powerful means for identifying complex mixtures of electro-optical materials used to form liquid crystals for specific applications. Library spectral searches are an additional important element in the instances discussed. Future applications involving complex dyes in the liquid crystal mixture are shown to be feasible.

Synthetic rubbers and plastics are complex mixtures varying widely in composition, molecular weight, polarity, and additive content. "**Analysis of Rubber and Plastic Chemicals by LC/Spectroscopy**" features the use of LC/FDMS (field desorption mass spectrometry) for the exact identification of individual compounds. The major limitation of the method is that a small percentage of compounds are poor desorbers. Examples are also shown where ATR and NMR spectroscopy can be successfully used with LC for important analyses. Microgram LC fractions are sufficient in many cases.

Analytical Chemists at the Launch Pad

Helein D. Bennett

National Aeronautics and Space Administration, Microchemical Analysis Section, Kennedy Space Center, FL 32899

Originally published in ANALYTICAL CHEMISTRY, 1977, Vol. 49, No. 3.

Are your analytical samples small like the point of a pin or large like the side of the barn? Do you use statistical sampling, or might your method be described as "catch as catch can"? Are your analytical procedures well described in standard references, or must you adapt or devise methods to fit the problem? All these questions can be answered "yes" by the chemists at a unique, "on the spot", and sometimes "last minute" analytical facility at the nation's spaceport where the side of the barn is more likely to be the side of a rocket and statistical sampling and standard methods are the exception rather than the rule.

The Microchemical Analysis Section at Kennedy Space Center provides nonroutine chemical analysis to all the Center activities and, on request, to other NASA Centers, other federal government agencies, and state and local government agencies. Since the types of problems encountered and the material to be analyzed can be literally anything and the time available for analysis is sometimes extremely short, the laboratory is highly instrumented. Optical microscopy is used frequently during analysis of solids and liquids and sometimes is the sole or major technique used. The general analytical approach is illustrated in Figure 1.

In determining the course of analysis, there are several additional considerations that are not easily illustrated but may indicate a path different from that in Figure 1. These considerations, a series of questions, include: Where did the sample come from? What is it likely to be? What is nearby or in the same system? What was happening when someone decided an analysis was needed (funny odor, peculiar appearance, etc.)? Why is this analysis needed (safety, materials compatibility, hardware malfunction, etc.)? Is it like some sample analyzed previously? The answers to these questions may make it possible to use one or two analytical techniques and a minimum length of time to confirm a good deduction. Limited availability of the sample is one other consideration that will dictate a path involving nondestructive methods initially, followed by whichever destructive method will give as much of the information requested as possible.

Many of the analytical problems presented to the laboratory are associated with the identification, and sometimes the quantitative analysis, of materials involved in failures of spacecrafts, launch vehicles, or ground support systems at the Center. This often requires a multidisciplinary approach involving not only many of the techniques used in this laboratory but also analytical work by two other groups: the Malfunction Investigation Staff in the areas of metallurgical, mechanical, and electronic or electrical failures and the Materials Testing Branch in the areas of properties of materials, oxygen compatibility, and environmental testing.

Although the laboratory personnel are occasionally asked to perform analyses at the launch pad or to take samples directly from the pads and other areas, most materials are carried into the Microchemical Analysis Laboratory by the requester (usually an engineer) who may not know which measurements are needed and what they may mean. Consequently, one of the most important services of the laboratory is interpreting analytical chemistry for the customer.

Sample Size and Type

Samples arrive in all kinds of shapes, sizes, and containers. Large samples have included an Apollo 12 liquid hydrogen tank nearly 3 ft in diameter which was examined by transmission electron microscopy, electron diffraction, and electron microprobe x-ray analysis to determine the cause of a leak in the solid-state weld between the titanium alloy tank and the stainless steel inlet/outlet port (1). Also, a spacecraft lunar adapter (the cone around the lunar module during launch), measuring some 6.4 m high and 7.9 m in diameter at its base, was tested with pH paper to determine that all traces of a hypergolic oxidant (nitrogen tetroxide) propellant spill had been removed.

Examples of small samples, the type most often analyzed by the laboratory, are many. Particles on millipore filters have frequently been analyzed to determine the identity of contaminants in spacecraft fluid systems. Particles in the threads of screw-type fittings and on the pins or sockets of plug-type fittings are also frequent objects for analysis. Most gas and liquid samples arrive in standard containers, but in unusual circumstances almost anything may be used. On one occasion, when a nitrogen system was being disassembled, an engineer unexpectedly found some liquid in supposedly dry stainless steel tubing and grabbed the nearest container, an ashtray containing a few ashes, to catch the liquid for analysis. The liquid was identified as water; the ashes were not analyzed.

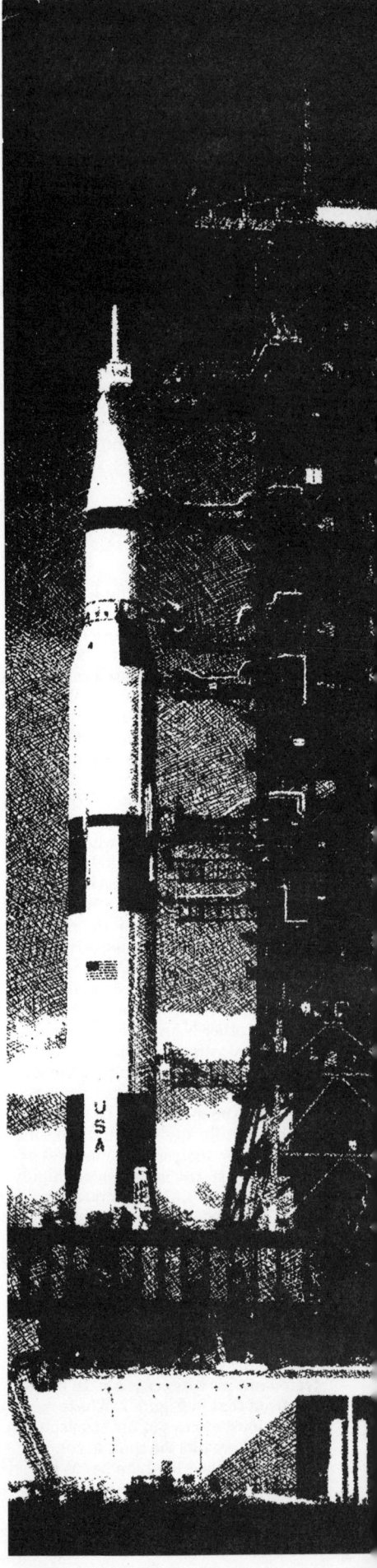

The following problems illustrate the analytical approach used at Kennedy Space Center.

Is There Enough Antioxidant in the Spacecraft Coolant?

During the Apollo program, ethylene glycol–water solutions were used as heat exchange media in the command module and lunar module environmental control systems and in the service module fuel cell system. The compositions of two of these solutions are shown in Table I.

Shortly before Apollo 11 was launched, thousands of tiny crystals were found in the ethylene glycol–water coolant of the lunar module, although few had been found in preceding Apollo missions. The crystals were identified as 2,2′-dithiobisbenzothiazole (Compound I) by the Microchemical Analysis Section using infrared spectroscopy, mass spectrometry, and x-ray diffraction. Compound I is the oxidation product of sodium mercaptobenzothiazole (Compound II) which functions as an antioxidant in the glycol solution.

Upon learning that sodium sulfite had been added as an antioxidant during the preparation of the 50% sodium mercaptobenzothiazole solutions used in previous Apollos, Johnson Space Center decided that sufficient sodium sulfite to give a final concentration of 3–6 mg/L should be added to the glycol solution on hand at Kennedy Space Center for use in the first spacecraft to land on the moon. The Microchemical Analysis Section was asked to confirm that "enough" sodium sulfite was present in the final solution to prevent the oxidation of the inhibitor to the resultant precipitate.

Several potential colorimetric and polarographic methods of analysis were considered and discarded because of erratic results. Initial attempts to use conventional oxidation–reduction titration failed because a voluminous white precipitate preceded and obscured the normal starch end point of the iodometric titration. When the precipitate was identified as 2,2′-dithiobisbenzothiazole, the titration procedure was modified to use a turbidimetric end point detected by a colorimetric accessory to a potentiometric recording titrator (2). A blank was obtained by bubbling nitrogen through the acidified solution to remove the liberated sulfur dioxide. The analytical results were of satisfactory precision, having a relative standard deviation of about 5%. Perhaps more important, the concentration determined in the on-board fluid was about 4 mg/L, a level judged to be adequate for the lunar landing mission. The successful flight of the Eagle in July 1969 ("Tranquility Base here. The Eagle has landed.") proved that judgment, along with thousands of others, to be correct.

What Are These Yellow-Green Crystals?

Identifying the precipitate and determining the sulfite concentration level of the lunar module coolant did not solve all the analytical problems posed by glycol solutions. The glycol–water solution used in the fuel cell system of Apollo 16 provided a very puzzling problem. Pale yellow-green needlelike crystals were collected on millipore filters during flushing of the system. The infrared spectrum of the crystals was surprisingly similar to that of 2,2′-dithiobisbenzothiazole, but was missing several sharp bands. This suggested the presence of some other compound similar to mercaptobenzothiazole, although this glycol–water solution (Table I) should have contained no compounds of this type. The electron microprobe x-ray analyzer showed major quantities of nickel and sulfur associated with the needles, suggesting the presence of nickel mercaptobenzothiazole. However, the x-ray diffraction pattern, which was of excellent quality, did not match that

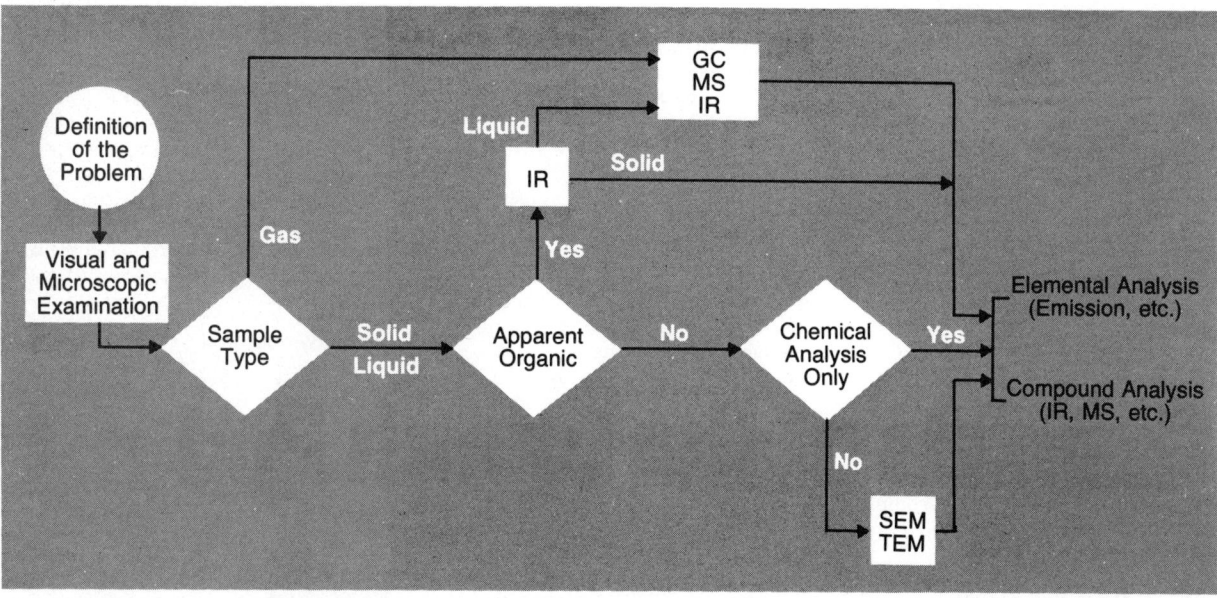

Figure 1. Identification of unknown material

of the nickel compound prepared in the laboratory as a reference, nor did it match any other pattern, either among the 20 000 in the Powder Diffraction File, searched both manually and by computer, or the reference patterns prepared in this laboratory. The few peaks found by mass spectrometry also failed to identify the material. A summary of the initial analytical results is given in Table II.

A comparison of the solution composition (Table I) and the analytical results (Table II) revealed no possible source for the yellow-green crystals whose identity was still unknown. The systems engineers were asked some questions about the materials used in the fuel cell system. "Is there any nickel in the system?" *Well, the tubing in the system is stainless steel which does contain some nickel—and, oh yes, there is a nickel frit in the system, too. But there shouldn't be any nickel in the solution because the pH is too high (7.8) to attack nickel.* (Wrong. The rate is slow, but traces of nickel could dissolve.) "Is there any organic material in the system?" *Yes. There is a rubber bladder which is necessary to pressurize the system so that the glycol solution can be pumped through the cooling loops of the fuel cell system while the spacecraft is in orbit.* "Do you have another bladder we can analyze?" *Yes.*

At this point, the course of the analysis was profoundly affected by the infrared spectroscopist, who had formerly worked for a rubber company. Knowing that various sulfur compounds, such as thiocarbamates and thiazoles are used as accelerators in the curing of rubber, he consulted with a friend in the rubber industry. With the friend's assistance the infrared spectrum was tentatively identified as that of an oxydiethylene dithiocarbamate salt. However, no reference spectrum was available, and no source of the material was known. Consequently, bis(4-morpholinecarbodithioato)Nickel-II (Compound III), the proper name for the compound, was synthesized in the laboratory. Comparison of the original sample material with the synthesized material by infrared spectroscopy, mass spectrometry, and x-ray diffraction showed the two materials to be the same.

It was theorized that 4,4′-dithiodimorpholine (Compound IV) had been used as an accelerator in the bladder, forming the zinc analog of III during vulcanization of the rubber. (Inorganic zinc compounds are frequently part of

Table I. Formulation (100 Parts by Weight) of Apollo Ethylene Glycol–Water Solutions

Lunar module environmental control		Service module fuel cell	
Ethylene glycol	35.00	Ethylene glycol	62.50
Water	Remainder	Water	Remainder
TEAP[a]	1.60	Na_2HPO_4	0.55
NaCap[b]	0.18	$K_2B_4O_7 \cdot 8H_2O$	0.70

[a] TEAP = triethanolamine phosphate. [b] NaCap = 50% solution of sodium mercaptobenzothiazole.

Table II. Analysis of Yellow-Green Crystals

Method	Results
Infrared spectroscopy	Similar to but not the same as Compound I
X-ray diffraction	Very sharp lines indicating highly crystalline material, probably very pure. Pattern not in Powder Diffraction File. Not same as lab x-ray pattern of Compound I
Emission spectrography	Major Ni, minor Cu, Al, Mg, Fe
Electron microprobe	Major Ni and S, highly correlated; low minor to trace Cu
Mass spectrometry	Molecular weight peak of 382 m/e. Weak peaks at 86, 130, 131, 196, and 260 m/e. (Molecular weight of Compound I is 332)

rubber formulations.) If the zinc compound then migrated to the surface of the bladder or "bloomed", it could react with the nickel ions which had been found to be present in the glycol solution to form Compound III.

$$O\diagup\hspace{-6pt}\diagdown N-\underset{\underset{S}{\|}}{C}-S-Ni-S-\underset{\underset{S}{\|}}{C}-N\diagdown\hspace{-6pt}\diagup O$$

III

$$O\diagup\hspace{-6pt}\diagdown N-S-S-N\diagdown\hspace{-6pt}\diagup O$$

IV

A new bladder that had a gray-white bloom was found. The infrared spectrum of the bloom matched that of the zinc analog of Compound III, completing the solution of the analytical problem.

The "Great Green Plague"

Unfortunately, not all analytical problems undertaken in the laboratory have neat answers. Corrosion products of stainless steel are frequently found on various samples, but the exact composition of the corrosion product depends on the type of alloy, the corrosive compound(s), and the time the alloy has been exposed to the corrosive compound(s). The product most frequently identified is a hydrated iron oxide, a composition which is consistent with environmental corrosion caused by the Florida coastal climate. However, not all corrosion is due to the environment.

One example of this was a stainless steel tubing contaminant found when a mobile launcher that had been downmoded after use in the Apollo program was being refurbished for use in the Skylab program. A brownish-to-dark olive drab material that occurred on the interior surface of some of the tubing was found by emission spectroscopy to contain major iron, manganese, and chromium; minor nickel and silicon; and traces of tin and five other metals. The numerous individual corrosion products, however, could not all be identified. The major ones identified by x-ray diffraction were iron oxide (Fe_3O_4), chromic oxide (Cr_2O_3), and manganese oxide (Mn_3O_4). These highly oxidized compounds were unusual compared to the hydrated iron oxides commonly found. Eventually, hundreds of feet of tubing were found to be contaminated with the "great green plague", and one office in the lab began to resemble a tubing farm as more and more samples were submitted for analysis. The cause of the contamination could not be definitely identified, but it was concluded that the olive drab contaminant most probably was formed during the final annealing stage of the fabrication process. The problem was solved by replacing most of the tubing. After cleaning which involved pickling, some of the contaminated tubing was later used in less critical systems.

Acknowledgment

The analytical contributions by coworkers L. G. Bostwick, R. Burton, W. R. Carman, W. A. Holden, J. F. Jones, R. M. Nichols, T. A. Schehl, L. D. Underhill, former coworkers K. Stevens, and A. C. Danielson of the Uniroyal Corp. are gratefully acknowledged.

References

(1) L. G. Bostwick and R. Burton, "Combined Electron Microscope, Electron Diffraction, and Electron Microprobe Analysis of Identical Microstructures in the Debond Area of a Dissimilar Metal Joint", NASA TN D-6327, August 1971.
(2) H. D. Bullard, L. D. Underhill, and G. L. Baughman, "Turbidimetric Determination of Sulfite Ion in Inhibited Ethylene Glycol–Water Solutions", KSC-TR-1073, September 1970.

Analysis of Orange Juice

Stuart A. Borman

American Chemical Society, Washington, DC 20036

Originally published in ANALYTICAL CHEMISTRY, 1980, Vol. 52, No. 12.

Sugar is cheaper than orange juice. And that is why the "100% Orange Juice" you buy may not always be as pure as all that. Adulteration, the substitution of cheaper ingredients for authentic products with the intent to defraud the buyer, occurs around the world with many commodities, and food products are no exception. Analytical chemistry is often called upon by the regulatory agencies to keep them one step ahead of those food processors who may want to cut a few corners here and there to reap higher profits.

Orange juice is easily adulterated by blending corn syrup or sugar syrup with orange juice concentrate before packing. Orange concentrate was priced a few years ago at about $0.40 per pound of solids, but the price today is more like $1.00 or more per pound of solids. Since sugar costs only about $0.15–0.20 per pound, the incentive to substitute is strong. A food processor can mix orange concentrate with corn or sugar syrup, then add a little more citric acid, orange oil, and, if necessary, a little β-carotene for color, and voìla—twice as much orange juice, or at least more.

Simple forms of adulteration are easy to detect, according to S.V. Ting, coordinator of processing and by-product research at the University of Florida Agricultural Research and Education Center in Lake Alfred. This is especially true if the amount of adulterant is over 30%. Adulteration by the addition of sugar syrup and corn syrup can theoretically be detected by an ashing procedure. Orange juice from any particular region normally has a very narrow range of ash content.

But there is a certain amount of biological variability in this value, and the determination itself has a certain amount of analytical uncertainty. The ash value can also be altered by the addition of appropriate inorganic compounds to simulate the proper composition.

If no inorganic materials are used and if the added syrup is high, there is no problem in its detection. But if the added adulterant is only 10 or 15% of the total ingredients, the detection cannot be made with certainty. When thousands of gallons of the product are involved, this amount of adulteration can still yield significant economic returns to the unscrupulous processor.

Another tool used to detect orange juice adulteration with sugar or corn syrup is sugar analysis, according to Dr. Ting. There are three main sugars in orange juice, glucose, fructose, and sucrose, and they are normally present in a ratio of approximately 1:1:2. If corn syrup, which contains mostly glucose, or cane syrup, which contains mostly sucrose, is added to orange juice, these ratios are changed. Ting measures these ratios by determining the three sugars with a colorimetric procedure that he himself developed. However, the availability of partially hydrolyzed cane syrup and the recently developed high fructose corn syrup makes it possible to produce an adulterated product with sugar composition very similiar to that of authentic orange juice.

Orange juice contains mainly citric acid, but it also has small amounts of malic acid and isocitric acid. Since there is a relatively constant proportion between citric acid and isocitric acid in citrus juices, determination of

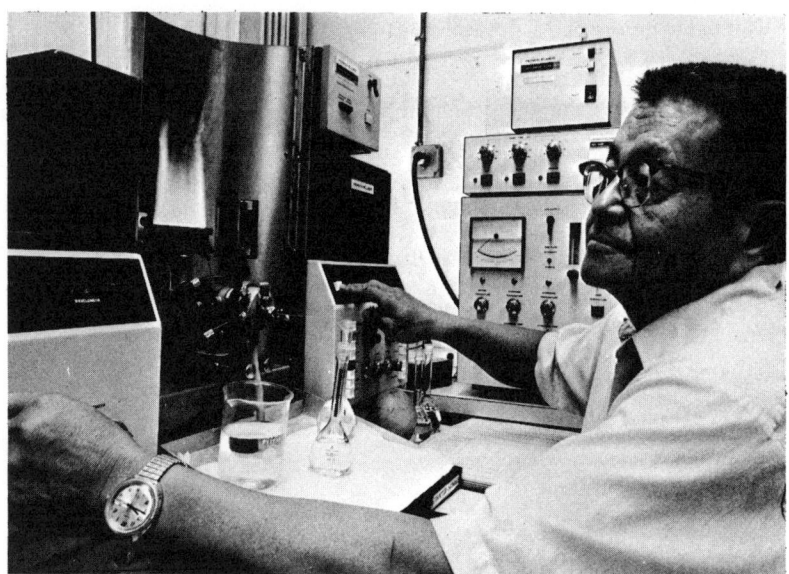

Jerry Ting measures mineral content of citrus juice samples. K, Na, Ca, Mg, and P are most often measured as indicators of juice adulteration. With the exception of P, which is usually determined colorimetrically, all are determined with the atomic absorption spectrophotometer, shown here

these minor acid components is another indicator of purity.

The total amino acid content and the relative amounts of the various amino acids are also normally in a certain range. The amino acid content of juice containing corn syrup or sugar syrup will be too low, since the added sugars do not contribute to the total amino acid content. Total amino acids can be measured by formalin titration. But there are countermeasures here too—the addition of inexpensive amino acids to increase the total content to the normal level. In this case the amino acids must be separated chromatographically to determine their distribution pattern.

Fighting food adulteration is very much an economic struggle. The tools and manpower needed to ensure food authenticity are expensive. The Food and Drug Administration (FDA) is responsible for waging this battle, but its resources are limited. "Defrauding the public is a crime. The FDA may expend a great deal of effort catching one guy and may succeed in having a fine of, say, $50 000 imposed. But it is not likely that a prison sentence will be meted out, and sometimes the case may be dismissed," explains Ting. "In the meantime, the agency has probably spent $50 000 or more just to bring this case to trial."

Ting himself is not concerned primarily with food adulteration. His knowledge of analyses useful in the detection of adulterated orange juice is just a spinoff from his research into the isolation, identification, and quantitation of the chemical constituents of citrus fruit. Ting's group hopes to learn more about how these constituents affect fruit quality and how their concentrations change during maturation, in processing and storage, and as a result of different horticultural practices.

Asked for a guess as to how many constituents might be present in an orange, Ting simply replies, "Nobody knows. That is the reason for the investigations. At least 100 constituents have been reported in orange oil alone. The number goes into the hundreds in orange juice."

A number of analytical techniques and instruments are utilized by Ting and his colleagues in their research. Nitrogen is determined with a classical Kjeldahl procedure. Gas chromatography is used to study essential oil composition, and to determine sugars, organic acids, and amino acids both qualitatively and quantitatively. Sugars are also measured by a colorimetric procedure.

Vitamin C is determined both titrimetrically and fluorimetrically. Fluorimetry is also used for the analysis of vitamin B_1. High performance liquid chromatography is used to determine vitamin B_2 and folic acid, and to study various other constituents of citrus juices, such as flavonoids and limonoids.

Where some constituents are too low in concentration to be determined by these methods, enzymatic procedures or microbiological assays have to be used. For example, some of the trace vitamins in citrus juices, such as vitamin B_6, folic acid, niacin, and pantothenic acid, are determined by microbiological assay.

As new methods and new instruments become available, they are quickly adopted as weapons in the struggle against adulteration. Certainly more support is needed for the development of these methods and for the purchase, maintenance, and operation of analytical instruments in this scientific game of hide-and-seek.

Ting's group, for instance, is investigating the adaptation of a $^{13}C/^{12}C$ ratio mass spectrometric technique for the detection of orange juice adulteration. This technique has been proved effective in the assurance of the quality of honey and apple juice, both of which have sometimes been subject to adulteration with high fructose corn syrup (see the February 1979 issue of ANALYTICAL CHEMISTRY, pp 224-232 A). Ultimately, success in this war against unfair profits and deception of the consumer depends on the sophistication of the analytical methodology used to detect it, and on the vigilance and skill of the analytical chemists on the front lines.

Search for the Elusive Phosphorus Source

Ralph L. Campbell and Ruth Zuback
Standard Oil Co., Cleveland, OH 44128

Originally published in ANALYTICAL CHEMISTRY, 1980, Vol. 52, No. 14.

As part of the Research and Development Division of Sohio, our Petroleum Tech Service Group can expect to receive samples and problems from nearly all of the operating petroleum divisions. Occasionally, we discover problems ourselves as a result of regular quality assurance audits; others are brought to our attention by alert field engineers engaged in special surveys. This was the case in February 1976, when Sohio was about to take over operation of Ohio Turnpike service stations. At that time, a quality audit conducted by our field engineers indicated that a number of unleaded gasoline samples from pumps at several of the plazas had phosphorus concentrations exceeding the EPA-regulated maximum of 0.005 g/gal. This situation was of considerable concern since a concerted and costly effort had been made early in 1974 to ensure that no lead or phosphorus, which might damage automobile catalytic converter antipollution devices, could find its way into the company's gasoline delivery system.

To track down the source of the contamination, the Sales Tech Division (Sohio's Scotland Yard) assigned one of its crack sleuths well versed in the ways of analytical chemistry. It didn't take him long to conclude that the phosphorus results were correct and the problem was real. Preliminary analyses of a group of selected samples (see Table I) indicated fuels from pumps in intermittent service had variable and high (0.05 g/gal) phosphorus concentrations while samples from high-service pumps were consistently low. Suspecting that the gasoline pump dispensing hoses might be the source of the phosphorus, our detective asked the lab for a series of leaching studies comparing hoses from the current supplier with those from other manufacturers.

A preliminary examination was made of two similarly identified hose samples from warehouse storage. We found that hose construction varied, that phosphorus was present in the hose materials, and that phosphorus could be extracted from the hoses by unleaded gasoline after an extended residence time. These results, shown in Tables II and III, confirmed the suspicion that the pump hose could be considered a prime source of high phosphorus.

A leach test was then designed to compare the weathering of gasoline pump hoses from a number of manufacturers. Our goals were to furnish our purchasing people with data to help them make knowledgeable hose selections in the future, and to compile evidence with which to confront the hose supplier. A quality control program for testing gasoline dispensing hoses was also recommended as a result of this study.

Experiment I

In our first controlled test, two samples of hose from the supplier (Codes "X-1" and "X-2") and three from other manufacturers (Codes "U," "V," and "W") were examined. Six-foot sections of each hose were filled with

Table I. Phosphorus in Unleaded Fuel Samples from Ohio Turnpike Plazas

Sample source	Operation	P, g/gal
Underground tank	—	0.004
Island 1	Closed	
Pump A		0.400
Pump B		0.072
Pump C		0.354
Island 2	Closed	
Pump A		0.186
Island 3	In use	
Pump A		0.0001
Pump B		0.0003
Island 4	Closed	
Pump A		0.140
Pump B		0.840
Pump C		0.154

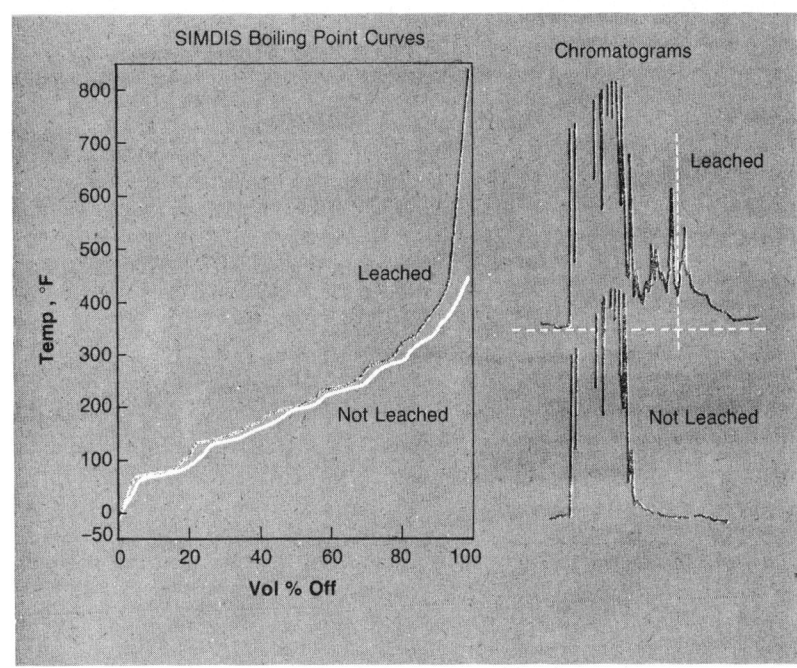

Figure 1. Comparison of leached fuel from hose "X" with base

unleaded gasoline and suspended in a U shape with both ends plugged. After a week at room temperature the fuel was removed and tested for API gravity, SIMDIS (simulated distillation by gas chromatography [GC]), P, Zn, Mn, Pb, S, and color. The used hoses were rinsed with unleaded gasoline, filled again with fresh fuel, and allowed to soak for a second week. This fuel was subjected to the same tests as the first week's leach fuel.

Gravity and SIMDIS were run to determine whether significant boiling range changes had occurred during fuel residence in the hose. Leaching of phosphorus, sulfur, and zinc was monitored because these elements had been found in the hose construction materials. Lead concentration was of interest because of the EPA-regulated maximum of 0.05 g/gal. Also, at that time an organo-manganese additive was being used in unleaded gasoline, and we were interested in its fate.

Lead and zinc were determined by atomic absorption spectrometry (AA), sulfur and manganese by X-ray fluorescence spectrometry (XRF), and phosphorus by the ASTM D3231 spectrometric procedure. The GC simulated distillation was done according to the ASTM D3710 procedure. Infrared spectrometry was used in making qualitative identifications of high boiling materials isolated during the GC runs. Hose materials were examined by scanning electron microscopy (SEM) using the X-ray emission mode.

The data we obtained indicated that S, Zn, P, and some high boiling materials were leached out of all the hose samples during the first test week and that lesser amounts were leached the second week. The "U"

Table II. Preliminary Gasoline Dispensing Hose Leach Test, March 1976

	Analysis of unleaded test fuel	
	Untreated fuel	After one week residence in new hose[1]
P, g/gal	.003	0.20
Distillation		
IBP, °F	95 (Typical)	95
10% Rec. at °F	118	118
90% Rec. at °F	339	420
FBP, °F	416	605

[1] "X" Soft Flex, sample 1
IBP = initial boiling point; Rec. = recovered; FBP = final boiling point.

hose sample was least affected, phosphorus levels in the test fuels increasing only to 0.002 g/gal. The highest levels of phosphorus, 0.4 g/gal, were leached from the two "X" hose samples. Sulfur, zinc, and high boiling materials followed the same pattern as phosphorus concentration. Discoloration of the test fuel also increased with phosphorus concentration and was an instant indicator of hose deterioration. No significant variations in lead and manganese levels were noted.

Experiment II

A second leach test was run two months later to confirm results from the first experiment, to standardize test conditions, and to examine a revised formula hose submitted by one of the manufacturers.

In addition to unleaded gasoline, a synthetic fuel (30% toluene, 70% isooctane) was used as a leach fuel. This was done at the request of one of the hose manufacturers who was concerned about possible effects of gasoline additives.

Duplicate leach tests using unleaded gasoline were run with fresh samples of "U" and revised-formula "X" hoses, and single tests were done with the synthetic fuel. A single test with unleaded gasoline was run on a

Table III. Preliminary Gasoline Dispensing Hose Leach Test, March 1976

	Phosphorus in gasoline dispensing hose	
	Unused hose[1]	Used hose[2]
Outside black rubber layer, wt % P	.0169	.0479
Inside black rubber layer, wt % P	.0207	.0359
Red rubber inner layer, wt % P	(no red layer)	.0505
Fibers from inner layer, wt % P	.0086	.0291
Total wt. % P	.0376	.1343

[1] "X" Soft Flex, sample 2
[2] "X" Soft Flex, sample 1

Table IV. Gasoline Dispensing Hose Leach Test, Experiment II

Hose	Synthetic fuel[1]					Unleaded gasoline									
	Base	"U"		"X"		Base	"U"				"X"				"Z"
Replicate							A		B		A		B		A
Week no.[3]		1	2	1	2	—	1	2	1	2	1	2	1	2	1	2
API Grav.	58.0	58.2	58.3	57.0	58.3	61.0	59.7	59.9	59.9	60.0	57.7	59.7	58.0	59.6	57.4	58.9
SIMDIS																
IBP, °F	193	193	195	192	193	−10	17	16	−10	−7	−10	−7	−10	33	−10	−7
96% Rec. at °F	242	243	242	675	241	399	437	408	424	405	674	419	674	431	755	435
98% Rec. at °F	243	797	243	678	407	423	672	441	793	437	678	456	678	479	758	752
FBP, °F	244	804	800	682	675	450	801	676	804	799	682	676	681	676	761	758
Est. % >450 °F	—	4	2	10	4	—	4	2	5	2	6	2	7	4	7	4
P, g/gal[4]	.001	<.001	<.001	.063	.027	<.001	<.001	<.001	<.001	<.001	.044	.027	.046	.023	.135	.060
Zn, ppm[5]	—	70	50	96	44	—	49	45	46	41	98	41	99	37	208	98
Mn, g/gal[6]	—	—	—	—	—	.053	.041	.047	.041	.048	.031	.044	.031	.043	.034	.044
Pb, g/gal[5]	—	<.01	<.01	<.01	<.01	<.01	<.01	<.01	<.01	<.01	<.01	<.01	<.01	<.01	<.01	<.01
S, Wt %[6]	<.002	.022	.018	.033	.013	.017	.037	.033	.033	.028	.045	.027	.046	.027	.046	.035
Color[2]	0	2	—	1	—	0	3	—	3	—	2	—	2	—	7	—
IR ID, >450 °F	—	—	—	—	—	—	Polyether-ester				Polyether compounds				Phthalate-ester	

[1] 70% isooctane, 30% toluene; [2] Scale: 0 to 10, colorless to dark; [3] Fresh feed charged to 6 ft of hose at beginning of each week; [4] ASTM D 3231; [5] AA; [6] XRF

IBP = initial boiling point; FBP = final boiling point; Est. % 450 °F = estimated % of fuel boiling at temperatures exceeding 450 °F; IR ID = infrared spectrometric identification of fraction boiling at temperatures exceeding 450 °F

third manufacturer's hose (Code "Z"). As in Experiment I, a second week leach test was also run. Both first and second week leach fuels were tested in the same manner as in Experiment I, as were the fresh stock fuels. In addition, high boiling materials from the SIMDIS runs were captured for identification by IR spectrometry.

In general, the data obtained in this experiment, shown in Table IV, confirmed the results of Experiment I. No significant differences between the effects of the synthetic fuel and unleaded gasoline were observed. The data showed the leaching test was repeatable enough to detect differences in hose performance.

Again, phosphorus concentration was not excessively increased by deterioration of the "U" hose samples. Tests with new-formula "X" hose samples showed phosphorus at 0.05 g/gal. Though this represented a tenfold improvement over the original test specimens (Experiment I), the phosphorus content still exceeded the EPA maximum of 0.005 g/gal. In this experiment the "Z" hose sample showed the greatest effect of deterioration, phosphorus concentrations increasing to 0.35 g/gal.

SIMDIS analysis showed varying amounts of high boiling materials were leached from the hose samples, with the "U" hose again showing the least deterioration. The high boilers, probably rubber plasticizers, were identified by infrared as polyether esters in the "U" hose, mixed polyethers in the new-formula "X" hose, and phthalate esters in the "Z" hose. In Figure 1, the SIMDIS chromatogram of fuel from the "X" hose is compared with that of the base fuel. The late-eluting mixed polyethers are quite evident.

As a result of this study, we learned that:
- Extended residence times for gasoline in the initially used ("X") pump-dispensing hoses caused leaching of phosphorus, zinc, sulfur, and plasticizer into the fuel;
- hoses made by different manufacturers varied in their resistance to deterioration;
- the type and quality of a hose cannot be determined from its markings; and
- a leaching test with phosphorus as an indicator can be used to reliably predict hose quality.

Recommendations

An immediate result of this investigation was the replacement of all "X" hoses in unleaded gasoline service. Also, it was recommended that fuel be completely drained from all limited-service pump hoses when not in use.

For the longer range, a quality control program to test new purchases of gasoline dispensing hoses was recommended to the purchasing department. This was accepted, and we have since tested several batches of hoses each year using the leaching test described here.

To the present time, no phosphorus has been detected in the many service station quality audits conducted over the past four years. The problem has been solved.

Acknowledgment

In addition to W. R. Turri, Sales Tech's master sleuth, who initiated this project and guided our efforts, we are indebted to many specialists from our Analytical and Petroleum Tech Service Groups for their expert assistance in solving the problem. They are: G. Alexander, T. Andrews, J. Boyko, L. E. Brown, R. H. Duff, M. A. Hazle, J. Hedger, R. Kollar, C. LaRocca, J. Laskowski, J. Lydic, G. Markos, A. Varnes, and J. Zelinko.

Adapted from "Repair, Replace, Sue, or Be Sued, Which?", R. L. Campbell and R. Zuback, Paper No. 26, ACS 179th National Meeting, Division of Analytical Chemistry, Houston, Tex., March 23–28, 1980.

Over-the-Counter Drug Analyses with HPLC

Margaret A. Carroll, E. Roderick White, and John E. Zarembo

Smith, Kline, and French Laboratories, Philadelphia, PA 19101

Originally published in ANALYTICAL CHEMISTRY, 1981, Vol. 53, No. 9.

The complexity of over-the-counter (OTC) drug preparations makes the identification of their active constituents extremely difficult. Unlike the single compound ethical drug often found in prescription preparations, OTC products are frequently formulations complicated by the numbers and relative quantities of active ingredients as well as by a large variety of excipients. Excipients include antioxidants, emulsifiers, stabilizers, flavoring agents, and buffers.

The most effective analytical approach followed in this laboratory for analyzing OTC products complements high performance liquid chromatography (HPLC) with various spectrometries, specifically the mass spectrometries (electron impact, chemical ionization, field desorption, GC/MS) and IR, UV, and NMR (^{13}C, ^{1}H) spectrometries. Application of these analytical techniques permits the conclusive identifications of compounds required for compliance with safety and efficacy regulations.

The use of HPLC obviates earlier complex, time-consuming, and less specific approaches, viz, liquid–liquid extractions and colorimetric assays. The knowledge obtained in analyzing OTC preparations also serves as a basis for solving other pharmaceutical problems.

Examples of these HPLC approaches to the analyses of OTC preparations are given below. Initially, HPLC was applied to the analysis of ethical products such as antibiotics. HPLC columns used for these assays were packed with medium-efficiency 37-μm pellicular materials. The introduction of high-efficiency 5- and 10-μm packings gave the increased resolution and greater sensitivity required for stability-indicating methods. The development of ion-pairing techniques for reversed phase systems permitted the analysis of many pharmaceutical compounds that were previously difficult to analyze with an ion-suppression reversed phase system because of the pH restrictions (pH 2–8) on the columns. The current method permits the separation of active compounds from their decomposition products, precursors, contaminants, and excipients. One major advantage of the technique is the simplicity of sample preparation, which is due to the high resolving power of HPLC for complex mixtures. Extensive pretreatment is seldom needed, most samples requiring only dissolution, dilution, and injection. The methods are precise ($RSD_{1\sigma} < 1.5\%$) and accurate.

OTC Choline Salicylate Product

One OTC product that was analyzed contained choline salicylate. Choline salicylate is an aspirin substitute and is stable in aqueous solution. The HPLC method developed for the salicylate used camphorsulfonic acid in the mobile phase to completely suppress salicylate ionization.

Since both the starting materials and formulations are viscous liquids, we wanted to establish a procedure for measuring samples volumetrically. This was done by filling a "to contain" pipet to the mark with sample, allowing it to drain into a volumetric flask, and then carefully rinsing the material adhering to the wall of the pipet into the flask with mobile phase liquid. A known amount of internal standard was added, and the sample diluted to volume. To prove that this technique was valid, the precision of the volumetric measurement was compared to that obtained with weighed samples. The average values obtained for both techniques were identical with RSDs of 0.8% and 0.7%, respectively. Thus, no problem appeared to occur with the volumetric measurement.

Attempts were made to decompose choline salicylate in controlled experiments. The sample was subjected to heat, UV radiation, and complexing

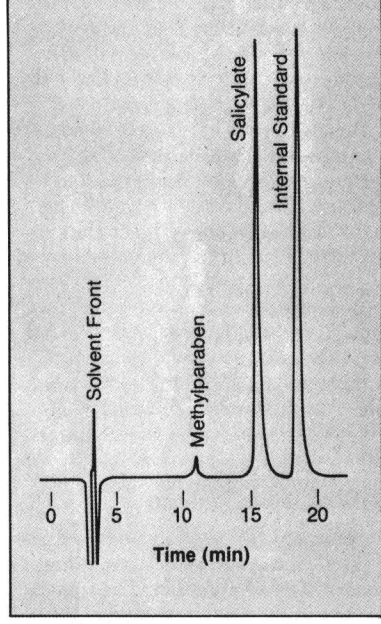

Figure 1. HPLC separation of salicylate in a choline salicylate formulation. Column, Dupont Zorbax C_8 (25-cm × 4.6-mm i.d.); AUFS, 0.04; flow rate, 1.0 mL/min; detector, 280 nm; injection, 10 μL; mobile phase (50:50) methanol: 0.025 M camphorsulfonic acid

and oxidizing agents. The HPLC and TLC (thin layer chromatography) chromatograms showed no decomposition products. Thus, salicylate appeared to be exceptionally stable under these conditions.

A chromatogram of the separation is shown in Figure 1. Methyl paraben is a preservative in the formulation. Ethyl paraben is the internal standard. The purity of the salicylate peak was investigated by an absorbance ratio technique (1). Absorbance ratios are determined by measuring the absorbance of the peak at several wavelengths and calculating the ratios of the absorbances. These ratios are constants for pure compounds. The ratios were determined on three samples—a reference standard; a 6-month, 50 °C sample of starting material; and a 6-month, 50 °C formulation. The absorbance ratios of the samples showed no significant variations. While this experiment did not completely eliminate the possible contribution of additional components to the peak, it was strongly indicative that a single entity was present, and we felt that the method had been validated.

Samples were analyzed for several months without any problems. Then, starting material samples that had been stored for 6 months were received. A sample stored at 50 °C assayed 1.9% higher in choline salicylate than did a sample stored at 5 °C. The assay on the 5 °C sample was in agreement with the result on the original sample. If decomposition had occurred at the higher temperature, it probably would have caused low, not high, results. Absorbance ratio measurements and TLC evidenced no decomposition products that might interfere. No inorganic material was present. IR and MS data showed no significant differences. Since the samples were in solution, the possibility of water loss was considered. Karl Fischer analyses for water solved the problem. A material balance was obtained on both samples—99.8% on the 5 °C sample and 99.2% on the 50 °C sample. The change in choline salicylate assay value was compensated for by the change in water content.

OTC Allantoin Product

Another OTC product analyzed was a skin healing and protecting lotion that contained allantoin. The complexity of the matrix precluded the use of frequently applicable colorimetric and titrimetric procedures, and the literature described no appropriate HPLC methods. We therefore developed an HPLC procedure specific for allantoin that included the use of a variable wavelength UV detector set at 220 nm.

The initial investigation included testing a variety of HPLC reversed phase columns, including Waters μBondapak C_{18}, Dupont Zorbax C_8 and C_{18}, and Whatman Partisil-10 ODS-2. Allantoin, however, was not retained on any reversed phase column—even those with very heavy carbon loading. Nor was ion pairing or ion exchange HPLC applicable because allantoin was not retained by any of these columns.

Since none of these approaches was successful, the next step was to investigate bonded normal phase systems. Bonded normal phase columns tolerate water well. The fact that allantoin is very water-soluble and only sparingly soluble in methanol made these columns appear promising for separation of compounds as highly polar as allantoin. Cyano columns, one of the most common types of normal phase columns, were tried first.

Column packings having cyano functionality only did not retain allantoin. One packing, however, Whatman PSX 10/25 PAC, has some amino functionality in addition to cyano, and it retained the compound slightly. We decided to try packings having amino groups.

Both Waters μBondapak and Dupont Zorbax amino columns gave satisfactory retention and peak shape. Figure 2 is a chromatogram of the separation of an allantoin formulation on a Waters column using a water–acetonitrile (11:89) mobile phase. Many components in the formulation also absorb at 220 nm, but these are well separated from allantoin.

The multiplicity of ingredients in the formulations caused other difficulties besides separation problems. Sample preparation is vital to the success of this type of analysis. Large differences in the solubility of components formulated with allantoin presented one of the most difficult problems encountered in the HPLC analysis of these lotions. To successfully assay these preparations, the sample must be completely solubilized or separated from insoluble material by filtration, centrifugation or other means to prevent column blockage by insoluble material at the head of the column. Allantoin lotions were particularly difficult to deal with.

It was observed that certain formulations did not dissolve in the mobile phase which was the preferred diluent. Attempts to solubilize the formulations by heating in methanol were unsuccessful because of decomposition detectable by the presence of extraneous peaks in the chromatogram. Tetrahydrofuran (THF) was a good solvent for the matrix, so mixtures of THF and water were tried. A 30:70 mixture of these solvents kept allantoin in solution and dissolved most of the remaining components. While the solution was slightly cloudy, any undissolved material settled on standing, and the supernatant liquid was suitable for injection. Samples of spiked placebo were assayed to validate the method. The recoveries averaged 99.6% with an RSD of 1.1%.

An additional problem arose when a component in one of the placebo mixtures was found to elute at exactly the same retention volume as allantoin. This interference would have caused allantoin results to be 10% too high. Attempts to modify the chromatographic conditions to solve this problem—first by changing the mobile phase and then by employing a completely new system on a silica col-

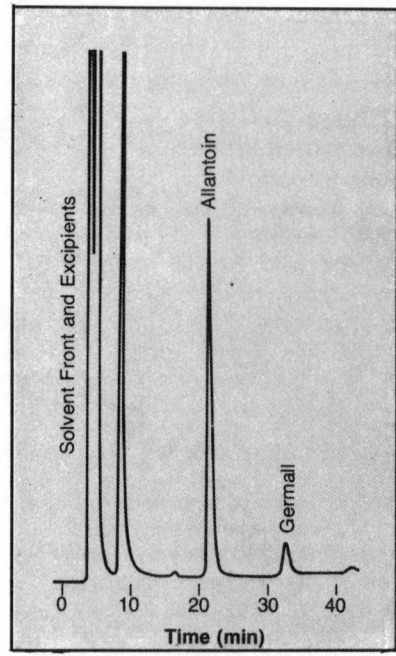

Figure 2. HPLC separation of allantoin in a lotion. Column, Waters μBondapak NH_2 (30-cm × 3.9-mm i.d.); AUFS, 0.08; flow rate, 1.0 mL/min; detector, 220 nm; injection, 25 μL; mobile phase, (11:89) water:acetonitrile

umn—were not successful. Extractions from acidic, neutral, and basic solutions were also unsuccessful in removing the interference. An attempt was made to identify the source of the problem by investigating the constituents of the placebo. Germall 115, a preservative in the lotion, is a condensation product of allantoin and formaldehyde (2). To determine if this excipient was the source of the HPLC interference, a sample of Germall 115 was injected into the instrument. A small peak eluted at the retention volume of allantoin. To identify the chemical composition of the unknown peak, the placebo was injected, and the unknown peak was collected and examined by mass spectrometry.

Electron impact (EI) mass spectrometry indicated that the spectrum was essentially identical to that of a reference sample of allantoin. To verify further that the peak was allantoin, a field desorption (FD) mass spectrum was obtained. Since FD is a soft ionization technique, many compounds that are easily fragmented by EI retain their molecular integrity. Field desorption mass spectrometry showed a molecular ion peak at m/e 158, which corresponds to that expected for allantoin. This particular lot of Germall 115 appeared to contain some unreacted precursor.

Many formulations had been prepared with this lot of material, and assays were required. The amount of interference in the placebo was found to correspond to the amount of allantoin expected from the theoretical content

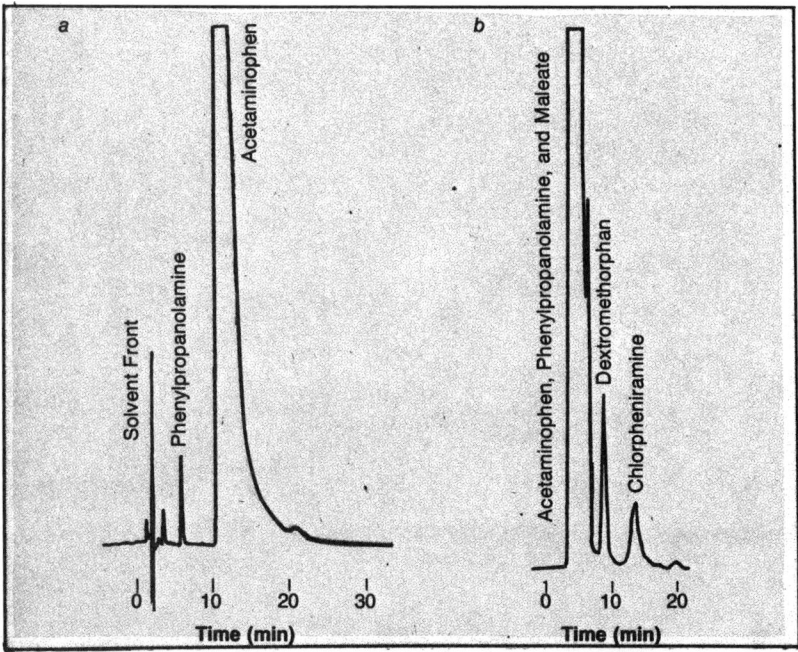
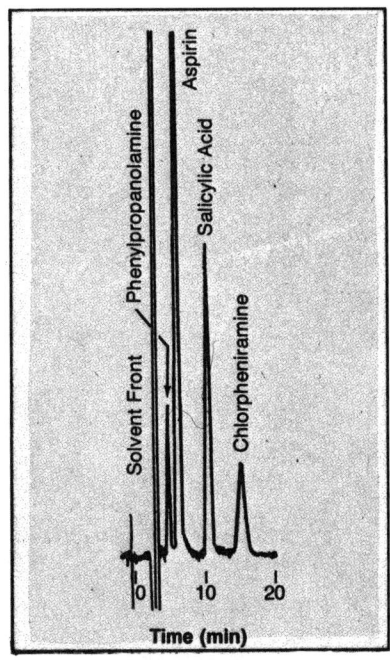

Figure 3. (a) HPLC separation of phenylpropanolamine hydrochloride and acetaminophen in a dissolution sample. Column, Whatman PXS-1025 ODS (25-cm × 4.6-mm i.d.); AUFS, 0.02; flow rate, 1.2 mL/min; detector, 256 nm; injection, 20 μL; mobile phase, (9:91) methanol:0.01 M methanesulfonic acid. (b) HPLC separation of dextromethorphan hydrobromide and chlorpheniramine maleate in a dissolution sample. Column, Waters μBondapak C_{18} (30-cm × 3.9-mm i.d.); AUFS, 0.02; flow rate, 1.0 mL/min; detector, 262 nm; injection, 100 μL; mobile phase, (75:25) methanol:H_2O (0.02 M dioctylsulfosuccinate sodium salt, 0.1% phosphoric acid)

Figure 4. HPLC separation of phenylpropanolamine, aspirin, salicylic acid, and chlorpheniramine in an allergy remedy. Column, Waters μBondapak C_{18} (30-cm × 3.9-mm i.d.); AUFS, 0.04; flow rate, 1.0 mL/min; detector, 256 nm; sample, 20 μL; mobile phase (40:60) methanol:H_2O (0.0125 M camphorsulfonic acid)

of Germall 115. The formulations were corrected by subtracting the equivalent values for allantoin calculated from the Germall concentrations from the total allantoin content of the samples. The placebo was monitored with the stability samples so that any change in allantoin content would be observed and the proper correction made for the samples.

OTC Cough–Cold and Allergy Products

To illustrate the uses of HPLC in the analysis of multicomponent products, ion-pairing reversed phase procedures used for some of these complex formulations are described. One product was a cough–cold remedy that contained phenylpropanolamine hydrochloride, acetaminophen, chlorpheniramine maleate and dextromethorphan hydrobromide. It was necessary to obtain assay and dissolution data on original and stability samples of the solid dosage forms. The extremely dilute (one capsule or tablet/900 mL) solutions of dissolution media were difficult to analyze. To define the solubility characteristics of the preparations, the dissolution medium was sampled at varying time intervals. A large number of samples of low concentration were generated. The HPLC instruments were found to have sufficient sensitivity and stability to assay these solutions successfully. The automated assay procedure permitted samples to be analyzed unattended overnight.

Because of the marked differences in polarity of the four compounds, two HPLC systems were used for the analysis of the cough–cold remedy. Figures 3a and 3b are chromatograms of a dissolution sample at $t = 15$ min. Depending upon the chain length of the ion-pairing reagent employed and the methanol content, it was possible to have phenylpropanolamine elute before or after acetaminophen. With a short chain reagent (such as methanesulfonic acid) and 9% methanol, phenylpropanolamine eluted before acetaminophen. This was desirable because of differences in concentration and UV response of the two compounds. Phenylpropanolamine and acetaminophen eluted at the solvent front when dioctylsulfosuccinate was used as the ion-pairing reagent in a mobile phase containing 80% methanol. In the same system, the less polar compounds, dextromethorphan and chlorpheniramine, were retained and quantitated.

Another multicomponent product, an allergy remedy, contained phenylpropanolamine hydrochloride, chlorpheniramine maleate, and aspirin. A single HPLC system with a camphorsulfonic acid mobile phase separated these compounds and salicylic acid, the decomposition product of aspirin. A chromatogram of the separation on a Waters μBondapak C_{18} column is shown in Figure 4. Aspirin hydrolyzes rapidly to salicylic acid in an aqueous system and cannot be quantitated in overnight runs. However, it can be quantitated if each sample is prepared shortly before injection. A manual UV assay for aspirin and salicylic acid was found to be more convenient than the HPLC system and required less overall operator time.

To summarize, the reliability of HPLC methods supplemented by other analytical techniques for the analysis of OTC products has been investigated. The examples illustrated in this paper show the utility and advantages of this effort. Large numbers of samples can be assayed automatically and complex mixtures separated into their individual components for qualitative and quantitative determinations. The simplicity of the technique permits its use by both chemists and trained technicians. Mass spectrometry, thin-layer chromatography, and ultraviolet and infrared spectrometries can all be used to complement the HPLC method and to aid in meeting the challenge of over-the-counter drug analysis.

References

(1) Yost, R.; Stoveken, J.; MacLean, W. *J. Chromatogr.* **1977**, *134*, 73.
(2) Berke, P. A. U.S. Patent 3 248 285, 1966.

Analysis of Products Used in the Electronics Industry

Peter Cukor

GTE Laboratories, Waltham, MA 02154

Originally published in ANALYTICAL CHEMISTRY, 1976, Vol. 48, No. 1.

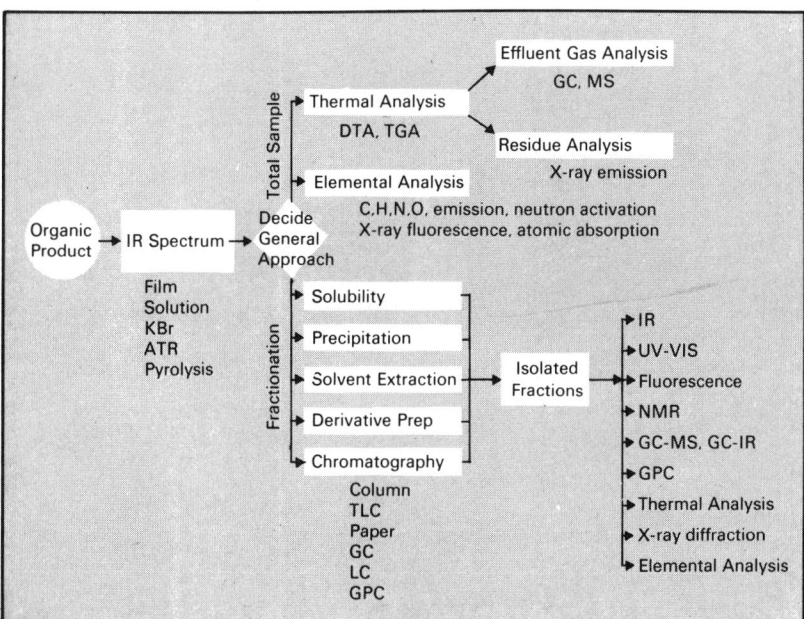

Figure 1. Analytical approach for characterization of organic products used in the electronics industry

The electronics industry has used a great variety of materials in a unique manner to accomplish the communication and home entertainment revolution of our times. The design and fabrication of integrated circuits (IC) which are at the center of this revolution consist of a series of complicated chemical processes. The purity of some of the materials used in IC manufacturing, such as silicon and boron, is extremely high. A trace amount of plasticizer extracted by a solvent from its plastic container can get deposited on a silicon wafer during its cleaning with solvent and may ruin several integrated circuits. Consequently, processing takes place under meticulously clean conditions in rooms with filtered laminar flow air and high-purity deionized water.

The integrated circuits are only one of many items produced by the electronics industry. General Telephone and Electronics, Inc., is a good example of a broad base technology company with business activities extending into areas of telephone service and equipment, lighting and home entertainment products, and chemicals and parts manufacturing. This diversified activity means that the analytical chemists at GTE's central research laboratory in Waltham work on a wide variety of analytical problems requiring different analytical approaches.

Although most of the analytical facility is devoted to the study of inorganic materials, there does exist within the Materials Evaluation Facility, a group whose primary responsibility is the analysis of organic materials. Members of this group provide the support required by scientists involved in research, development, and production and also spend a fair amount of time on the analysis of commercial products. The major reasons for the latter activity are summarized in Table I.

The analytical approach used for the characterization of organic products is shown in Figure 1. The examples that follow serve as specific illustrations of this general approach. The first one illustrates the determination of the chemical composition and probable polymerization mechanism of an experimental product. The second example shows how analytical work originally intended for troubleshooting led to the development of a raw material specification. Finally, an example is given from an area that is a new challenge to analytical chemists—namely, the analysis of insoluble polymers.

Analysis of Experimental Electron Beam Resist

Electron beam resists generally are soluble polymers which cross-link and are rendered insoluble by exposure to an electron beam. It is possible to print an image on a substrate by coating it with a polymer and exposing the polymer coating to a beam of electrons which trace the desired pattern. Those portions of the coating which are not bombarded with electrons are readily washed away, and the exposed substrate is available for chemical treat-

Table I. Reasons for Analyzing Commercial Products

Problems with vendors
 Batch-to-batch variations in products supplied
 Inability of vendor to supply sufficient quantity of product requiring either alternate vendors or in-house preparation of material
 Products may be used for purposes other than intended by vendor, and their use may lead to production problems

Materials for evaluation
 Evaluation for a proposed use by manufacturer or by own technical personnel
 Evaluation in connection with management decision, i.e., acquisition

Analysis of competitive products to confirm patent claims

83/4463-0065 $07.00 © 1983 American Chemical Society

ment such as etching. The cross-linked polymer pattern, on the other hand, will be retained on the substrate, protecting it from the etchant. The advantage of an electron beam resist over the more commonly used photoresist is the ability to produce much finer lines in the image. An experimental electron beam resist was evaluated for a potential supplier who also requested a detailed chemical analysis.

General examination of this sample (odor, heating a small portion on a hot plate) indicated that the resist was dissolved in a solvent. Infrared spectra identified the volatile portion as 2-propanol and the residue as a silicone. With this preliminary information, it was possible to map a strategy for more detailed characterization as shown in Figure 2. The volatile fraction amounted to 80% of the sample, and its evaporation in the thermogravimetric analyzer under nitrogen was completed around 200°C. The residue was thermally stable up to 650°C, beyond which it exhibited a gradual weight loss. By GC–IR it was shown that the 2-propanol was of high chemical purity.

The nonvolatile portion of the beam resist was studied by IR and by proton NMR in $CHCl_4$ solution. In addition, the elemental composition of the nonvolatile was determined by several techniques: X-ray fluorescence, optical emission spectrography, neutron activation, and C, H, and N organic microanalysis. These data are summarized in Table II.

The IR band in the vicinity of 1100 cm^{-1} definitely suggested the Si—O—Si type structure observed in SiO_2, silicones, and other siloxanes. The exact location of this band (1080 cm^{-1}) further indicated a cyclic-siloxane, and its position and shape suggested a tetramer $(Si—O)_4$:

This structure was further inferred by comparison of the intensity of the Si—O—Si stretching band with the C—H band intensities. The relatively strong Si—O—Si band indicated a minimum number of groups attached to the silicon atoms. The NMR spectrum indicated that the —CH_3 and —OH groups present were on the same carbon atom. The spectral data are consistent with the structure shown at the top of the next column.

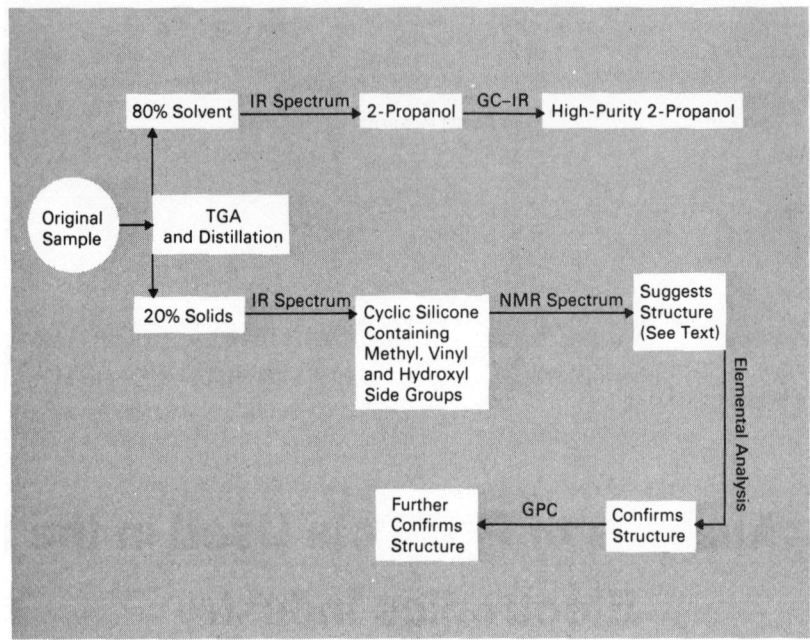

Figure 2. Analysis scheme of electron beam resist

Elemental composition calculated on the basis of this structure is in good agreement with elemental analysis data as shown in Table II. The low value shown for carbon probably is due to SiC formation.

Gel permeation chromatography was used to measure weight average molecular weight of the polymer. A calibration curve constructed from n-paraffins yielded an average molecular weight of 360 for the polymer; a calibration curve based on polyglycols gave 660 as the average molecular weight. The curve was skewed toward the high molecular weight side, and resolution of the curve into two Gaussian peaks showed that the molecular size of the two species was roughly in a ratio of 2:1, indicating the possible presence of a dimer.

The calculated molecular weight of the proposed structure is 633 which is in good agreement with the molecular weight obtained by GPC (660) when the polyglycols calibration curve was used.

Besides determining the structure of the resist material, the analytical investigation also yielded information about how the resist functioned.

Upon exposure of a film of dry resist to a stream of electrons, IR showed a noticeable decrease in the intensities of the C—H and OH bands in relation to the Si—O—Si band intensity. The appearance of a new band at 1700 cm^{-1} suggested the formation of C=O, and this correlated with a pronounced decrease in the vinyl group absorption at ~960 cm^{-1}, indicating that the cross-linking was accomplished through the vinyl groups. This was further supported by the decrease of the vinyl overtone at 1950 cm^{-1}. The spectra also showed a shift in the Si—O—Si band to lower frequencies which suggested an opening of the tetramer ring and the formation of an open chain polymer.

Table II. Characterization of Electron Beam Resist

IR DATA

Absorption band, cm⁻¹	Rel intensity	Conclusion
3300	Med	OH stretch (also see 880 cm⁻¹)
3000–3100	Weak	CH stretch in olefins
2960	Weak	CH in methyl groups under influence of Si are much weaker than C—CH₃
1930	Weak	Possible overtone of lower frequency C=C
1600	Med	C=C stretch
1400	Med	CH in plane bending in H₂C=C + SiCH₃
1265, 760	Med	SiCH₃ single methyl. 1265 one band, no Si(CH₃)₃ groups
1080, 760	Strong	Si—O—Si cyclic tetramer, open chain is single or double peak at different wavelengths
1000, 960	Weak	—C=CH₂
880	Med	OH deformation in SiOH

NMR DATA

ppm	Rel intensity of protons	Assignment
1.25	2	2 OH groups
2.20	6	2 CH₃ groups
6.06	24	8 CH=CH₂ groups

ELEMENTAL ANALYSIS

Element	Method	Found, %	% Theoretical in proposed structure
Si	Neutron activation	34.5	33.8
O	Neutron activation	29.7	28.9
C	Elemental analyzer	Greater than 26.8	32.5
H	Elemental analyzer	4.43	4.8
N	Elemental analyzer	Not detected	
S, Cl, Br	X-ray fluorescence	Not detected	
Others	Emission spec.	Si only	

Analysis of Cemented Carbide Binder

This example illustrates the problems that may occur when a change in raw material suppliers takes place. In the preparation of cemented carbides, an organic binder is used to hold the powdered material together prior to firing. During firing, the binder is decomposed and is given off as low molecular weight volatile matter while the powdered carbide is cemented together.

One of the most popular binders has been a certain paraffin wax. Due to a shortage of petroleum products, the supplier was unable to continue providing this material. No obvious alternative supplier could readily be found, and an assortment of possible binders was evaluated. A screening procedure based on the use of the thermogravimetric analyzer (TGA) and firing conditions as close as possible to those used in the manufacturing process was set up, and the thermograms of the candidate binders were compared with that of the original paraffin wax. This rapid screening procedure eliminated all but three paraffin waxes as possible substitute binders.

Next, small batches of carbide cements were prepared with each of the three waxes. In these tests only one of the three performed satisfactorily. The question then arose as to what other property besides the profile of the thermal degradation influences the behavior of the binder. Accordingly, the three waxes which passed the TGA screening test were further characterized. Molecular weight distributions were obtained by GPC analysis, and the distribution curves corresponded very closely to the TGA curves, suggesting that the decomposition temperatures were a direct function of the molecular weights of the components. Another, more careful look at the manufacturing process revealed that the problem occurred not during, but prior to, firing. That is to say, the decomposition temperatures of all three waxes were satisfactory, but their wetting characteristics were vastly different. Infrared analysis of the waxes was carried out, and the one with the highest CH₃/CH₂ absorption ratio displayed the best wetting characteristics. These findings explained why only one of the three waxes proved to be a usable binder. The investigation resulted in the establishment of specification for paraffin wax binders in terms of TGA and IR data.

Analysis of Plastic Wire Coating

One of the more recent challenges to analytical chemists has been the analysis of insoluble or partially soluble polymers. In analyzing these materials, the most efficient tools of characterization, transmission infrared spectrophotometry and various forms of chromatography, usually are not applicable without modification. Hence, the increasing interest in reflectance and pyrolysis infrared spectrophotometry and in pyrolysis gas chromatography. Also differential thermal analysis may be used to generate an identifying fingerprint thermogram of many insoluble plastics. Neutron activation and X-ray fluorescence have proved to be effective ways of analyzing plastics for metals, halogens, and oxygen without the need for cumbersome wet ashing procedures.

A cable containing a plastic wire coating was analyzed. A flow diagram of the analysis is shown in Figure 3.

With an ATR spectrum and DSC curve, the principal constituent of the coating was identified as polyvinyl chloride. The presence of an ester-type plasticizer was also detected. The sample was extracted with methanol to remove the plasticizer. The methanol extract, which was equal to 20% of the total sample, was subjected to liquid chromatographic separation, and the material of the major peak was collected and identified by its IR spectrum as diethylhexyl phthalate. Analysis for C, H, N, O, and Cl confirmed the finding that the coating was plasticized PVA, but it also indicated that not all of the material was accounted for, since the total added up to only about 92%. TGA yielded a 10% residue. Emission spectrographic analysis of the residue showed antimony as the principal constituent. Neutron activation analysis of the original plastic confirmed this finding. The composition of the wire coating may be summarized as follows: 70% polyvinyl chloride, 20% diethylhexyl phthalate plasticizer, and 10% antimony trioxide fire retardant.

The examples cited illustrate the

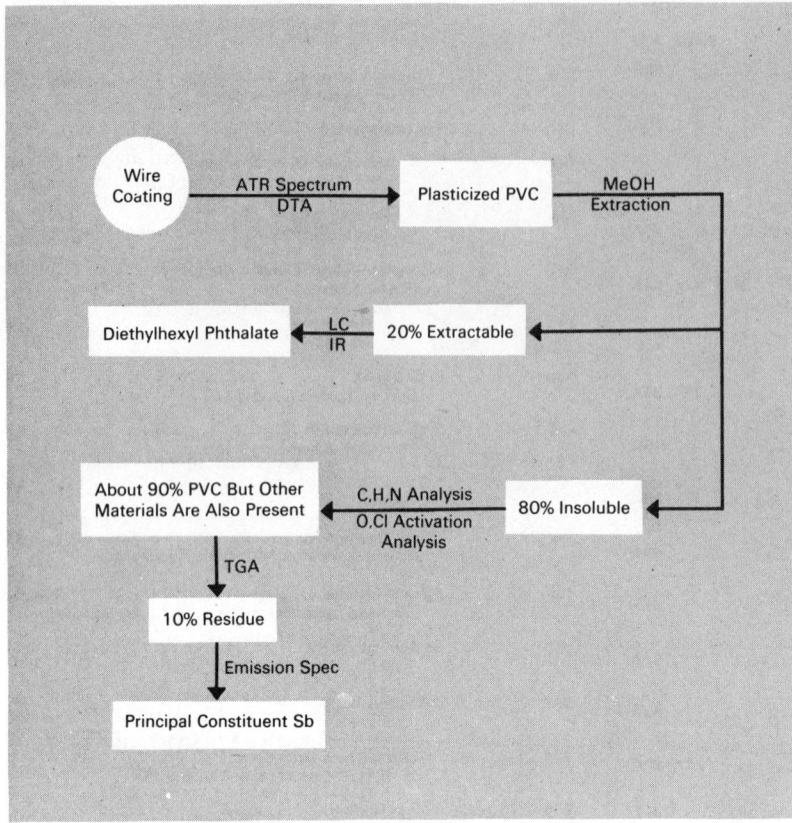

Figure 3. Analysis of wire coating

complexity of the analyses required for the characterization of commercial products. It should be obvious that a variety of techniques have to be utilized for rapid and effective analysis of these samples. It is also important to point out that in most cases only the major and the minor components are found and identified. Additives which may be present in trace quantities are only recognized if their presence is suspected and specific effort is made to detect them.

Acknowledgment

The author acknowledges the contributions of many people to the work discussed here, particularly those of Carmine Persiani and Edward Lanning, also Amy Fermin, Michael Rubner, Arthur Russell, Betty Orofino, Frank Mason, James Kranick, and Jorge Flores. He is further grateful for the contributions of all other members of the Materials Analysis Department of GTE Laboratories.

Assuring the Quality of Honey: Is It Honey or Syrup?

Landis W. Donor, Irene Kushnir, and Jonathan W. White, Jr.

U.S. Department of Agriculture, Agriculture Research, Science, and Education Administration, Eastern Regional Research Center, Philadelphia, PA 19118

Originally published in ANALYTICAL CHEMISTRY, 1979, Vol. 51, No. 2.

Honey, a natural product of limited supply and relatively high price, traditionally has been a target for adulteration. As a result, bulk honey markets are being lost to mixtures and substitutes in the form of sugar cane and corn-derived syrups. When such mixtures are appropriately labeled, there is no legal violation. However, the evidence of widespread mislabeling of these materials as pure honey represents a fraud to the consumer. The adulteration of honey with various sweet syrups without fear of detection is a great threat to the integrity of the honey markets, to the economic resources of beekeepers, and to the production of the more than 11.5 billion dollars worth of agricultural crops that depend on the honeybee for pollination.

The corn processing industry now produces a new low-cost sweetener, high fructose corn syrup (HFCS), in great amounts by making use of advances in bound enzyme technology. It has been estimated (1) that the food industry will use four billion pounds of HFCS annually by 1980, with per capita consumption reaching 18.2 lb, 14% of the total nutritive sweetener consumption. HFCS production involves the enzymatic isomerization of a portion of the glucose in conventional corn syrup to the sweeter sugar, fructose. The resulting syrup is then refined with activated carbon and ion-exchange treatment, and yields the two monosaccharide sugars, glucose and fructose, with only slight amounts of other materials.

A research program designed to develop methods for the detection of honey adulteration by HFCS was initiated at the Eastern Regional Research Center early in 1975. We considered several approaches with the hope that some would result in methods for HFCS detection in honey, regardless of improved production methods of this sweetener by the corn processing industry.

The Analytical Approach

Two general analytical approaches can be taken in attacking a problem such as the detection of mixtures of honey and HFCS.

Approach One: Identification of a constituent or property of the adulterant and detection of its presence in suspect honeys. Pure honeys would be shown either not to possess the chosen characteristic or to possess it at a much lower level.

Approach Two: Identification of a constituent or property of honey that is always present at a certain level. The addition of an adulterant without the characteristic would lower the concentration of the constituent or the value of the property.

Tests are available to detect the traditional adulterants of honey, including conventional corn syrup and commercial invert syrup (2). Approach One was used in the development of the official methods for identifying each syrup in honey. Conventional corn syrup (primarily glucose) contains appreciable amounts of higher molecular weight glucose polymers not present in honey, and these are detected by paper and thin-layer chromatography. Inverted cane syrups (consisting primarily of glucose and fructose) are detected in honey through the presence of hydroxymethylfurfural (HMF), which is present at significantly higher levels in acid-inverted cane syrups than in pure honeys.

Approach Two is of no value for the detection of adulterated honey because of the wide variability among the known constituents of honey. The most authoritative study of United States honey composition (3) illustrates the variability encountered. In that study, all commercially significant domestic honey types and blends were analyzed, and the results can be considered truly representative of domestic honey composition with regard to the components determined. The concentration ranges for the major constituents of honey (490 samples analyzed) and of HFCS samples from a major commercial source are given in Table I. The ranges indicate that a considerable amount of a sweetener like HFCS, with its sugar composition resembling honey, could be added without the mixture exceeding normal honey concentration limits. The increased moisture content can be compensated for easily by a clever manipulator. In the survey (3) the presence of other honey constituents over a wide range precluded the use of dilution of a honey property for detecting adulteration. Approach Two was used for just two possibilities examined in our research, unsuccessfully both times, as indicated in Table II.

The composition of honey depends upon two most important factors, the floral source and the composition of the nectar. Less important are certain external factors, including climate and differences in processing. All factors contribute to the variability of honey composition and to its enormous complexity [22 minor di- and trisaccharides have been identified (4)] and make most approaches to the detection of adulteration unworkable.

Early in our research program we had to learn as much as possible about the properties and composition of HFCS from the various commercial sources. Our goal was to find a constituent or property common to the various HFCS samples but not found in any honeys. Little information regarding the minor HFCS constituents that may come through the rigorous refining process was available in the literature, and examination of significant numbers of representative honeys with respect to any candidate HFCS components was required. It was our expectation that no single test would serve the objective of this search, and the need to screen large numbers of suspect samples would require relatively simple indicator tests; confirmatory analysis could be more complex. Furthermore, tests, to obtain legal standing, would have to be subjected to formal collaborative testing by independent laboratories, under the auspices of the Association of Official Analytical Chemists.

Establishing tests for HFCS in honey proved more challenging than earlier methods for the traditional adulterants, since HFCS is simpler, more closely resembles honey composition with regard to major components, and is more highly refined. Complicating the problem is the fact that methods of HFCS production are continuing to evolve, and trace constituents found to be unique in present syrups may be eliminated by new refining processes. Accordingly, we applied Approach One to the problem of detecting HFCS in honey in eight separate investigations (Table II); three proved useful.

Evaluation of Various Approaches

Three approaches to the problem involved determining lower molecular weight ions and molecules and were not valid for judging honey purity. A suggestion had been made (5) that examination of the sodium/potassium ratio was useful because HFCS is refined by ion-exchange treatment, and the original cations present in HFCS are replaced by sodium. Honey has long been known to be relatively poor in sodium but rich in potassium. A paper evaluating literature data (6), however, demonstrated that the sodium/potassium ratio is of little use as the sole parameter because of the extreme variability of these elements in honey. The convenience of atomic absorption analysis would have made this a very attractive approach, but this variability precluded its use.

Early in our research, we were optimistic that a useful test might be the determination of the monosaccharide sugar psicose, which in early production lots of HFCS had been reported (7) to be present at levels to 1% because of base-catalyzed isomerization of fructose. This sugar is not present in honey, and HPLC methods were available for its detection in standard mixtures with the major honey sugars, fructose and glucose. The presence of psicose in HFCS, however, indicated to the corn sweetener manufacturers that the process was not ideal, and they were successful in producing syrups free of this sugar and other by-products considered detrimental to the quality of their product. This was our first experience with a potential method being eliminated because producers were obtaining better (from their viewpoint), more highly refined syrups. This was not our last such experience because, in a sense, we were shooting at a moving target.

The amino acid proline is present in unusually high levels in honey and absent from HFCS. Consequently, proline was measured in 740 honey samples (8) to determine whether it is present over a sufficiently narrow concentration range to permit the use of Approach Two. A convenient colorimetric method is available for such a test, but the wide range of values

Table I. Concentrations of Major Constituents of Honey and High Fructose Corn Syrup (HFCS)

Constituents	Honey		HFCS	
	Range (%)	Mean (%)	Range (%)	Mean (%)
Moisture	13.4–22.9	17.2	23–29	26
Fructose	27.2–44.3	38.2	30–42	36
Glucose	22.0–40.7	31.3	31–35	33
Di-, tri-, and higher saccharides	4.6–23.3	10.1	3.1–5.7	4.4
Other		3.2		0.6

Table II. Approaches for Detection of Mixtures of High Fructose Corn Syrup and Honey

Method	Constituent determined	Found to be applicable
Approach		
Differential scanning calorimetry	Organic compounds	No
Gas-liquid chromatography	Isomaltose/maltose ratio	Yes
Gel filtration, affinity chromatography	Polysaccharides	No
High-pressure liquid chromatography	Monosaccharide (psicose)	No
Immunodiffusion	Polysaccharides, proteins	No
Stable carbon isotope ratio analysis	$^{13}C/^{12}C$ ratio	Yes
Thin-layer chromatography	Dextrins, polysaccharides	Yes
Turbidimetry (with concanavalin A)	Polysaccharides	No
Approach Two		
Atomic absorption spectroscopy	Sodium/potassium ratio	No
Colorimetry	Proline	No

found precluded the use of proline determination as an indicator of honey purity.

Several methods were designed to reveal differences between the minor macromolecular (polysaccharide and protein) fractions of honey and HFCS. In one approach, we constructed an autoanalyzer to determine the molecular weight distribution of polysaccharides from honey and HFCS. Initially, this approach appeared to be very promising. In a second method based on the differences in the polysaccharide fractions from honey and from HFCS, we took advantage of the unusually high degree of branching present in polysaccharides from some HFCS samples. The soybean lectin, concanavalin A, associates with terminal glucose units in branched polysaccharides through a multivalent interaction and leads to precipitation and quantification by turbidity. Interesting results were obtained (9) but, as with the other macromolecular approaches, were rendered useless for the adulteration problem when HFCS containing no high molecular weight carbohydrate polymers became commercially available.

An immunochemical approach wherein rabbits were injected with HFCS materials and tested for elicitation of antibodies was attempted. HFCS polysaccharides were conjugated with bovine serum albumin and keyhole limpet hemocyanin and, after being administered to rabbits, produced immune sera, which were isolated by scientists at the Western Regional Research Center of the U.S. Department of Agriculture. Unfortunately, interaction between the immune sera and the injected materials was not inhibited by HFCS polysaccharides. Efforts to prepare a protein concentrate from HFCS for preparation of an immune serum in rats and rabbits were unfruitful. Considering the standard immunodiffusion techniques available, we were disappointed that these sensitive and convenient approaches to the problem did not work.

We also attempted to apply differential scanning calorimetry to the adulteration problem. However, the virtually identical profiles obtained for honey and HFCS again reflected the similarity in composition of the two products.

Stable Carbon Isotope Ratio Method

Clearly, we needed to identify a property of HFCS that would not be affected by new refining processes but would be characteristic of products derived from the corn plant. The breakthrough came when we evaluated the ratios of the stable isotopes of carbon in representative samples of honey and HFCS. This approach had been used (10) in detecting the illegal addition of cane sugar to Israeli citrus juice, and it was suggested (10) that the $^{13}C/^{12}C$ ratio might be useful in detecting fraudulent substitutions of various types of plant products with corn-derived materials. Methods for detecting the adulteration of maple syrup with cane sugar (11) and the adulteration of natural bean vanillin with synthetic vanillin (12, 13) had utilized $^{13}C/^{12}C$ analysis.

The reasons for differences in $^{13}C/^{12}C$ ratios among members of the plant kingdom are now beginning to be understood and result primarily from the three different pathways by which carbon dioxide is fixed into organic compounds via photosynthesis. In plants using the classic Calvin (C_3) cycle, the primary photosynthetic product is the 3-carbon acid, 3-phosphoglycerate. In plants using the Hatch-Slack (C_4) cycle, the initial products of carbon dioxide fixation are the 4-carbon acids oxalacetate, malate, and aspartate. Plants in a third category, crassulacean acid metabolism, have the enzymatic capability for initially fixing carbon dioxide by the C_4 pathway and then shuttling it into the C_3 system. All plants are slightly lighter in ^{13}C than the carbon dioxide of the atmosphere, and Calvin (C_3) plants discriminate to a greater extent than do Hatch-Slack (C_4) or crassulacean acid metabolism plants. The $^{13}C/^{12}C$ ratios of a sample are reported as per mil (‰) deviations from a limestone standard and are defined as:

$$\delta^{13}C \text{ (per mil, ‰)} = \left(\frac{^{13}C/^{12}C \text{ sample}}{^{13}C/^{12}C \text{ standard}} - 1\right) \times 10^3$$

These values are determined by isotope ratio mass spectrometry after complete sample combustion to carbon dioxide. A flow diagram of the procedure is shown in Figure 1. Calvin (C_3) plants have $\delta^{13}C$ values of -22 to -33‰, and Hatch-Slack (C_4) plants have values from -10 to -20‰ (14, 15). Crassulacean acid metabolism plants have intermediate $\delta^{13}C$ values.

Corn, sugar cane, sorghum, and other grasses native to the tropics fall into the Hatch-Slack category with $\delta^{13}C$ values in the upper range. Pure HFCS then would have a similar value. It was our hope that all honey samples would be in the range expected for Calvin plants and that this might be a characteristic of all flowering, nectar-bearing honey sources. To test this, we selected 84 samples from our collection of pure honeys to represent all commercially important United States honey sources from wide geographical areas. We analyzed 35 imported honey samples from 15 countries, with geographical latitude the primary consideration in their selection. Details of the method and results of $\delta^{13}C$ analysis of these samples are described elsewhere (16–18); the distribution of values are diagrammed in Figure 2. The coefficient of variation for all honey samples was 3.86%, the smallest yet encountered for any constituent or physical property of honey. Mixtures of HFCS and honey would be expected to have $\delta^{13}C$ values equal to the sum of the fractional contribution of each. This was found to be the case.

A collaborative study of the method was conducted (17) by seven laboratories, each testing four prepared honey–HFCS mixtures and a pure honey. Because of the excellent agree-

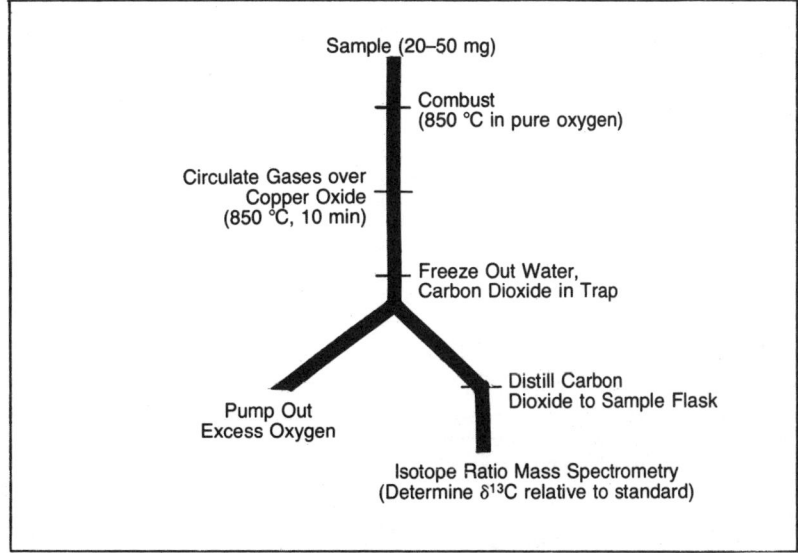

Figure 1. Flow diagram of method for $^{13}C/^{12}C$ determination

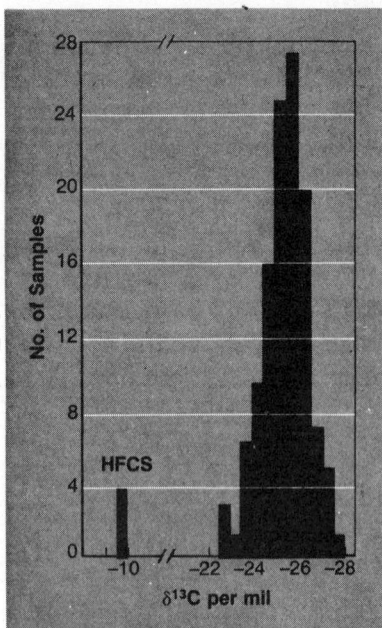

Figure 2. Distribution of $\delta^{13}C$ values for honey and HFCS

Reprinted with permission from ref. 17. Copyright 1978 Association of Official Analytical Chemists

ment among the collaborators, the Association of Official Analytical Chemists has adopted the method as official first action for handling cases of honey adulteration by HFCS. The beauty of this method is that it is noncircumventable and will apply regardless of new refinements in HFCS production. The only requirement is that these syrups continue to be produced from corn or other Hatch-Slack (C_4) plants. This procedure is now being used by regulatory agencies and by the honey industry for self-policing. The $\delta^{13}C$ values indicate to them whether samples are pure or adulterated honeys. The upper (least negative) limit for authentic honey may be set with any desired degree of certainty; Table III indicates the confidence with which a honey of a certain $\delta^{13}C$ value may be considered pure. A sample with a value less negative than $-21.5‰$ can be classified as adulterated.

Screening Methods Developed

The carbon isotope ratio test is moderately expensive. Only a few laboratories possess the required instrumentation, and none is in the regulatory field. A need existed for more routine tests that could be conducted in ordinary laboratories for selecting samples sufficiently suspicious to justify the confirmatory isotope ratio test. Recently, we developed two such methods, one using thin-layer chromatography and the other gas-liquid chromatography.

The thin-layer chromatographic method (19) has been subjected to successful collaborative testing (20) and recommended for adoption as an official method of analysis. A flow diagram of the procedure is given in Figure 3. This very sensitive procedure involves isolation of a fraction containing oligo- and polysaccharides from both honey and HFCS by column chromatography on charcoal-Celite. After concentration, these fractions were examined by silica gel thin-layer chromatography; consistent differences between honey and HFCS fractions were revealed. Whereas pure honeys yielded only one or two blue-grey or blue-brown spots of R_f greater than 0.35, a series of spots or blue streaks extending from the origin characterized adulterated samples. The method detects HFCS and the traditional honey adulterants, even when present as 10% or less of the total mixture. An added advantage is that this procedure should detect in honey the presence of all starch-derived sugar syrups tested, regardless of the plant source. Figure 4 shows the differences in the chromatographic profiles of honey, HFCS, and adulterated mixtures. This procedure is being used routinely to screen samples, not only for HFCS but also for other adulterants of honey, including conventional corn syrup and inverted sucrose syrups.

A gas-liquid chromatographic method (21) based on the determination of maltose and isomaltose has been useful in our laboratory. However, in view of the small number of successful collaborative tests (20), it could not be recommended for adoption as an official method. The results of the maltose and isomaltose determinations in honey and in HFCS samples are given in Table IV. A discriminatory equation was developed from these data, and 81% of authentic honey samples and 78% of adulterated honey samples (as determined by $\delta^{13}C$ analysis) were correctly classified.

Society Benefits

We are optimistic that awareness of these convenient new methods for detecting honey adulteration will minimize the threat to the integrity of honey markets. This will help protect the many thousands of beekeepers whose economic resources depend on confidence in the purity of their product. As a result, the population of honeybee colonies will be maintained at the high level so essential for the pollination of billions of dollars in food, feed, and fiber crops.

Since its advent about 30 years ago, isotope ratio mass spectrometry has been a powerful tool, particularly in

Table III. Probability of $\delta^{13}C$ Value of Authentic Honey Sample Being Lower Than a Stated Limit

Probability of a sample lower than limit	(%)	Limit $\delta^{13}C$ (per mil, %)
5 of 6	84.1	−24.4
43 of 44	97.7	−23.4
769 of 770	99.87	−22.5
24 999 of 25 000	99.996	−21.5

Reprinted with permission from ref. 17. Copyright 1978 Association of Official Analytical Chemists.

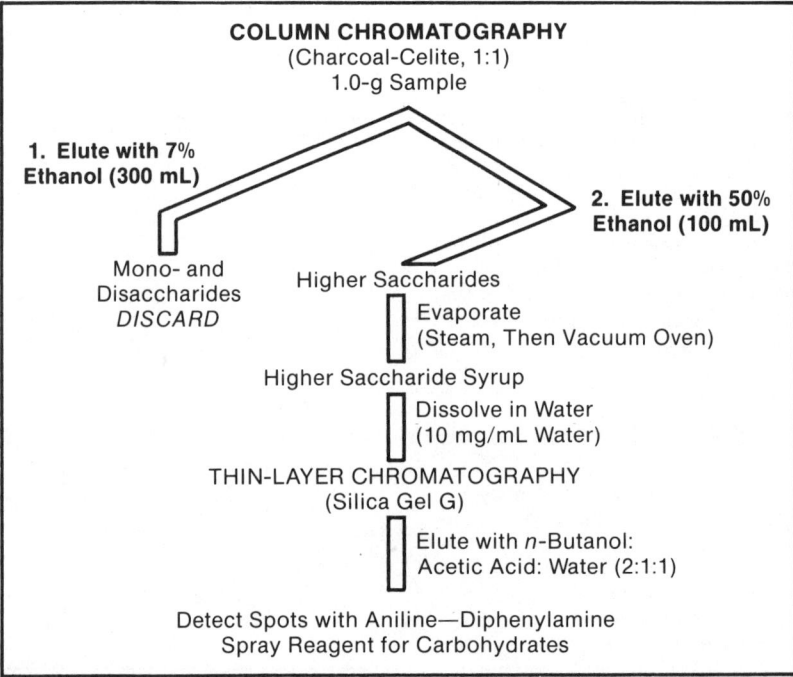

Figure 3. Thin-layer chromatographic test for honey adulteration

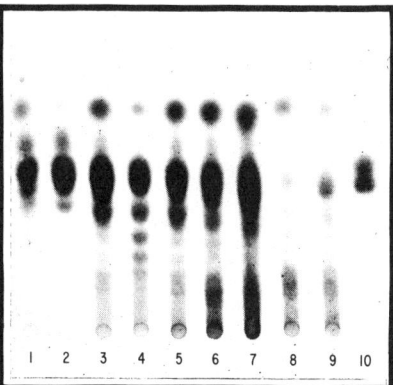

Figure 4. TLC plate of oligo- and polysaccharide fractions from honey, honey–HFCS mixtures, HFCS and standard trisaccharides

1, 2: Pure orange and clover honeys. 3–6: Mixtures of honey with 5, 10, 25, and 50% HFCS, respectively. 7: Mixture of honey with 5% conventional syrup. 8, 9: HFCS samples from two manufacturers. 10: Mixture of trisaccharides raffinose and melizitose

the realm of basic research. Now it has been applied to a major problem for the food and agricultural industries. Undoubtedly, numerous applications will be forthcoming as more is learned regarding natural variations in $^{13}C/^{12}C$ ratios and ratios of other stable isotopes among plants and their derived products.

The adulteration of natural vanilla extract with synthetic vanillin has been revealed by $\delta^{13}C$ measurements (12, 13); detection will hopefully result in this practice being discouraged. More recently, U.S. Customs authorities have been confronted with the problem of determining whether shipments of imported candied pineapple and papaya are processed with honey or inexpensive syrups from C_4 plants. A method was developed (22) to determine the nature of the processing syrup by $\delta^{13}C$ analysis. The method also has been applied by the apple juice industry to determine whether HFCS has been mixed with apple juice before production of apple juice concentrates.

The thin-layer chromatographic method (19) is both highly sensitive for the detection of honey adulteration by HFCS and convenient for use by regulatory agencies and the honey industry. Added advantages of this method are its detection of inexpensive C_3 plant-derived syrups in honey and its potential for further development and application to future honey adulteration by new sweeteners. It has been recommended that this method replace the old paper chromatographic method (2) for detecting the presence of commercial glucose, one of the traditional adulterants, in honey.

References

(1) *Chem. Eng. News*, **54** (17), 13 (1976).
(2) "Official Methods of Analysis," 12th ed., sections 31.134–31.136, 31.138–31.139, Association of Official Analytical Chemists, Washington, D.C., 1975.
(3) J. W. White, Jr., M. L. Riethof, M. H. Subers, and I. Kushnir, "Composition of American Honeys," Tech. Bull., U.S. Dept. Agric. No. 1261, 1962.
(4) L. W. Doner, *J. Sci. Food Agric.*, **28** (5), 443–56 (1977).
(5) R. S. Shallenberger, W. E. Guild, Jr., and R. A. Morse, *N.Y. Food Life Sci.*, **8** (3), 8–10 (1975).
(6) J. W. White, Jr., *Bee World*, **58** (1). 31–5 (1977).
(7) J. M. Newton and F. K. Wardrip, "Symposium: Sweeteners," G. E. Inglett, Ed., Chap. 8, pp 87–96, Avi Publ., Westport, Conn., 1974.
(8) J. W. White, Jr., and O. N. Rudyj, *J. Apic. Res.*, **17** (2), 89–93 (1978).
(9) L. W. Doner, *J. Agric. Food Chem.*, **26** (3), 707–10 (1978).
(10) A. Nissenbaum, A. Lifshitz, and Y. Stepek, *Lebensm. Wiss. Technol.*, **7**, 152–4 (1974).
(11) C. Hillaire-Marcel, O. Carro-Jost, and C. Jacob, *J. Inst. Can. Sci. Technol. Aliment.*, **10** (4), 333–5 (1977).
(12) J. Bricout and J. C. Fontes, *Ann. Falsif. Expert. Chim.*, **716**, 211–5 (1974).
(13) P. G. Hoffman and M. Salb, *J. Agric. Food Chem.*, in press, 1979.
(14) M. M. Bender, *Phytochemistry*, **10**, 1239–44 (1971).
(15) B. N. Smith and S. Epstein, *Plant Physiol.*, **47**, 380–4 (1971).
(16) L. W. Doner and J. W. White, Jr., *Science*, **197**, 891–2 (1977).
(17) J. W. White, Jr., and L. W. Doner, *J. Assoc. Off. Anal. Chem.*, **61** (3), 746–50 (1978).
(18) J. W. White, Jr., and L. W. Doner, *J. Apic. Res.*, **17** (2), 94–9 (1978).
(19) I. Kushnir, *J. Assoc. Off. Anal. Chem.*, in press, 1979.
(20) J. W. White, Jr., I. Kushnir, and L. W. Doner, *ibid*.
(21) L. W. Doner, J. W. White, Jr., and J. G. Phillips, *ibid*.
(22) L. W. Doner, D. Chia, and J. W. White, Jr., *ibid*.

Table IV. Gas Chromatographic Determinations of Maltose and Isomaltose in Honey (80 U.S. Samples, 35 Imported Samples) and in HFCS (21 Samples)

	Maltose		Isomaltose	
	Mean (%)	SD	Mean (%)	SD
Domestic Honey	1.93	0.51	0.64	0.37
Imported honey	2.17	0.53	0.87	0.50
HFCS	0.72	0.26	1.50	0.82

Reprinted with permission from ref. 21. Copyright 1979 Association of Official Analytical Chemists.

Snap, Crack, Blister, and Peel: An Analytical Approach

George L. Fix

Raytheon Co., Equipment Development Laboratories, Sudbury, MA 01776

Originally published in ANALYTICAL CHEMISTRY, 1980, Vol. 52, No. 12.

Each of us has experienced anger and frustration from an elastomer, adhesive or paint that failed to perform to our expectations. The knowledge that such failures are a common experience may provide a topic of conversation to share with others, but it offers little consolation when a laborious personal project is suddenly doomed to failure. Material failures are far more important to a corporation, however, for its continued business depends on the reliability of its products. This is particularly true for manufacturers of military hardware.

The combination of stringent performance requirements with state-of-the-art technology, electronics and materials makes military hardware the most challenging commodity a corporation could elect to manufacture. When a material fails, and some do, the demand for high reliability in military hardware requires that the "root cause" of the failure be accurately determined and reproduced. The personnel assigned to study the root cause obviously vary with the nature of the failure—mechanical, metallurgical, electronic, or chemical. Failure analysis of organic materials is one of many activities undertaken by materials engineers, many of whom are actually chemists cleverly disguised in a white shirt, brown shoes and a fully outfitted pocket saver.

An extremely important search for a root cause occurred in the early sixties because of serious problems with polyurethane encapsulants. The encapsulants were employed to isolate electronic components electrically and support them during shock and vibration. Many materials performed admirably in the laboratory but failed in the stifling heat and humidity of South Vietnam. The root cause: Hydrolytic degradation transformed the polyurethane encapsulants into a soft, sticky and relatively conductive ooze. This root cause determination identified some basic types of organic materials as "4-F" and led to hydrolytic stability testing of materials for use in military electronics.

The light weight and high density of today's electronics have significantly increased the application of organic materials. At the same time the performance capabilities of organic materials have increased so dramatically that some plastics are actually stronger than steel. Fortunately, the analytical instrumentation available to the materials engineer has kept pace with the advancement of other technologies. Surprisingly to some within the electronics industry, a well-equipped materials engineering laboratory includes an ESCA/Auger spectrometer, a differential scanning calorimeter, a high pressure liquid chromatograph, and a Fourier transform infrared spectrometer.

Adhesive and Cohesive Failures

In analyzing a failed adhesive, encapsulant or elastomer, the type of failure mechanism determines the selection of the analytical techniques applied to the problem. For example, an adhesive failure occurs at the interface between the adhesive and the substrate. Adhesive failures are generally best investigated by surface-sensitive techniques such as multiple internal reflectance infrared spectrometry, ESCA and Auger spectrometry.

A cohesive failure occurs within the adhesive. Cohesive failures require evaluation of bulk properties and most often utilize solvent extraction, liquid chromatography, thermal analysis and infrared spectrometry. Also, a very powerful tool for evaluating both types of failures is the scanning electron microscope (SEM) equipped with an X-ray fluorescence analyzer. Semi-

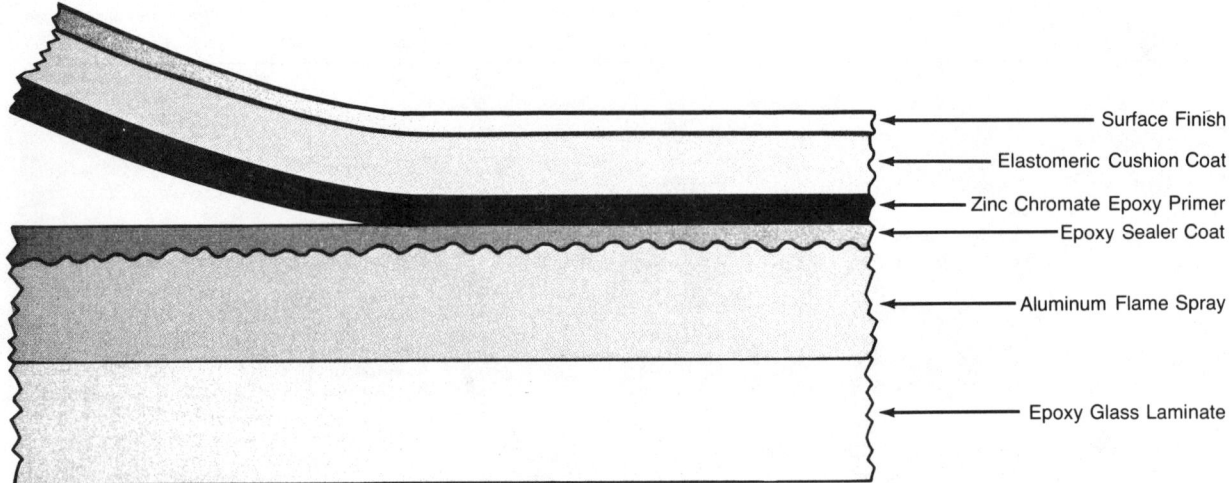

Figure 1. Cross section of the failed paint system of a large radar antenna

quantitative information on the inorganic constituents of adhesive failures is easily obtained and often provides a basis for the rest of the investigation. SEM photographs of the fracture surface of cohesive failures readily reveal voiding and noncharacteristic modes of fracture propagation.

Paint Adhesive Failure

An interesting example of a catastrophic adhesive failure occurred last winter with the polyurethane paint system of a large radar antenna. As the photograph shows, the paint stripped from the antenna surface like shelving paper, all 250 square ft of it. Beneath the peeled paint was a yellow-brown liquid, visible in the photograph on the exposed section of the antenna surface. Figure 1 shows the cross section of the antenna, epoxy sealer coat and polyurethane paint system. Samples of the paint film, the yellow-brown liquid and scrapings of the epoxy sealer coat were collected for analysis. Understandably, we did not section samples of the epoxy glass laminate of the antenna, since it was in excellent condition.

The free paint film was found to have excellent adhesion between layers. Solvent extraction, liquid chromatography, infrared spectrometry and differential scanning calorimetry revealed no problem with the paint system. ESCA analysis of the primer surface disclosed only the normal constituents of an epoxy-based zinc chromate primer.

The yellow-brown liquid was an aqueous solution containing two discrete constituents, one yellow and one brown. Infrared spectrometry, SEM X-ray fluorescence, and wet chemical methods were all used for its analysis. The yellow solute was identified as water-soluble chromates, a normal constituent of primers for aluminum surfaces. Their presence was not surprising since their function is to bleed, should moisture contact the primer surface. An excess of water soluble chromates can cause similar failures, however, and the primer was initially suspect. The brown solute was found to be an amine. Since both the primer and the sealer coat are amine cured liquid epoxy based systems, the origin of the water soluble amine was uncertain.

The scrapings of the epoxy sealer coat were analyzed by solvent extraction, liquid chromatography, and infrared spectrometry. These analyses revealed that the sealer coat had contained too much amine curing agent.

To check the cause of the failure, the conditions were reproduced. The improperly formulated sealer coat was duplicated, applied to aluminum test panels, cured and overcoated with the paint system. No exposure to humidity, however, could duplicate the catastrophic failure of the antenna. Since the facilities of the vendor who applied the sealer coat and paint system were as cool as 50 °F during the winter months, the misproportioned sealer coat was applied to cold aluminum test panels. Once again humidity exposure would not cause duplicate catastrophic failure in the test panels.

Although a discrete problem with the epoxy sealer coat had been identified analytically, reproducing the failure mechanism seemed impossible. Amidst all of the theories and speculation, one technician's observation surfaced. The unused mass of the adhesive, which was used to prepare test coupons, really "steamed and smoked" when it gelled. Examination of the unused resin batches that frequently litter the laboratory during the fervor of a root cause determination showed a dull occlusion on their surface. This surface film was indeed quite water soluble. The high heat generated by the exothermic reaction of the large mass of the unused material was known to cause the steam and smoke phenomenon, but the resultant surface film had previously escaped observation. Excessive heat applied during the curing of the sealer coat was indeed found to duplicate the failure. When the vendor was contacted, the source of the problem became apparent. The root cause: To compensate for the coolness of the building, the vendor had positioned too many infrared lamps too close to the antenna. This action resulted in a rapid exothermic reaction which caused this epoxy to exude a resinous film that was quite water soluble. When the antenna was exposed to the environment, water permeated the paint system and dissolved the exudate. The dissolution of the adhesive interface resulted in the release of the large free film of paint.

Pin Connector Failure

In addition to adhesive and cohesive failures, electrical failure is also encountered with polymeric materials used in the electronics industry. The electrical failure may be attributed to one or more electrical properties such as leakage current, breakdown voltage, dielectric constant or dissipation factor. Similarly, one or more characteristics of the material may affect its electrical performance. Filler content, type and size; entrapped air; degree of cure; hydrolytic stability; crystallinity; stoichiometry and oxidation resistance are but a few examples.

An interesting example of an electrical failure occurred recently with an epoxy adhesive used to dam off wire wrap pin connectors. The epoxy, as shown in Figure 2, was applied to prevent leakage of the module encapsulant into the contact area of the connector. The completed electronics assembly gave satisfactory tests at the

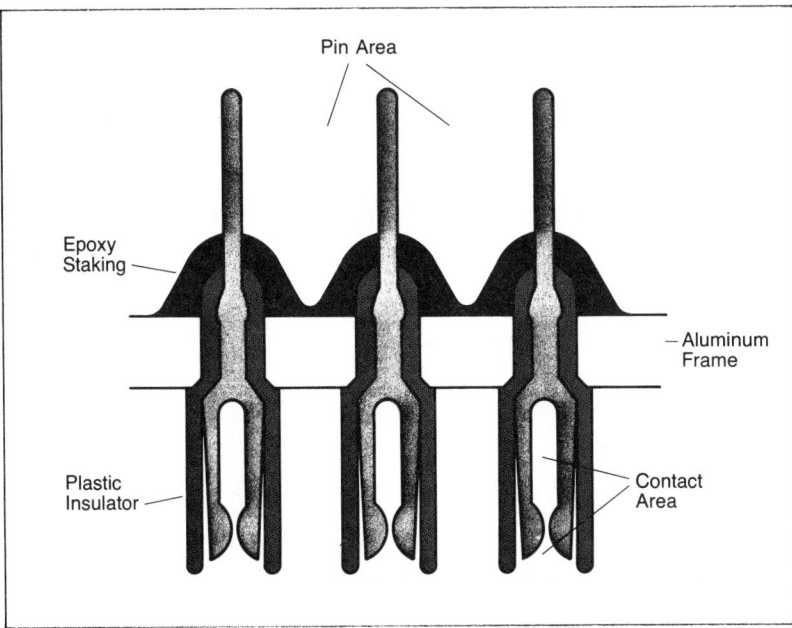

Figure 2. Cross section of a wire wrap connector Epoxy staking is applied to prevent the subsequent encapsulation of the pin area from reaching the contact surfaces of the connectors

factory but when put into service, it leaked electrically.

The sample quantity available for analysis was too small for Soxhlet extraction or other macro physical analyses. Infrared analysis, however, showed considerable differences among the relative intensities of many of the peaks of the epoxy when compared to the spectrum of the properly formulated and cured adhesive. Further infrared analysis of adulterated mix ratios of the epoxy resin and curing agent enabled the mistaken proportion to the determined by standard peak ratioing techniques. The proper proportion for this adhesive is four parts of resin to one part curing agent. The proportion of the failed adhesive was determined to be one part resin to four parts of curing agent—quite a difference!

As part of the root cause determination, one fact remained to be explained: The electronic assembly had initially passed electrical leakage testing but failed in service. One of the suspects was humidity because of the amine curing agent employed. So the incorrect mix ratio was prepared and electrically tested just after being cured and again following humidity exposure (7 days @ 60 °C, 95% RH). Indeed, as can be seen in Figure 3, humidity dramatically affected the resistivity of the misproportioned adhesive. Only the dry adhesive could be tolerated. The initial electrical testing had been performed shortly after an oven curing cycle, but by the time the assembly was placed in service, moisture had affected the marginal electrical properties of the misproportioned adhesive.

Cable Jacketing Failure

Experience has shown that the most common cause of crazing in electronic cable jacketing is environmental. Most frequently, cable jacketing failures are the result of the inadvertent substitution of a lesser-grade material (which cannot withstand the environment) for the specified MIL-SPEC material.

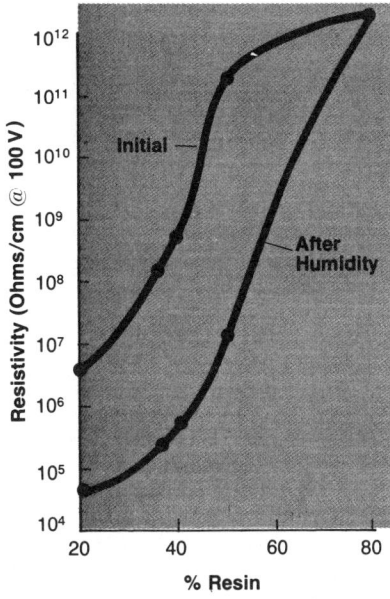

Figure 3. Resistivity of various proportions of an epoxy adhesive. Initial values were measured shortly after the formulations were cured for 4 h at 70 °C. Following humidity exposure, 7 days @ 60 °C and 95% RH, only the proper formulation, 80% resin, was unaffected.

This was the instance in a recent failure of a neoprene cable jacketing. The jacket was severely crazed after only six months exposure to the comparatively temperate summer of New England. Neoprene and most other rubbers obtain many of their physical properties from the plasticizers, antioxidants, UV stabilizers and other performance additives from which they are compounded. These additives are usually easily solvent extracted from the rubber, and their presence, absence or relative concentration often provides the most meaningful information as to the root cause of the failure. In this particular instance GPC analysis showed that the only performance additive in the failed neoprene rubber was a low molecular weight plasticizer, even though the jacketing was supposed to possess UV and oxidation resistance. The lack of performance additives easily explained the root cause of the jacketing failure.

In another instance, the jacketing of a high voltage cable assembly was found to be badly crazed in localized areas after a brief period of service. The same comprehensive analyses performed on this jacketing revealed that it had, indeed, been properly formulated. The service environment of high voltage and its associated production of ozone was suspect, yet a depletion of the antiozone additive was not observed in the bulk analyses. ESCA analysis, however, showed an extraordinarily high concentration of oxygen in the crazed surfaces. Further examination of the service environment revealed that inadequate ventilation of ancillary structures permitted high concentrations of ozone to develop in the areas of localized crazing.

Root Cause Determination: The Analytical Approach

No one technique or combination of techniques is always applicable to the root cause determination of a material's failure. Sample size is often the determining factor, but more frequently the type of failure or even the failure environment will determine which techniques and analytical routines are applied. Like other occupations involving applied analytical chemistry, investigations by the materials engineer into the root cause of a material's failure are systematic. The sequence and types of analytical techniques applied to disclose the origin of a failure are based upon known facts, prior experience, observations, and judgment. The complex and diverse nature of electronic materials' failures necessitates multiple techniques and a well-directed and dedicated analytical approach to determine the root cause.

Multitechnique Approach Solves Construction Materials Failure Problems

William G. Hime

Erlin, Hime Associates, Northbrook, IL 60062

Originally published in ANALYTICAL CHEMISTRY, 1974, Vol. 46, No. 9.

Microscopists and analytical chemists at the Erlin, Hime Associates Laboratories are regularly confronted with problems involving failures of construction materials: cement, concrete, metals, paints, and coatings. The general approach to the solution of these problems involves initial study of the material by the techniques of petrographic microscopy to discover the mechanism of the failure, followed by the application of analytical techniques to ascertain the causative agents. The techniques typically used and the kinds of information obtained are listed in Table I.

It is estimated that over 90% of construction materials failure problems can be solved by the approach suggested in Table I, provided the microscopist–chemist team is expert enough in the chemistry and behavior of construction materials to know what to look for. For example, the presence of very large quantities of many substances has little effect on the properties of cement or concrete, but very small quantities of others cause enormously deleterious effects. To illustrate, silica in the form of quartz can be present as the major concrete component. But silica in the form of opal must be limited to a few percent. Even more powerful in their immediate effect are certain organic substances which at a thousandth of a percent level affect the setting, workability, or strength of concrete. The following three examples of "failure analyses" illustrate the approach.

"Unset" Concrete— Getting the Lead Out

When concrete forms were removed on a large construction project in New York, everyone held their breath. Occasionally, the concrete came pouring out. The uncertainty finally dictated

Table I. Techniques Used for Hardened Concrete

Technique or method	Information obtainable
Light microscopy	Air-void system Aggregate-composition, texture, classification, reaction rims Proportions of aggregate and paste Cracking patterns Identification of solid admixtures Extent of cement hydration Composition, fineness, and dispersement of relic cement particles Identification of hydrated cement compounds Identification and location of secondary compounds Detection of "unaccommodative" chemical reactions Physical properties of the paste such as hardness, granularity, porosity, density
Atomic absorption	Quantitative analyses of "oxides" present in cement and concrete
Infrared spectroscopy	Identification of organic admixtures (air-entraining, set-retarding, and workability agents)
Wet-chemical analysis	Cement content Chemical composition of aggregate Chemical composition of paste Chemical composition of secondary compounds Detection of some organic substances
X-ray diffractometry	Aggregate mineralogy Identification of secondary compounds Identification of hydrated and unhydrated cement compounds
X-ray fluorescence	Identification and relative proportion of elements present in aggregate

that the multimillion dollar project be halted until the cause for the failure-to-set problem was determined and corrected. By the time a sample of the concrete was received in the laboratory, the "unset" concrete had already hardened. Microscopical analysis revealed unusual, thin rims on the cement particles, suggesting an excessive amount of a cement set-retarder.

Since sabotage had been suspected, the concrete was analyzed for sugar—a known set-retarder. (A cup of sugar can delay the set of yards of concrete for weeks.) Colorimetric methods did not detect sugar; therefore, other extracts of the concrete were then analyzed by infrared and ultraviolet spectroscopy for known cement hydration retarders, such as other polysaccharides and lignosulfates. These results also were negative.

X-ray fluorescence measurements were then made. Trace quantities of lead and zinc were detected. Since experience has indicated that quantitative analyses of quite varied materials are made more accurately by atomic absorption, AA determinations for lead and zinc were performed, and about 0.03% of each was found. Such quantities, when present as alkali-soluble compounds, are known to delay severely cement hydration.

Further work resolved the mystery. A dredged river gravel was being used as the aggregate for the concrete. A thin band of the "New Jersey" lead and zinc deposit passed across the river. The dredging operation thus accounted for the sporadic occurrence of these elements in the concrete.

Holey Concrete

The quality of concrete is usually monitored by compression tests of samples taken during the "pour." Unfortunately, a lot of construction may take place before the initial (usually three-day) results become available. Thus, when tests on a large road paving project indicated strengths of 50% below requirements, all work was stopped while the laboratory team worked on the problem. The general approach taken in a problem of this type is illustrated in Figure 1.

The microscopists quickly determined the failure mechanism—15% air in the concrete. About 5% air by volume is frequently specified because it provides great protection against freeze-thaw deterioration. Such a quantity does not significantly affect strength, but each additional percent of air leads to a loss of about 5% strength. A photomicrograph of a polished section of the holey concrete is shown in Figure 2.

Samples of the concrete were extracted with a number of solutions, and the extracts were prepared for

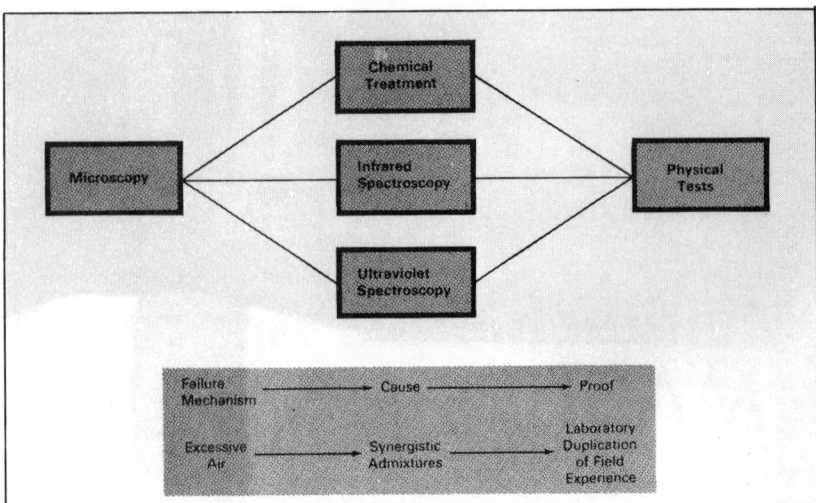

Figure 1. General approach to study of admixture problem

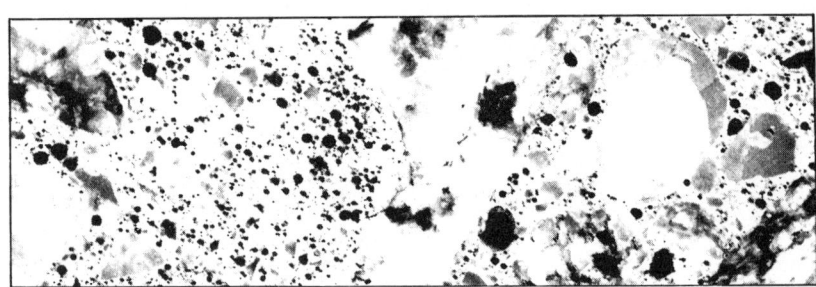

Figure 2. Photomicrograph of polished section of highly air-entrained concrete. Black "holes" are sectioned microscopic bubbles of air

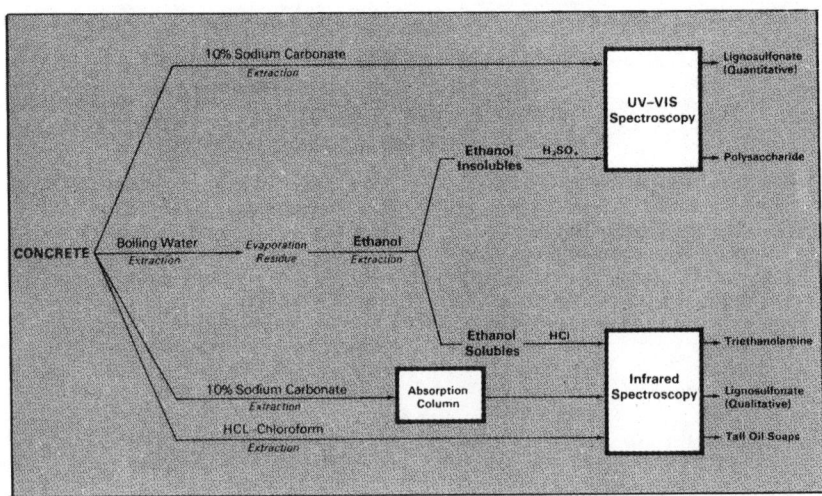

Figure 3. Analytical methods and results for admixture problem

analysis by absorption spectroscopy as detailed in Figure 3. Infrared revealed the presence of two commercially available admixtures, materials added to concrete to produce special properties. One admixture, a triethanolamine salt and tall oil soaps, is added to cause entrainment of air. The other admixture identified (triethanolamine, polysaccharides, and lignosulfo-

nate) is sold to increase the "workability" of plastic concrete. The concentrations of these admixtures were determined by infrared and ultraviolet-visible spectrophotometry by comparison of the sample extracts with extracts from "known" concretes containing the identified admixtures.

But one admixture singly was known to entrain only 5% air, and the other only 2%. A concrete mix containing both admixtures in the determined dosages was prepared, and this particular combination proved synergistic—over 15% air had been entrained, and 50% loss in strength had resulted. Unfortunately, there was no way to save the mile of concrete pavement (about 10 million lb) that had been placed, but succeeding pavement set satisfactorily by controlling the amounts of the admixture in the concrete.

Concrete Scales

Many owners of wood frame houses recognize that the peeling of paint is a major economic headache. Concrete may experience the same distress. But when a concrete block building began losing not only its paint but also a $\frac{1}{4}$-in. layer of underlying concrete (Figure 4), consternation really abounded.

Microscopic analysis of the received "scales" revealed thin delamination layers within the spalled section received at our laboratory. This effect is characteristic of freeze–thaw damage to critically saturated concrete.

Analysis of the paint by infrared spectroscopy, solvent extraction, and pyrolysis techniques disclosed an alkyd type of paint. Such a paint is classified as "nonbreathing."

With this analytical data, the team speculated that moisture entering the building walls was prevented from escaping by the paint. The concrete near the paint surface became saturated and during winter froze because of its exposure at outside temperatures.

A site visit revealed a flashing detail error that allowed entrance of rain water into the walls. The corrective measures suggested were the elimination of the flashing detail error by properly redoing the flashing and the replacement of the nonbreathing paint with a breathing paint.

Figure 4. Underside of spalled concrete "flake" that originally extended across three masonry blocks. Section is about $\frac{1}{4}$-in. thick. Opposite side is painted

Chromatopyrography for Polymer Characterization

John Chih-An Hu

Boeing Aerospace Co., Quality Assurance Laboratories, Seattle, WA 98124

Originally published in ANALYTICAL CHEMISTRY, 1981, Vol. 53, No. 2.

An urgent problem arose in the Air Launched Cruise Missile (ALCM) project of Boeing Aerospace Company during the crucial competition for a major defense contract. During the final inspection before the scheduled delivery of the missiles to the U.S. Air Force, the quality assurance engineers discovered that one rubber part of the missile was not stamped with identification marks and its composition was unknown. The delivery was held up. The competition between Boeing and another aircraft company had reached a decisive stage, and it was critical to meet the delivery schedule. The part in question was an expensive item, and the nondestructive sampling analysis necessary allowed only a tiny amount of sample to be obtained.

Conventional mechanical and instrumental analytical methods were undesirable because (a) they were not specific enough for characterization; (b) they were too slow to meet the delivery schedule; or (c) they required a large sample size, which would result in destruction of the part. The urgency of the problem prompted the ALCM management team responsible for delivery to make an unprecedented visit to the laboratory to personally request a miraculous analysis within the hour to meet the delivery schedule. A new analytical approach—a technique called chromatopyrography (CPG), which had been developed in our Boeing Quality Assurance Laboratories—made this speedy analysis a reality. This article will describe the new approach and some of its applications to actual problem-solving.

Chromatopyrography, a one-step two-shot analytical technique, is a new adaptation of pyrolysis gas chromatography. Boeing's analytical chemists had originally developed it to modernize the quality control test methods

Figure 1. Chromatopyrography

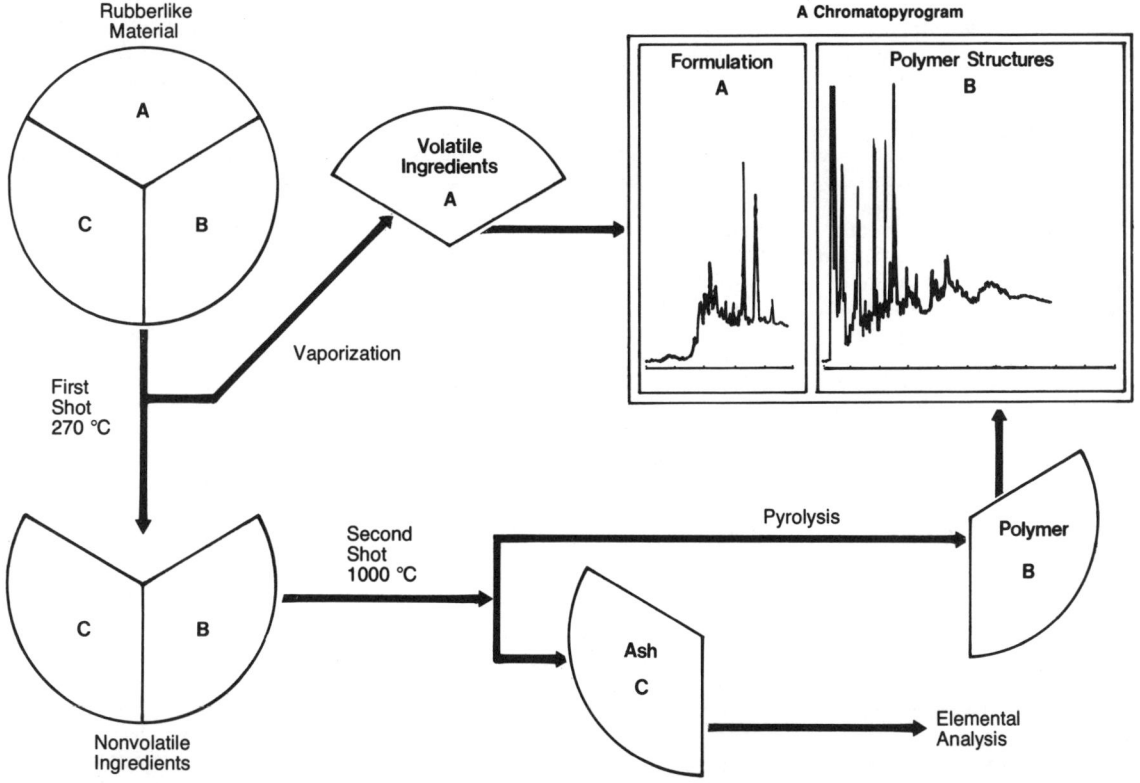

used in the inspection of incoming nonmetallic industrial goods and polymeric materials.

This modernization was necessary because the ever-increasing number of test samples from incoming materials mandated faster and more economical test methods. In the aerospace industry today, polymeric materials are widely used, especially those with relatively low density and consequent light weight.

Traditional Methods of Analysis

Traditional mechanical tests (i.e., tensile, elongation, etc.) are not satisfactory for the complete characterization of polymeric materials since they do not define a unique chemical identity of the material being tested. Also, they can be time-consuming, costly, and they often require large sample size and test specimens in specific forms. Instrumental methods such as chromatography, mass spectrometry, infrared spectrometry, and thermal analysis are faster than mechanical tests, and the data are more suitable and reliable for quality control purposes.

But academic instrumental methods often have to be modified for the solution of industrial problems. New approaches are sometimes required because industrial materials tend to be diverse and complex in composition, often making sample preparation troublesome. It was a new approach to a conventional instrumental technique—pyrolysis gas chromatography—that resulted in the development of chromatopyrography. It was used first for quality control purposes, but later it was applied to the solution of problems such as the unidentified missile part.

Polymer analysis by conventional pyrolysis gas chromatography suffers from problems with interlaboratory reproducibility and standardization. Results of cooperative studies among many laboratories in the U.S. (1) and in Europe (2–5) indicated that reproducible pyrograms could be obtained from pure polymers, but that commercial polymeric materials give poor results because of their chemical complexity. Unfortunately, the majority of samples received in industrial laboratories fall into this latter category. Most efforts to improve reproducibility in recent years have been concentrated on the design of an effective pyrolyzer with a fast pyrolysis temperature rise time (6). But little attention has been directed toward other weaknesses in the method.

The first of these weaknesses involves the solid sample introduction technique. The traditional procedure required a prepyrolysis waiting period after the sample was inserted into the injection port, but this waiting period resulted in sample losses. The amount of sample lost depended on the injection port temperature, boiling points of the volatile constituents of the sample, length of the waiting period, and the carrier gas flow rate.

Also, compounded polymeric materials contain definite amounts of volatile ingredients. Some volatile ingredients were lost during the waiting period, and the remaining volatiles caused irreproducibility problems. In some procedures, attempts were made to remove the volatiles by solvent extraction, but in addition to being time-consuming, this created problems of its own since residual solvent and impurities from the solvent complicate the analysis.

The analysis of these volatile ingredients is just as important as analysis of the polymer because, frequently, it is the formulation containing volatile ingredients that is unknown rather than the polymer itself. A large number of polymeric materials can be manufactured from a single type of polymer by varying the ingredients in these formulations.

We felt that an on-line process, which could be carried out in the carrier gas flow system, would be more desirable than elimination of the volatiles by extraction or use of a waiting period. Chromatopyrography (7–11) is such a process.

The CPG Approach

CPG, as shown in Figure 1, involves

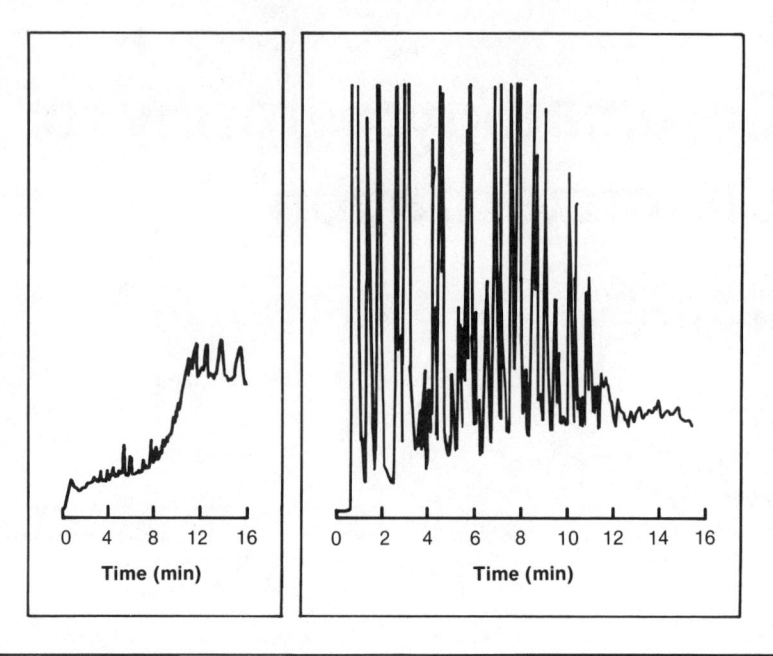

Figure 2. Chromatopyrogram of silicone rubber A

only one step but is actually a two-shot process. The first shot is designed to determine the volatile ingredients and the specific formulation. The second shot identifies the polymeric structure.

The procedure is actually quite simple. The injection port is preheated to 270 °C and maintained at that temperature. This heat along with a carrier gas flow of 30 mL/min causes an effective flash vaporization. The analysis begins as soon as the sample is inserted. This is the first shot, which immediately drives out all of the volatile ingredients and results in chromatogram A, which serves as a fingerprint of the formulation. After chromatogram A is complete, the second shot is fired simply by pushing the pyrolysis button. The thermally purified polymer is then pyrolyzed at 1000 °C for 15 s to develop pyrogram B, which serves as a fingerprint of the polymer. After pyrogram B is complete, the inorganic residue can be isolated for elemental analysis if desired. The combination of chromatogram A and pyrogram B forms what we call a chromatopyrogram. These two shots are more effective in completely characterizing a polymeric material than either shot alone.

The CPG approach effectively meets a number of requirements. First, the labyrinth-type design of the injection port is such that sample losses do not occur due to back flush of the carrier gas (9). During sample insertion, the carrier gas is momentarily released into the atmosphere along the outermost peripheral edge of the injector barrel and the space inside the injector liner where the sample is located is in a static condition without back-flush. As soon as the pyrolysis probe is sealed, the normal carrier gas flow is resumed and gas chromatographic analysis begins. All volatile ingredients of the sample are thus subjected to CPG analysis. Another factor is that the dynamic percolating conditions in the heated injection port substantially lower the boiling temperatures of the monomeric ingredients to such an extent that practically all volatile ingredients are vaporized instantaneously at 270 °C while the high polymers are not affected (11). The requirement of only a minute sample size in the range of sub-milligrams to micrograms also favors a rapid and complete vaporization of the volatile ingredients. And the fast vaporization results in sharp peaks and a stable recorder baseline during sample insertion.

Meeting all of these conditions means CPG can also be used for the direct introduction of heterogeneous viscous liquid samples for GC analysis (9), simplified headspace-type analysis, and splitless sample injection for capillary GC (9).

CPG Applications

CPG has modernized industrial quality control test procedures, and it also has problem-solving applications. Many real-world analyses involving rubberlike polymeric materials, which in the past required long research efforts, can now be completed by CPG within an hour.

It is CPG that solved the urgent identification problem we described earlier concerning the unstamped missile part. We knew the part should have been fabricated from an approved Boeing proprietary formulation that was based on a specific siloxane polymer. Figure 2 is the chromatopyrogram of the sample, which was removed from the missiles by a pseudo-nondestructive technique. It is identical to that of the specified Boeing proprietary material (silicone rubber A). The unmarked missile part had been unambiguously identified within the time limit and the missiles could be delivered on schedule! Figure 3 shows the chromatopyrogram of a different silicone rubber (silicone rubber B) for comparison purposes.

A second case that required a CPG solution occurred in the Boeing AWACS project. The AWACS (Airborne Warning and Control System) is the aircraft with a flying-saucerlike radome mounted on top of a Boeing 707 air frame. During a routine inspection of the prototype AWACS aircraft it was discovered that some rubber sheaths covering high voltage electric cables showed signs of crazing while other similar sheaths didn't show such failure signs. Failure of electric insulation might result in serious difficulties in operating the AWACS system, so the problem was immediately investigated. The approach was to first find out what materials were involved by chemical analysis and then to deduce the cause of failure.

The sample of the failed sheath was subjected to our former conventional rubber analysis: extraction with solvents, then pyrolysis and infrared (IR) spectrometry, a time-consuming procedure. The IR results of the failed material indicated a nitrile type of rubber, but the formulation of the sample was not positively identified. Therefore crazed and noncrazed rubber sheaths were reexamined with CPG, which immediately characterized both the formulation and polymer structure of the materials. From

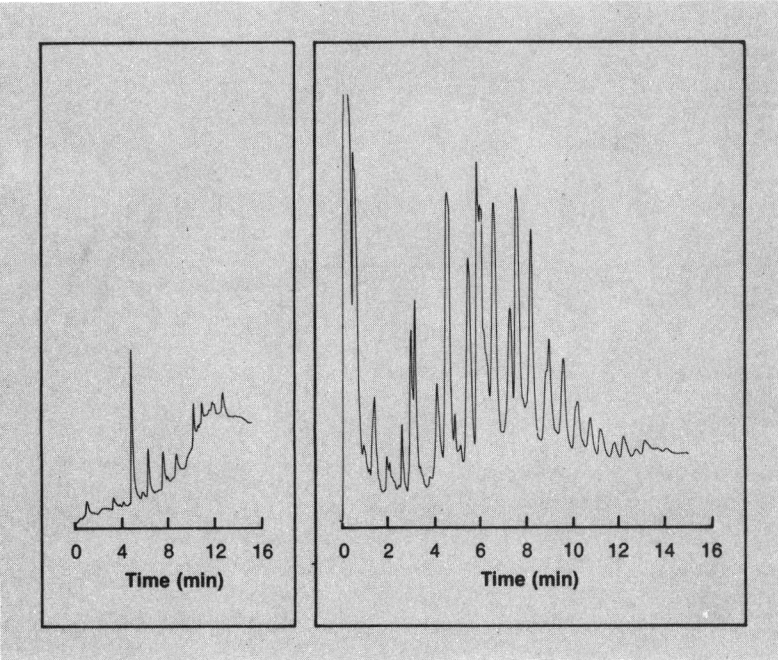

Figure 3. Chromatopyrogram of silicone rubber B

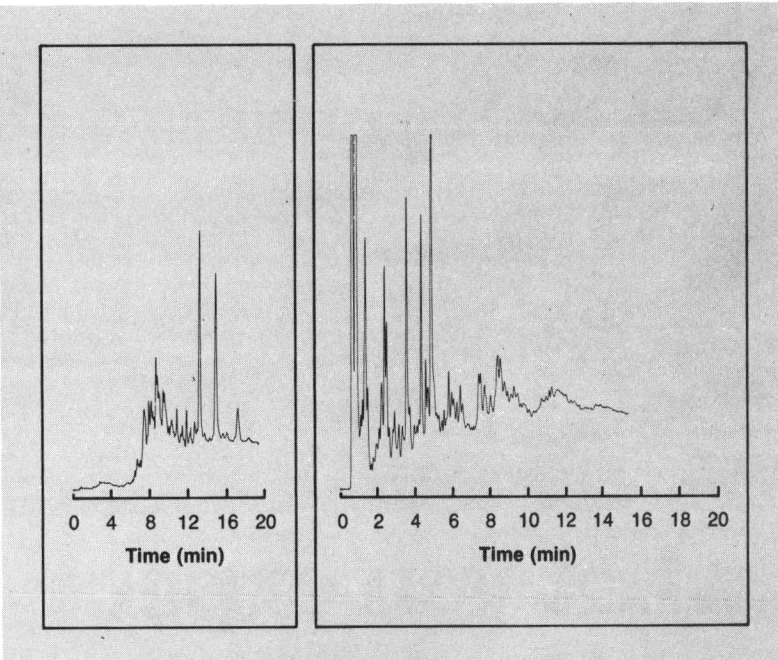

Figure 4. Chromatopyrogram of failed nitrile sheath

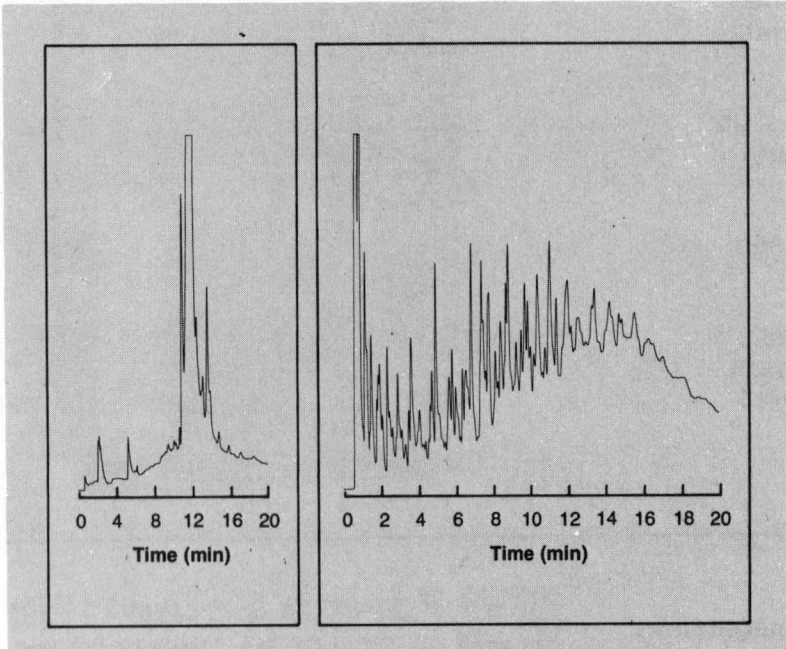

Figure 5. Chromatopyrogram of intact neoprene sheath

the "fingerprints" of the polymer structure the failed sheath was positively identified as a nitrile rubber and the good sheath as neoprene. The chromatopyrogram shown in Figure 4 is from the failed nitrile sheath, and Figure 5 is that of the intact neoprene sheath. The analyses were fast and the results were unambiguous.

It was deduced that the high voltage environment had apparently resulted in a high ozone concentration that attacked the nitrile rubber and caused the crazing of the sheath. The neoprene rubber was more ozone resistant and was not adversely affected. The problem was solved by replacing the failed nitrile materials with neoprene rubber sheaths. The CPG technique, when compared to the conventional method, was much faster, less costly and more specific.

A third case was a problem encountered in the hammer shop, where airplane metal parts are shaped in various forms. The rubber forming pad is one of the essential tools needed in the shaping process. One batch of rubber forming pads was found defective. The defective pads had torn and cracked after only three days in use while the previous pads had been used for months without cracking. The shop supervisor wondered why this had suddenly happened, so both the cracked new material and a used good pad were analyzed by CPG. The results revealed that the used good pad was made of a natural rubber while the cracked one was made of a synthetic nitrile rubber. The problem was solved by replacing the nitrile rubber forming pads with natural rubber materials.

These three cases exemplify how a new analytical approach can modernize quality control test methods and then be used for other industrial problem-solving applications. CPG has already proven itself useful in characterizing rubberlike materials and undoubtedly will be used to solve additional real-world problems in the future.

After all, two shots at any analytical problem are better than one!

References

(1) J. Q. Walker, *J. Chromatogr. Sci.*, **15**, 267–74 (1977).
(2) N. B. Coupe, C. E. R. Jones, and S. G. Perry, *J. Chromatogr.*, **47**, 291–6 (1970).
(3) C. E. R. Jones, S. G. Perry, and N. B. Coupe, in "Gas Chromatography 1970," R. Stock, Ed., Elsevier Publishing Company, Amsterdam, 1971, pp 399–406.
(4) N. B. Coupe, C. E. R. Jones, and P. B. Stockwell, *Chromatographia*, **6**, (11), 483–8 (1973).
(5) T. A. Gough and C. E. R. Jones, *Chromatographia*, **8**, 696–8 (1975).
(6) E. J. Levy, 26th Pittsburgh Conference on Analytical Chemistry and Applied Spectroscopy, Cleveland, Ohio, March 3–7, 1975, Paper No. 19.
(7) J. C. A. Hu, 30th Pittsburgh Conference on Analytical Chemistry and Applied Spectroscopy, Cleveland, Ohio, March 5–9, 1979, Paper No. 092.
(8) J. C. A. Hu, U.S. Patent 4 159 894, 1979.
(9) J. C. A. Hu, *Anal. Chem.*, **51**, (14), 1295–7 (1979).
(10) W. Worthy, *Chem. Eng. News*, **58**, (22), 26–27 (1980):
(11) J. C. A. Hu, *Anal. Chem.*, **49**, (4), 537–40 (1977).

Analysis of Liquid Crystal Mixtures

Trevor I. Martin

Xerox Research Centre of Canada, Mississauga, Ontario, Canada L5L 1J9

Werner E. Haas

Xerox Webster Research Centre, Webster, NY 14580

Originally published in ANALYTICAL CHEMISTRY, 1981, Vol. 53, No. 4.

It's fairly well-known in today's electronic age that the display device in your digital watch, calculator, desk clock, digital meter, and instrument panel contains liquid crystals. But have you ever considered the problems involved in analysis of these electrooptically sensitive, unusual chemical mixtures? Gas chromatography (GC) or on-line gas chromatography–mass spectrometry (GC/MS) have been used with some success in the past (1, 2), but we have found on-line liquid chromatography–mass spectrometry (LC/MS) to be the most useful technique in the analysis of these liquid crystal mixtures.

Before discussing in detail our approach to this analysis, however, it might be beneficial to review the nature and types of liquid crystals.

Liquid crystals are substances which have some of the properties of liquids (they flow, pour, and take the shape of their containers) and have some of the optical properties of solid crystals (such as birefringence and optical activity).

Most of the liquid crystals used in display devices are of the twisted nematic (meaning threadlike) type. They are long, thin organic molecules which, under the influence of an electric field, undergo a transition from the helical state to an aligned homeotropic state.

The practical application of this effect requires a room-temperature nematic liquid crystal having positive dielectric anisotropy; chemical, electrochemical, and photochemical stability; high resistivity; low viscosity for enhanced response times; low birefringence to prevent undesirable optical effects; and absence of color.

In order to be useful in electrooptic applications, the material must also be in the liquid crystalline (mesomorphic) state, which only exists in a certain temperature range. Above the upper limit of this temperature range the liquid crystal becomes an isotropic liquid and below it a crystalline solid.

But all these demands cannot be satisfied by a single liquid crystal component. They require the formulation of carefully balanced mixtures.

The mixtures may contain many different liquid crystal families—biphenyls, phenylcyclohexanes, cyclohexylcyclohexanes, Schiff bases, and others.

The Analytical Approach

LC/MS seemed a perfect technique to identify the components in liquid crystal mixtures since individual components of such mixtures are not unambiguously identifiable by ultraviolet–visible, infrared, or nuclear mag-

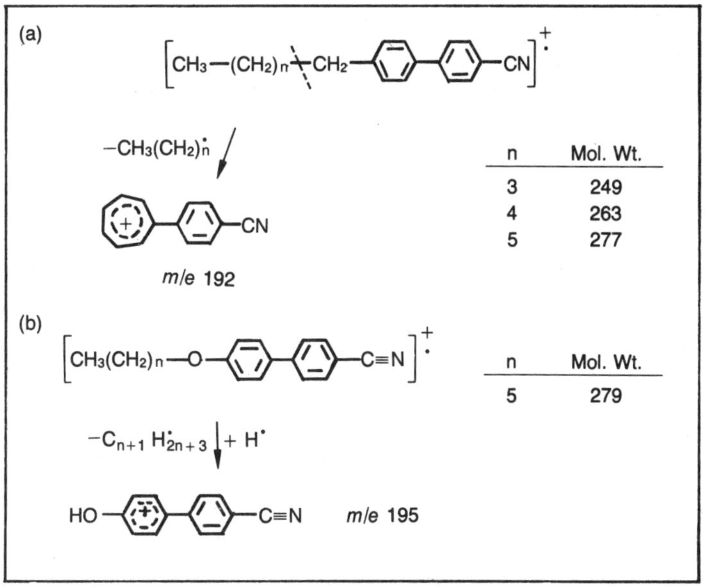

Figure 1. Major MS fragmentation patterns. (a) 4-*n*-alkyl-4'-cyanobiphenyls. (b) 4-*n*-alkoxy-4'-cyanobiphenyls

netic resonance (NMR) spectrometry without prior separation, isolation, and purification. Nor could GC separations for on-line GC/MS always be readily achieved, especially with the azoxy or Schiff base types of liquid crystals. The inherent sensitivity of the LC/MS technique seemed invaluable considering the small quantities of compounds contained in display devices, typically a total of 8–10 mg. Also, it was desirable to develop an analytical procedure that would be applicable to future problems of this type. One such anticipated problem was the analysis of pleochroic dyes in the presence of liquid crystal materials. Typically, these dyes may be highly polar substituted aminoanthraquinones or trisazodyes, which are not suitable for GC separations because of low volatility at normal working pressures or poor thermal stability at elevated temperatures. However, LC/MS offers a unique and convenient solution.

Very little information has been published concerning the mass spectra of liquid crystals or their fragmentation upon electron impact. To our knowledge, only two very recent papers have appeared (3, 4), and they describe the mass spectra of only a few compounds.

Since we had on hand in our laboratories a number of pure single component liquid crystals, as well as a selection of commercial mixtures of 2–6 unknown components, our analytical approach was as follows:

The mass spectral fragmentation pathways were first obtained for several typical members of various structural classes, using pure single component liquid crystals whose structures were verified with additional spectroscopic techniques. Thus, a standard libary of mass spectra was compiled and stored for comparison purposes. The major electron impact (EI) mass spectrometric fragmentation patterns for two representative classes of liquid crystal are shown in Figure 1. While not all-encompassing, the accurate assignment of molecular structures to unknowns depends on differences in the fragmentation patterns for different classes of compounds.

Next, the on-line LC/MS procedure was checked using simple mixtures prepared from single components. During this phase of the program, the MS parameters were optimized. In all cases, satisfactory LC separations were obtained using acetonitrile/water as the mobile phase at flow rates of 1–2 mL/min. The columns selected for the analysis were packed with either Partisil 5 ODS or Ultrasphere ODS-5 μm. Both columns were 25 cm × 4.6 mm id and possessed greater than 50 000 plates/m for the test compounds. Column eluant was presented via a split device to the Finnigan LC/MS belt interface. Generally, the split ratio was adjusted to allow 0.1–0.5 mL/min to pass to the belt surface. The belt interface was connected to a Finnigan 4000 quadrupole mass spectrometer continuously scanned from m/e 45 to m/e 500 in the EI mode at 3 s per scan. Flash evaporation of solute from the belt into the source of the mass spectrometer was achieved by heating the belt to 300 °C.

When scanning this mass range, virtually no interfering mass fragments arising from the mobile phase were observed. Backgrounds in general were very low with no change in level during complex gradient elution programs. During acquisition of data files the mass spectrometer was under computer control, using the INCOS 2000 system interfaced with a Tektronix terminal and cathode ray tube (CRT). For all the "synthetic" mixtures analyzed, satisfactory EI mass spectra were obtained for the components eluting from the liquid chromatograph.

Analysis of a Commercially Available Liquid Crystal Mixture

Next, several mixtures of "unknowns" obtained from various commercial liquid crystal supply houses were run. For brevity, only one example will be discussed here. Although extensive data on electrical, optical, and physical specifications of the liquid crystal mixture were supplied by the chemical manufacturer, the sample was identified simply as a mixture of biphenyls and pyrimidines. An isocratic separation of this six-component mixture was accomplished in 60 min on a Partisil 5-ODS column employing 65% acetonitrile/35% water as the mobile phase at a flow rate of 1.7 mL/min. The LC/MS run for this mixture is illustrated in Figure 2. The reconstructed liquid chromatogram (RLC) for this mixture is shown in addition to the mass chromatograms of several typical base peaks, ions m/e 192, 194, 195, 268, and 270.

Three methods could then be used to unambiguously identify the individual components. First, a search of the existing library of mass spectra of authentic liquid crystal structures could be carried out. If the compound was in the library, a direct comparison of the two spectra for fit and purity could be made. This was done for component 4 (Figure 2), which was confirmed as 4-n-pentyl-4′-cyanobiphenyl. Second, a computer-assisted

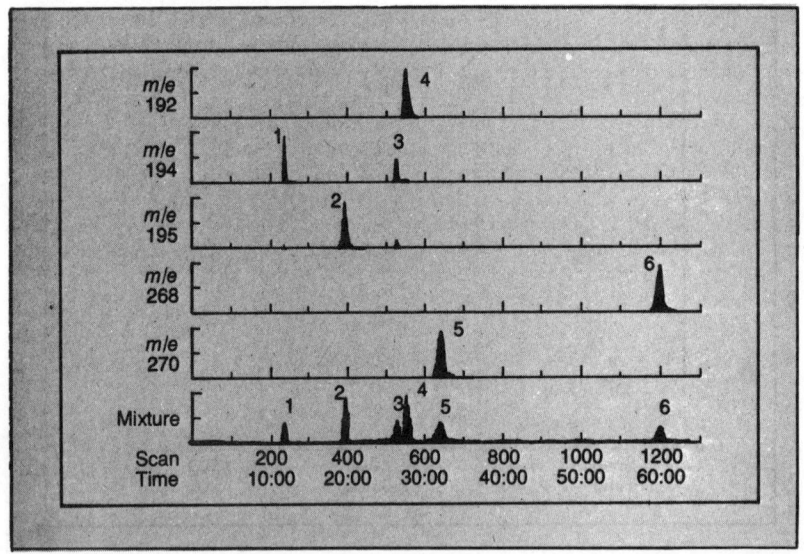

Figure 2. RLC for commercial liquid crystal mixture with selected mass chromatograms for characteristic base peaks

Table I. Structures of Components Present in Commercial Liquid Crystal Mixture

Component no.	Structure	Nominal mol. wt.	Base peak
(1)	$CH_3(CH_2)_4$—pyrimidine—phenyl—CN	251	194
(2)	$CH_3(CH_2)_4$—O—phenyl—phenyl—CN	265	195
(3)	$CH_3(CH_2)_6$—pyrimidine—phenyl—CN	279	195
(4)	$CH_3(CH_2)_4$—phenyl—phenyl—CN	249	192
(5)	$CH_3(CH_2)_3$—phenyl—pyrimidine—phenyl—CN	313	270
(6)	$CH_3(CH_2)_4$—phenyl—phenyl—phenyl—CN	325	268

search for selected molecular ions indicative of the individual homologous members of the various structural types of liquid crystals could be initiated. The net results of this selected technique could be displayed on the CRT of the INCOS terminal as a set of selected mass chromatograms. An examination of these selected mass chromatograms, along with those for the base peaks, often permits rapid identification of the individual components. Third, when the first two methods were not successful, the mass spectrum of each component could be displayed on the CRT or printed, and interpreted using basic MS knowledge.

With a combination of these three methods, it was possible to assign the structures depicted in Table I for the six components present in the liquid crystal mixture. The assumption was made that no branched alkyl groups were present. This assumption almost always holds true, since most liquid crystals useful for displays are long rodlike molecules. Occasionally, a small quantity of a compound with a chiral branched alkyl group is deliberately added to induce a uniform direction of "twist" in the device. If this type of component is suspected in a mixture, it must be isolated and the exact nature of the branching confirmed with ^1H NMR.

Analysis of a Mixture Isolated from a Commercial Display Device

Finally, several commercial display devices were broken apart, and the liquid crystal mixtures were extracted, concentrated, and subjected to analysis by on-line LC/MS. The general procedure for extraction first involves removal of the reflecting polarizer from the cell surface. The adhesive used to bond the reflecting polarizer is removed by gentle swabbing with dichloromethane. The intact cell is placed in a Teflon beaker and carefully broken into small pieces with a stainless steel rod to expose the inner surfaces. Acetonitrile is added to the beaker, which is then heated to dissolve the liquid crystal mixture. The acetonitrile solution is decanted and the solvent is removed under reduced pressure to give the liquid crystal mixture (usually about 10 mg) as an oily film. The mixtures then are subjected to LC/MS analysis. The results obtained for one of these mixtures will be discussed briefly.

The RLC for this liquid crystal mixture is shown in Figure 3. Eight components, labeled a–h, are well-resolved. Separation was achieved on an Ultrasphere ODS-5 μm column, employing a mobile phase gradient elution program from 73% acetonitrile/27% water to 100% acetonitrile. An on-column injection of 10 μg of liquid crystal mixture was made with approximately 10% of the column eluant passing to the belt interface. The efficiency of the belt flash evaporation system has been shown to be 40–50% (5). Thus, approximately 450 ng of total components were reaching the source of the mass spectrometer. It is evident that the mass spectrometer is capable of EI detection of 1–2 ng or less of a compound reaching the ion source, as component b is estimated to be present at 0.2–0.5 wt% of the mixture.

Using the base peak criteria described previously (see Figure 1), component a was identified as a member

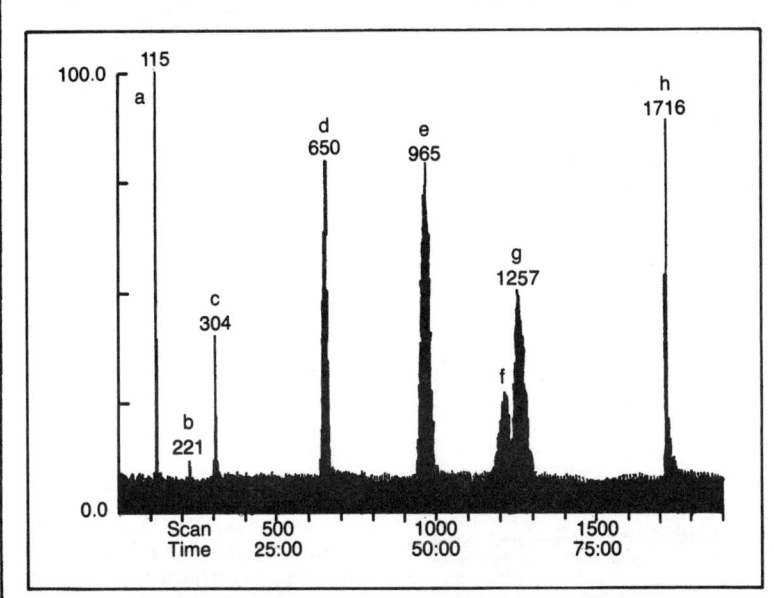

Figure 3. RLC from LC/MS run of liquid crystal mixture isolated from the commercial display device

of the 4-n-alkoxy-4'-cyanobiphenyl class (base peak m/e 195) and components c, d, and e as members of the 4-n-alkyl-4'-cyanobiphenyl class (base peak m/e 192). Their complete structures, determined through examination of their individual mass spectra and spectrum matching (where possible), are shown in Table II. Component b was readily identified from its mass spectrum as a 4-pentoxy-4'-cyanobiphenyl. However, it was present at such a low level that we suspected it might be an optically active additive containing a branched alkoxy group. Its identity was confirmed, after isolation from the mixture, as 4-(2''-methylbutoxy)-4'-cyanobiphenyl using ^{1}H NMR spectrometry at 250 MHz. Components g and h proved rather unusual and interpretation of their mass spectra proved more challenging. The mass spectrum of component h is shown in Figure 4, together with the assigned molecular structure and a rationalization of the major fragment ions observed. The intense ions at m/e 295 and m/e 175 are due to the stable acylium ions resulting from the usual ester cleavage mechanism, whereas the characteristic ions at m/e 121 and m/e 118 are derived as shown. The molecular ion, although weak, is clearly visible at m/e 458. The mass spectrum of component g is illustrated in Figure 5. The compound exhibits a weak molecular ion at m/e 332, but the spectrum is dominated by the intense ions at m/e 180 and m/e 110.

Table II. Identity of Components Present in Liquid Crystal Mixture Isolated from Commercial Display Device

Component, 10 µg total (Injected)	Structure	Nominal mol. wt.	Base peak in mass spec
(a)	n-C$_3$H$_7$—O—⌬—⌬—CN	237	195
(b)	(CH$_3$)(C$_2$H$_5$)CH*—CH$_2$—O—⌬—⌬—CN	265	195
(c)	n-C$_5$H$_{11}$—⌬—⌬—CN	249	192
(d)	n-C$_7$H$_{15}$—⌬—⌬—CN	277	192
(e)	n-C$_8$H$_{17}$—⌬—⌬—CN	291	192
(f^1)	CH$_3$, OH-benzoxazole-C(CH$_3$)$_3$, Cl substituted phenyl	315	300
(g)	n-C$_3$H$_7$—cyclohexyl—C(O)—O—⌬—O—n-C$_5$H$_{11}$	332	180
(h)	n-C$_5$H$_{11}$—⌬—C(O)—O—⌬—C(O)—O—⌬—n-C$_5$H$_{11}$	458	175

* denotes chiral center

^{1} The structure of component f could not be unambiguously assigned from the mass spectrum obtained from the LC/MS run. The structure shown, although proposed tentatively, is consistent with data obtained from ^{1}H NMR, UV, IR and MS.

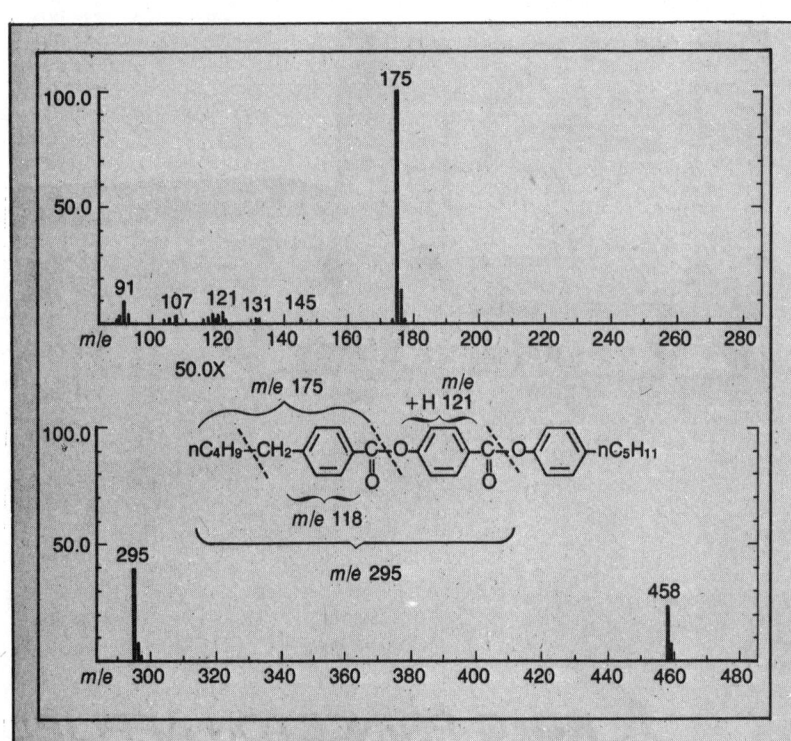

Figure 4. Mass spectrum of component h (Figure 3) of liquid crystal mixture isolated from the commercial display device

These ions are typical of alkoxyphenols and serve to define half of the structure as shown. Surprisingly, only a weak ion is observed at m/e 153 due to the cyclohexyl acylium ion. Evidently, this ion readily loses CO to give the substituted cyclohexyl secondary carbonium ion at m/e 125. The latter ion also defined the chain length of the alkyl group attached at position 4 of the cyclohexyl ring.

Anticipated Future Problems

Recently there has been considerable interest in the use of pleochroic dyes as additives to liquid crystal mixtures for use in color display devices operating in the cholesteric–nematic phase change mode. Some of the more promising dyes, in terms of high contrast, suitable solubility, and excellent photostability, are those based upon the amino- or diamino-anthraquinone structures (6, 7). LC/MS thus far offers the only useful and convenient technique for analysis and identification of liquid crystal mixtures containing these dyes. An LC/MS run was made for a liquid crystal mixture in which approximately 2% of 1,5-bis-(2'-phenylethylamino)-anthraquinone, a

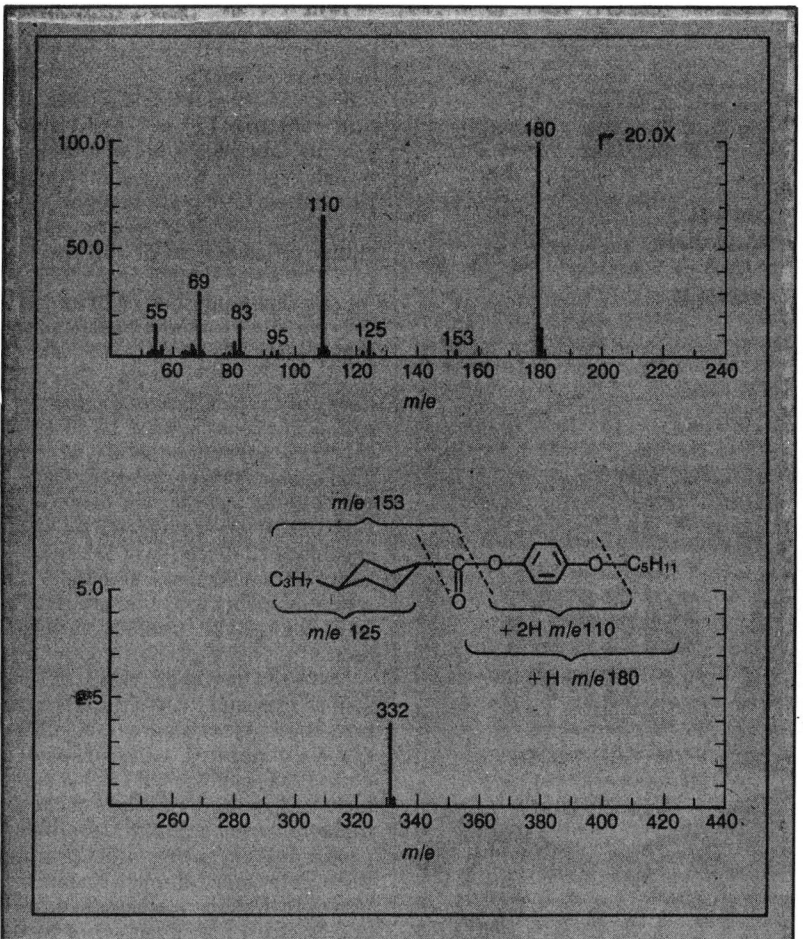

Figure 5. Mass spectrum of component g (Figure 3) of liquid crystal mixture isolated from the commercial display device

magenta dye, was dissolved. An isocratic separation was achieved in 12 min on a Partisil 5-ODS column using 95% acetonitrile/5% water as the mobile phase. The EI mass spectra of each component, as determined from the LC/MS run, matched perfectly the corresponding mass spectra of the individual pure components acquired using solid probe techniques, and the structures of the five components were readily determined.

References

(1) A. Boller, H. Scherrer, M. Schadt, and P. Wild, *Proc. IEEE*, **60,** 1002 (1972).
(2) H. Kelker, B. Scheurle, R. Hatz, and W. Bartsch, *Angew. Chem. Int. Ed. Engl.*, **9,** 962 (1970).
(3) R. L. Hubbard, "Proceedings of the Symposium on the Physics and Chemistry of Liquid Crystal Devices," Plenum Press, New York, N.Y., 331 (1979).
(4) R. V. Popanova, B. M. Bolotin, and M. S. Chupakhin, *J. Anal. Chem., USSR Engl. Transl.* **33,** 1413 (1978).
(5) W. H. McFadden, *J. Chromatogr. Sci.*, **17,** 2 (1979).
(6) M. G. Pellatt, I.H.C. Roe, and J. Constant, *Mol. Cryst. Liq. Cryst.*, **59,** 299 (1980).
(7) R. J. Cox, *Mol. Cryst. Liq. Cryst.*, **55,** 1 (1979).

Analysis of Rubber and Plastic Chemicals by LC/Spectroscopy

Jerry B. Pausch

BF Goodrich Research and Development Center, Brecksville, OH 44141

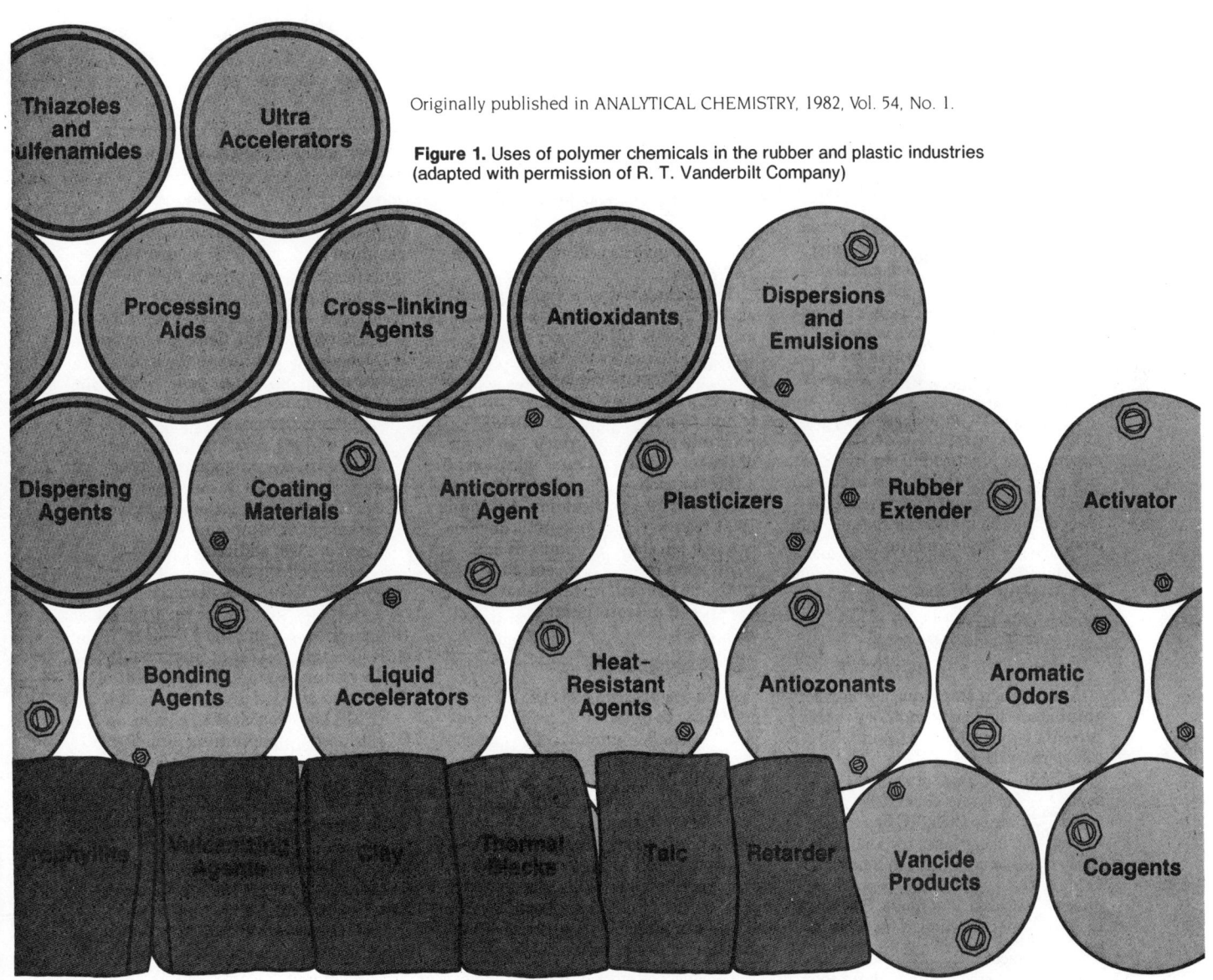

Originally published in ANALYTICAL CHEMISTRY, 1982, Vol. 54, No. 1.

Figure 1. Uses of polymer chemicals in the rubber and plastic industries (adapted with permission of R. T. Vanderbilt Company)

Natural rubber was first introduced to the European continent by Christopher Columbus after he observed American native boys playing with strange bouncing balls. However, the unique elastic property of rubber remained virtually untapped until rubber vulcanization was discovered by Charles Goodyear in 1839. From 1850 to 1950 numerous chemicals were developed and added to rubber to overcome two serious problems—its fast deterioration rate and its slow vulcanization rate. Aniline was developed as the first organic accelerator in 1906; antioxidants based on aromatic amines were discovered in 1921; alkylphenylenediamines became commercial antiozonants in the 1950s.

Today there are many thousands of these chemicals that play a major role in protecting and developing desired physical and chemical properties, not only in rubber (1) but in plastics or other synthetic polymers. Figure 1 illustrates a number of uses for these chemicals in the rubber and plastic industries.

The wide variety and complexity of these chemicals make analysis very difficult. They usually have high boiling points and molecular weights ranging from 200 to >2000 amu. Many are quite polar and contain several heteroatoms. Also, many contain more species than originally anticipated—as many as 20 to 50 components. These mixtures occur because the starting materials are not purified, and the manufacturing conditions can lead to the production of various by-products and/or oligomers. From a chemical analysis standpoint this task is obviously in the "supersleuth" category (2).

In this article we will describe the joint chromatographic–spectroscopic approach as practiced in the BFGoodrich Analytical Laboratories both for isolating microgram samples and for identifying the individual components from these complex mixtures.

The Analytical Problem

Measurement is the basis of all knowledge.
Lord Kelvin

A separation step using chromatographic methods is mandatory in this type of analysis. Majors (3) in 1970 was perhaps the first to show the utility of liquid chromatography (LC) for separating polymer chemicals like antioxidants. Since then, LC has found widespread acceptance for analyzing these types of chemicals in many fields such as agriculture, petroleum, pharmaceuticals, and foods. But since LC is not specific enough for identification, we decided to use mass spectroscopy (MS), infrared spectroscopy (IR), and nuclear magnetic resonance (NMR) to give us information on the LC peaks.

The complexity of polymer chemical analysis with reversed-phase, microparticulate LC columns is illustrated in Figure 2 with the resolution of about 40 components from a typical rubber antioxidant, a reaction product of aniline and acetone. Each peak probably represents from 0.1 to 10.0% of the total. For a 10-mm³ injection of a 10% solution, a 1% peak would represent about 10 µg on-column. Assuming 100% recovery, this amount is obviously much too small to obtain IR and NMR spectra using conventional methods. Instead, we use either MS or specialized microtechniques for IR and NMR.

Mass Spectroscopy

Mass spectroscopy has the inherent capability to obtain spectra on microgram or smaller amounts of material. Even for probe analysis, the rule of thumb has generally been, "if you can see it, there is too much." Scientists have been working for several years to develop interfaces joining the liquid chromatograph to the conventional mass spectrometer. This type of approach offers limited applicability in analysis of most polymer chemicals at this time, since polymer chemicals do not give good molecular ion spectra from either the electron impact (EI) or chemical ionization (CI) modes. Requirements in the polymer field dictate that ion formation must result from a low-energy process. Therefore, we decided to try the field desorption (FD) mode. In most cases the FD spectrum gives only one peak, which is usually the molecular ion [compounds having labile protons often give the (M + 1) ion] (4). After the molecular ion is known, accurate mass analysis from EI spectra can be much more definitive in pinpointing the exact identification of the compound.

There are two additional advantages with FDMS: Direct qualitative analysis of mixtures is often possible on a sample without prior isolation of the components, and the FDMS results give the number of components in the sample to guide development of the subsequent LC procedure.

FDMS has been close to a panacea for our needs in supporting liquid chromatography. Just the molecular weight can be a very big factor in identifying the overall structure. The major limitations of FDMS have been that a small percentage of compounds are poor desorbers. Sodium cation attachment and fast atom bombardment have been useful in some instances to circumvent this problem. With some polymer chemicals there may also be a mass range limitation unless a high

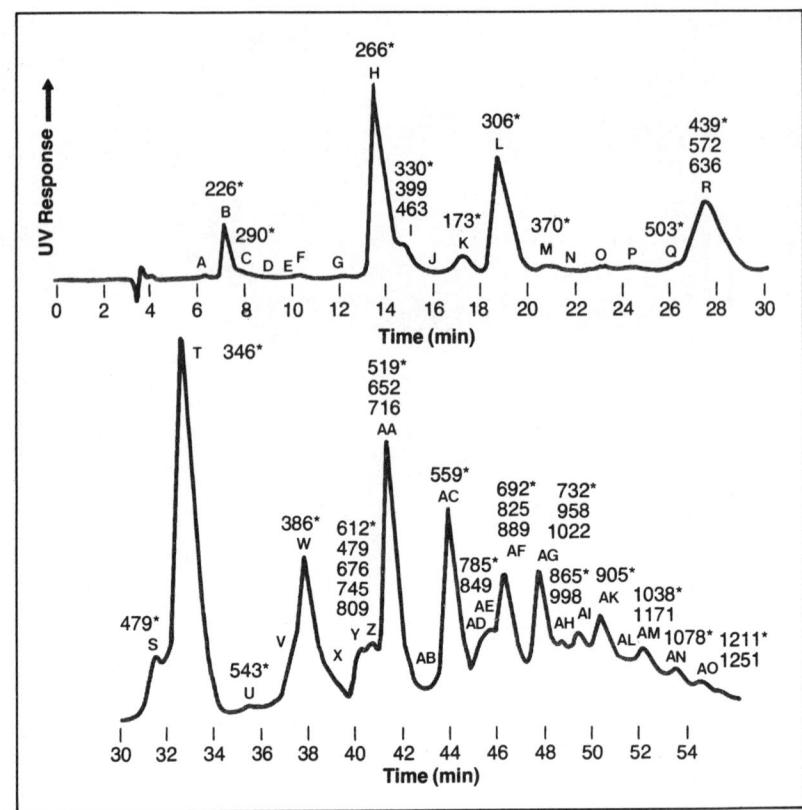

Figure 2. Reversed-phase liquid chromatogram for aniline–acetone reaction products (8). An asterisk indicates the major components

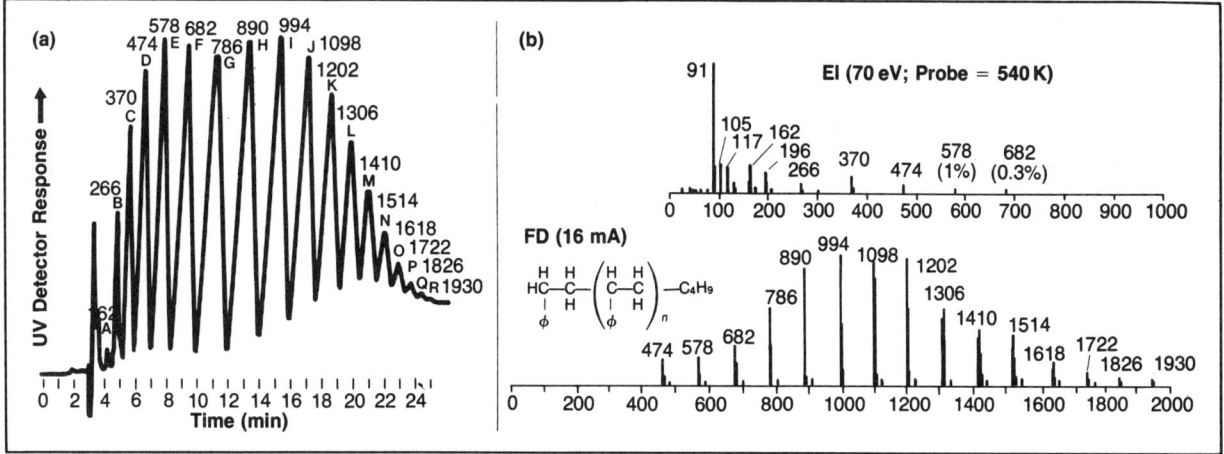

Figure 3. (a) Reversed-phase liquid chromatogram of low molecular weight polystyrene; (b) electron impact and field desorption mass spectra of polystyrene

field instrument is available.

A typical application where FDMS molecular weight values yielded straightforward interpretation of the data concerned a low molecular weight polymer analysis (5). Figure 3a shows the LC curve for a commercial polystyrene standard. Isolated fractions analyzed by FDMS gave essentially only molecular ions, so each polystyrene oligomer was easily verified. The utility of this approach is demonstrated in Figure 3b by comparing the EI and FD spectra on the total sample. It readily shows that EI ionization could not be applied for identifying the exact nature of the individual fractions. A side benefit of these FDMS data on oligomers is that a molecular weight distribution (6) can also be derived from the same experiment. Polystyrenes up to ~5300 amu have been characterized successfully. For these materials and other typical products like tackifying resins and polyglycols (7), a mass spectrometer with extended mass range is obviously necessary.

A second example was a polyurethane sample that was extracted for characterization of low molecular weight oligomers. FDMS again showed its unique ability to directly identify molecular species in a complex mixture. The three series of cyclic polyesters/polyurethanes shown in Figure 4 were readily identified (8).

Infrared Spectroscopy

One of the chief drawbacks to applying IR spectroscopy to problem solving in liquid chromatography is that traditional IR has never been regarded as a trace analytical technique. We have sometimes used FTIR for this purpose, but have relied more on the micro-attenuated total reflectance (ATR) approach (9) using a grating spectrometer. The ATR crystal is a parallel-piped, 0.5-mm-thick KRS-5 crystal with 45° end angles. It is very convenient to add the trapped effluent to the crystal dropwise and let the LC solvent evaporate. This permits some enrichment of the sample compared to examining the sample in solution. The spectra obtained are usually signal averaged four times for a 10-μg sample and further enhanced by subtracting out background and interfering components. These data are used routinely to complement the other spectroscopic results and are most valuable in the total analytical picture for making structural determinations.

AgeRite White is an aromatic amine antioxidant produced from p-phenylene diamine and β-naphthol. Under conditions of the manufacturing process, numerous tars may be formed as shown in Figure 5. Proposed structures were determined as usual from the FDMS and accurate mass data. For complementary data, IR spectra were obtained, for example, on the peak labeled mass 279 in Figure 6. The strong band at 1565 cm^{-1} is typical of the novel C=N vibration, which is conjugated with a C=C vibration in a ring structure. This band was first observed in previous analysis of ozonation products (10) of another aromatic amine. Further agreement was gained from the absence of the secondary N—H stretch at 3420 cm^{-1}. Finally, the 800 cm^{-1} region shows the presence of 2,3-disubstitution of naphthalene.

$$[-(CH_2)_4-O-\underset{\underset{O}{\|}}{C}-(CH_2)_4-\underset{\underset{O}{\|}}{C}-O]_n$$

MW = 200n

$$[-(CH_2)_4-O-\underset{\underset{O}{\|}}{C}-\underset{H}{N}-\underset{}{\bigcirc}-CH_2-\underset{}{\bigcirc}-\underset{H}{N}-\underset{\underset{O}{\|}}{C}-O-[-(CH_2)_4-O-\underset{\underset{O}{\|}}{C}-(CH_2)_4-\underset{\underset{O}{\|}}{C}-O]_n$$

MW = 340 + 200n

$$[-(CH_2)_4-O-\underset{\underset{O}{\|}}{C}-\underset{H}{N}-\underset{}{\bigcirc}-CH_2-\underset{}{\bigcirc}-\underset{H}{N}-\underset{\underset{O}{\|}}{C}-O-(CH_2)_4-O-\underset{\underset{O}{\|}}{C}-\underset{H}{N}-\underset{}{\bigcirc}-CH_2-\underset{}{\bigcirc}-\underset{H}{N}-\underset{\underset{O}{\|}}{C}-O-[-(CH_2)_4-O-\underset{\underset{O}{\|}}{C}-(CH_2)_4-\underset{\underset{O}{\|}}{C}-O]_n$$

MW = 680 + 200n

Figure 4. Cyclic polyurethane oligomers identified from extract of high molecular weight polyurethane

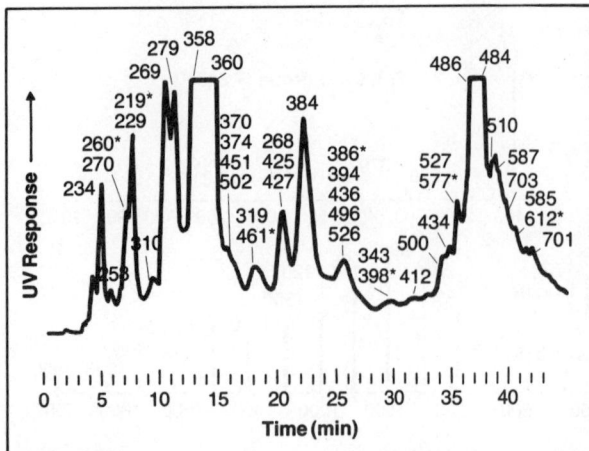

Figure 5. Reversed-phase liquid chromatogram of tar by-products from antioxidant production process

Figure 6. Micro-ATR infrared spectrum of ~20-μg tar by-product with molecular weight 279 (shown in Figure 5), KRS-5 parallel-piped crystal with 45° end angles, and ~25 internal reflections using Perkin-Elmer 180 infrared spectrometer

Nuclear Magnetic Resonance

As with conventional IR, iron-magnet NMR instrumentation does not possess the necessary sensitivity to examine samples routinely from analytical LC units. However, superconducting magnet NMR instruments can obtain useful proton spectra after averaging a number of pulsed scans. In this work a standard probe with 5-mm sample tubes has proven satisfactory to obtain the desired sensitivity, although we do concentrate the sample to the smallest possible volume. The biggest problem has been the presence of trace amounts of water. Most reversed-phase LC experiments use water as part of the solvent system, and it is difficult to remove completely. The customary approach is to evaporate the sample to dryness, add D_2O and reevaporate, then dilute in an appropriate organic solvent. "Solvent suppression" has also been valuable in some instances.

The usefulness of NMR is shown in an example involving trimethylolpropanetriacrylate (TMPTA), which is a commonly used cross-linking agent for polymers and adhesives. Proton NMR spectra on LC cuts confirmed the major component in the LC curve as the expected product from the various assignments shown in Figure 7a. An IR spectrum also verified this structure. NMR analysis of the primary impurity indicated two components. The identifications (even the ratio of amounts) were determined from Figure 7b as the incomplete (diacrylate) and overreacted products from the starting materials of trimethylolpropane and acrylic acid.

Rubber Antioxidant

The multicomponent nature of a typical rubber antioxidant was already illustrated by the liquid chromatogram shown in Figure 2. This resinous material is an acetone–aniline condensation product whose principal structure is shown in Table I as the oligomeric series with a monomer molecular weight of 173. The detailed composition of this product was generally unknown before liquid chromatography became available. But a total of eight oligomeric series containing 42 components has now been identified. These structures were deduced initially just from FDMS molecular weights and logical chemical reactions of the reactants. Atomic composition data and IR spectra were obtained on several fractions to confirm assignments of various structures.

Summary

The combined application of chromatography and spectroscopy is mandatory for analyzing polymer and rubber chemicals—chromatography for

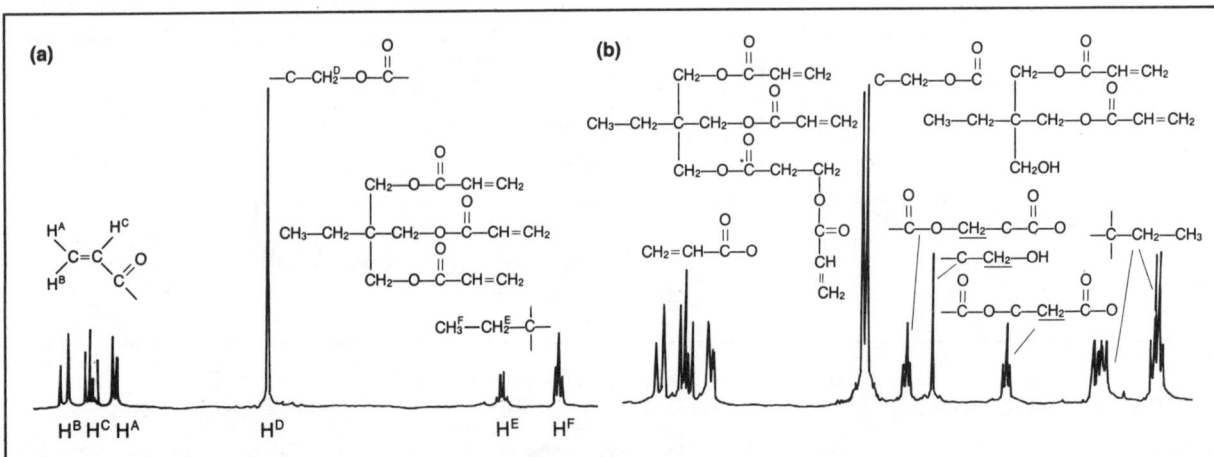

Figure 7. 200-MHz ^{1}H NMR spectra of (a) trimethylolpropanetriacrylate and (b) two impurities. Spectra obtained on ~50 μg in 5-mm tube on Bruker WH-200 NMR with quadrature detection; 16 scans averaged

Table I. Components Identified in Aniline-Acetone Resin

Series composition	n	Molecular weight[1]	LC peak[2]
$MW = 226 + 173n$	0	226	B*
	1	399	I
	2	572	R
	3	745	Y,Z
$MW = 93 + 173n$	1	266	H*
	2	439	R*
	3	612	Y,Z*
	4	785	AD,AE*
	5	958	AG
$MW = 290 + 173n$	0	290	C*
	1	463	I
	2	636	R
	3	809	Y,Z
$MW = 306 + 173n$	0	306	L*
	1	479	(S;Y,Z)[3]
	2	652	AA
	3	825	AF
	4	998	AI
	5	1171	AM
$MW = 157 + 173n$	1	330	I*
	2	503	Q*
	3	676	Y,Z
	4	849	AD,AE
	5	1022	AG
$MW = 173n$	1	173	K*
	2	346	T*
	3	519	AA*
	4	692	AF*
	5	865	AI*
	6	1038	AM*
	7	1211	AO*
$MW = 370 + 173n$	0	370	M*
	1	543	U*
	2	716	AA
	3	889	AF
$MW = 386 + 173n$	0	386	W*
	1	559	AC*
	2	732	AG*
	3	905	AK*
	4	1078	AN*
	5	1251	AO

[1] Nominal mass determined by FDMS.
[2] Peaks are lettered in Figure 2. An asterisk (*) indicates this is the major component of an LC peak.
[3] Apparently two isomers are separated by LC.

separation and spectroscopy for identification. Although the molecular weight is unquestionably the single most important piece of information for identifying components, atomic composition determinations from accurate mass data and functional groups information from NMR and IR provide vital spectroscopic support. Very high resolution accurate mass data are not required on compounds containing heteroatoms. Specific isomer configurations or conformations are not usually necessary for problem solving in polymer chemicals.

This analytical approach of utilizing spectroscopy as an off-line technique for identifying *microgram* amounts of isolated LC fractions has proven quite valuable in the BFGoodrich laboratories for determining material balances on production processes and for understanding the performance of these chemicals. It is also invaluable for FDA and TSCA approvals.

Acknowledgment

The work of several BFGoodrich scientists forms the basis of this article. It is my pleasure to recognize their analytical team achievements in analyzing polymer chemicals during the past few years: Robert P. Lattimer (mass spectroscopy); E. Ray Hooser (liquid chromatography); Dale J. Harmon (size exclusion and liquid chromatography); Jerome C. Westfahl (proton nuclear magnetic resonance spectroscopy); Hugh E. Diem (infrared spectroscopy); Paul M. Zakriski (liquid chromatography–mass spectroscopy); and C. K. Rhee, E. T. McDonel, J. C. Andries, P. P. Nicholas, and R. W. Layer (ozonation studies). Finally, the support of the BFGoodrich Company management for funding these laboratories and for permission to publish this article is appreciated.

References

(1) Rubber World 1979 Blue Book, "Materials, Compounding Ingredients and Machinery for Rubber," Hartman Communications, Inc.: New York.
(2) Grasselli, J. G. *Anal. Chem.* **1980**, *52*, 30 A.
(3) Majors, R. E. *J. Chromatogr. Sci.* **1970**, *8*, 338.
(4) Beckey, H. D. "Principles of Field Ionization and Field Desorption Mass Spectrometry"; Pergamon Press: New York, 1977.
(5) Lattimer, R. P.; Hooser, E. R.; Zakriski, P. M. *Rubber Chem. Technol.* **1980**, *53*, 346.
(6) Lattimer, R. P.; Harmon, D. J.; Hansen, G. E. *Anal. Chem.* **1980**, *52*, 1808.
(7) Lattimer, R. P.; Hansen, G. E. *Macromolecules* **1981**, *14*, 776.
(8) Lattimer, R. P.; Welch, K. R. *Rubber Chem. Technol.* **1980**, *53*, 151.
(9) Perkin-Elmer Product Bulletin, Accessory Data Sheet L-525.
(10) Lattimer, R. P.; Hooser, E. R.; Diem, H. E.; Layer, R. W.; Rhee, C. K. *Rubber Chem. Technol.* **1980**, *53*, 1170.

Environmental

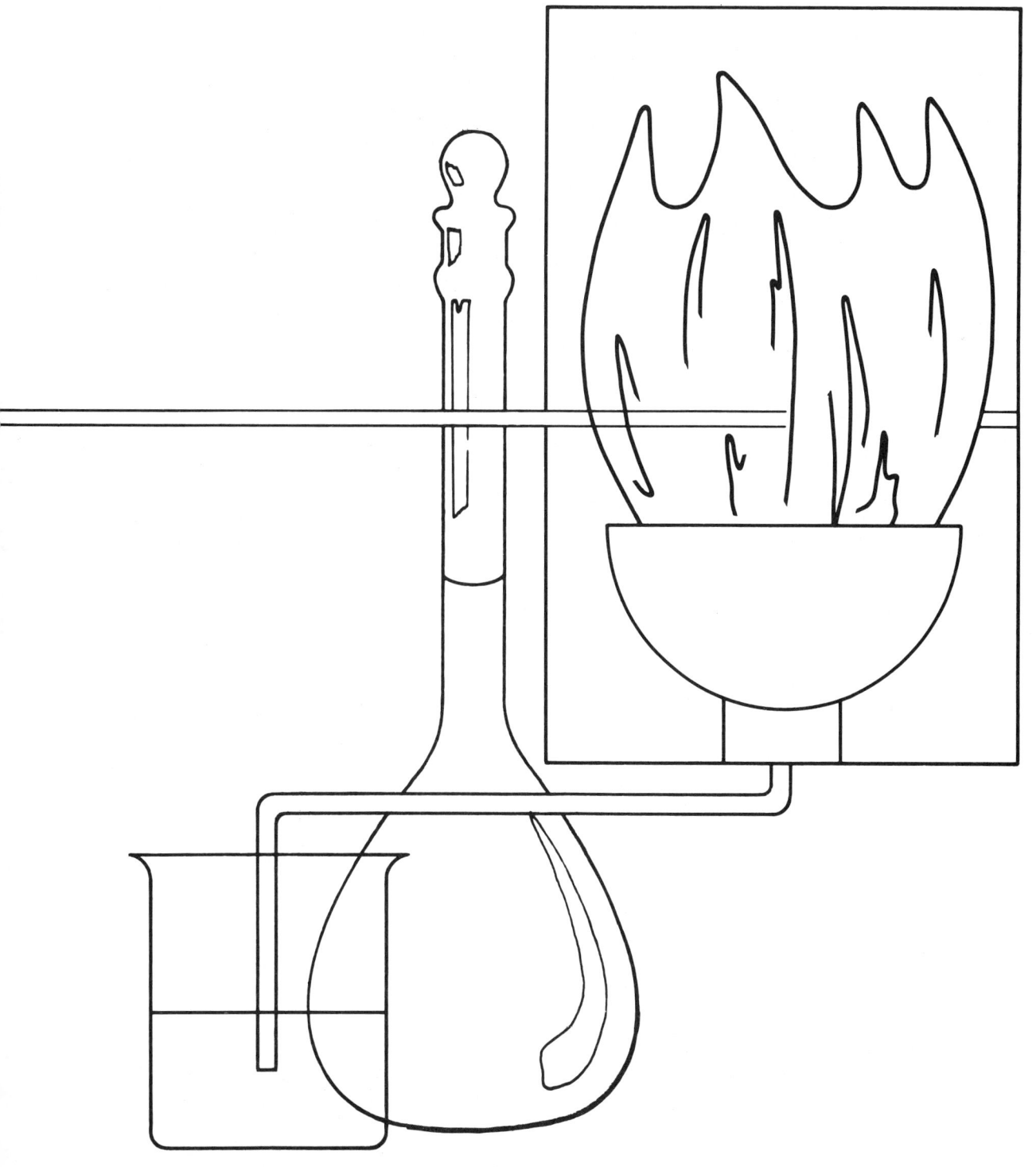

Since Rachel Carson's book "The Silent Spring," environmental activities have consumed increasingly large amounts of time for analytical chemists. Environmental problems are particularly difficult because they involve trace amounts of materials in air, effluent, or waste streams that are not themselves well understood. In addition, there is also a sense of urgency involved in the solution of such problems. The analyst, in all cases, must have a good understanding of the basic chemistry, which may furnish a clue to resolving the problem. Important positive results of the rapid evolution of environmental analysis are the lower thresholds of detection for many compounds and the improved sensitivities and precisions that have been developed. High demands are placed on the reliability and credibility of the data, and decisions based on the work of the analyst have far-reaching impacts on regulations and on society.

Oil spills occur naturally and from actions of humans. In "**Who Spilled the Oil**," the author describes efforts funded by the Coast Guard to determine the guilty parties in significant spills. Difficulties encountered by the analyst include the weathering of the pollutant and, in some instances, the presence of pollutants other than oil. Facilities to study test methods in a controlled environment are discussed, and cases of successful identification, satisfactory for use in a court of law, are described.

In 1978, President Carter signed a bill authorizing the "Federal Plan for ocean pollution research, development, and monitoring, fiscal years 1979 to 1983." The analytical chemistry elements of that plan are described in "**Ocean Pollution**." Measurement technology is often the limiting factor in developing information for defining ocean pollution problems. A multiagency approach is described for determining where pollutants arise and in what volumes, how they are transported and transformed, and how they affect marine organisms and humans.

Glaciers exist over very long periods of time. Each layer of a glacier is a record of the precipitation which fell at the time the layer formed. The layers can be dated by geochronology. In "**Plutonium in Glaciers**," studies of plutonium concentrations in glaciers from Greenland and Antarctica, which correlate with atomic testing events, are described. Attention is now being focused on carbon measurement as an indicator of the impact of fossil fuel burning on our environment.

Silicosis can be a severe health problem. The eruption of Mt. St. Helens spread ash over a wide area of the nation. In the series of articles on this subject, including "**Crystalline Silica in Mt. St. Helens Ash**" and two Focus features, the heated controversy over the silica content of the ash is discussed. Geological laboratories report little silica in ash samples, while government chemists report levels as high as 10%. The controversy rages over two questions: (1) Are the samples different? and (2) Do the preparation methods used modify the composition? The lack of communication between the two camps is highlighted.

The "**Identification of Organic Compounds in an Industrial Wastewater**" discusses the difficulty in isolating organic pollutants in industrial waste and rivers. Based on survey analyses, the analyst is more concerned with "what is present," rather than "how much." Because of the low concentrations of pollutants present, GC/MS methods have been the predominant wastewater analysis techniques. Examples are given which show that much information about an industrial process can be obtained from analyzing its effluent.

Vinyl chloride monomer (VCM) is suspected as a carcinogen initiating angiosarcoma, a rare form of liver cancer. In "**Industrial Analytical Chemists and OSHA Regulations for Vinyl Chloride**," the authors describe the urgent need for a portable sampler that monitors worker exposure to VCM. The method issued by NIOSH had the potential for sample loss over time. A new, much improved system for more accurate analysis was developed in a short time.

The Texas Air Control Board asked its analytical staff whether the general public exposure to toxic elements could be determined economically, whether the analytical results could identify pollution sources, and whether there were other ways analytical techniques could benefit the agency. In "**Detecting and Defining Air Pollutants: One Laboratory's Experiences and Approaches**," the answers to all questions proved to be "yes." This group has developed new cost-effective methods, isolating polluting sources from air analyses, and progressing to new applications with wide usage. Multi-instrumental approaches are described.

In analyzing spent fuel rods, the analyst is faced with handling highly radioactive materials containing many long- and short-lived fission products, unused fuel, and plutonium. In "**Sampling and Analysis of Radioactive Solutions**," government scientists describe the development of a new sample loading technique using anion resin beads which obtained good samples as small as 1–3 ng of each element. In addition to serving as an excellent bead sample for analysis by mass spectrometry, there are many shipping cost savings in sending samples to Vienna (International Atomic Energy Agency) for analysis.

Who Spilled the Oil?

Alan P. Bentz

U.S. Coast Guard, Research and Development Center, Groton, CT 06340

Originally published in ANALYTICAL CHEMISTRY, 1978, Vol. 50, No. 7.

In and around U.S. waters in 1976, there were 10 660 oil spill incidents involving more than 23 million gal of oil (1). The Federal Water Pollution Control Act, as amended in 1972, gave a public mandate to law enforcement agencies, such as the U.S. Coast Guard and the U.S. Environmental Protection Agency, to protect the United States waterways from pollution. Oil spills were identified with special provisions and funding for cleanup. The contingency fund for cleanup costs in "mystery" spills has amounted to $45 million. During the same period, $49 million was contracted in cleanup costs; $18 million has already been recovered. Whenever the source of a "mystery" spill can be identified, the spiller is required to repay the money from the fund.

The identification of the source of an oil spill requires a forensic chemical analysis, which means that it must be persuasive in a court of law. Since cleanup costs are civil penalties, only a preponderance of evidence is required, unlike criminal cases where the evidence must convince the judge or jury beyond a reasonable doubt. Oil identification differs from most forensic analyses, such as blood typing, ballistic determinations, or street drug identification, all of which involve relatively stable samples. The major difference is that from the moment an oil is spilled, it starts to change in composition due to evaporation of light ends, solubility losses, oxidative changes, or other altering processes. The problem then is to convince a judge that an unweathered oil can be the source of a spilled weathered oil despite obvious differences caused by weathering.

The Analytical Approach

The analytical approach required first of all selections from all of the methods most feasible for weathered oil analysis, i.e., those least affected by weathering. Of those most suitable for development, the Coast Guard selected thin-layer chromatography (TLC), gas chromatography (GC), fluorescence, and infrared spectroscopy (IR). Other suitable methods include liquid chromatography (LC), low-temperature luminescence, and emission spectroscopy. Several methods were used to give statistically independent parameters. When these independent methods agree, the statistical confidence is overwhelming (>99%). Also, if one method is affected by contamination, the others may well not be. There is, for instance, no guarantee that spills will occur in unpolluted waters. For example, in New York Harbor, several spills had accumulated carbonyl-containing components before being deposited on Long Island beaches. These contaminants adversely affected the infrared "fingerprint" without being detectable by fluorescence. Conversely, there have been instances where bilge cleaners used by tankers caused spurious fluorescence peaks in the resulting slick, with no noticeable effect on the infrared spectrum. Chromatographic methods (2) can remove polar impurities, and a saponification technique removes esters, acids, and alcohols (3).

The selected analytical methods had to be developed, refined, and tested to determine their value and limitations. Testing required the analysis of oils with known weathering histories. Schemes were devised to simulate natural weathering. Exxon (4), under Coast Guard contract, devised a carefully controlled simulated indoor weathering tank. Clean seawater was circulated through the tank from a 55-gal reservoir under controlled rate and temperature conditions. A fan

vidual peaks. Table I shows the number of peaks sorted into the ranges 0–5, 5–10, and >10%. When comparing identical unweathered oils, all peaks lie within the 0–5% range. Generally, if one allows a "weathering window" of 10%, all peaks lie within that range when comparing a weathered with an unweathered oil. However, for the No. 4 and No. 6 fuel oils shown in Table I, 2 and 7 peaks lie outside that range, respectively. To compensate for that, Anderson et al. (7) applied weighting factors (WF) as coefficients to each of the peaks. The coefficients range from 0.3 to 1.0; the higher coefficients are applied to those peaks that change the least with weathering, and vice versa. Table I shows that when the coefficients are applied, none of the peaks varied by more than 10%. The same coefficients were tested for different No. 4 fuel oils to see whether they were able to differentiate between unlike oils. In fact, the coefficients were found to make the differences even more pronounced.

Another pattern recognition concept is illustrated in Figure 5. In this scheme, a partially weathered suspect oil is analyzed along with the unweathered oil by a vector method which defines a plane in 18-space representing the oil in question. The computer can then select in this plane that vector which most closely approximates the spill, i.e., a computer-simulated weathering. Figure 5 shows a plot of the 18 points used for a spill and suspect, with a computer-simulated weathering curve superimposed as the "best fit" to the spill. It is immediately apparent that this simulated curve is a close match to the spill. Different oils would deviate considerably at one or more points along the curve.

Applications

These techniques have been applied successfully to a large number of oil spills on the navigable waters of the U.S. The Coast Guard has also been asked to apply the techniques in other situations. A case in point is a recent criminal investigation in which the police were trying to place an automobile at the scene of a crime. Analysis of oil-soaked soil from the crime scene did not match samples of oil from the leaking automobile transmission, thus eliminating any further consideration that the auto under investigation might have been the "getaway" car.

It is important for the Coast Guard to exonerate a suspect as well as to identify an oil spiller. For this reason, the techniques are applied to spills of known origin, such as the *Argo Merchant*. Then when tarballs show up on Cape Cod or Nantucket, it can be demonstrated whether they did or did not originate from the *Argo Merchant*. Also, the *Grand Zenith* disappeared a short time after the *Argo Merchant* grounding. The Coast Guard Cutter *Dallas* sampled a slick about 300 miles off the coast and flew the sample by helicopter to the laboratory. It matched the *Argo Merchant* even though the slick had been on the water two to three weeks. The very next day, a second slick was sighted and sam-

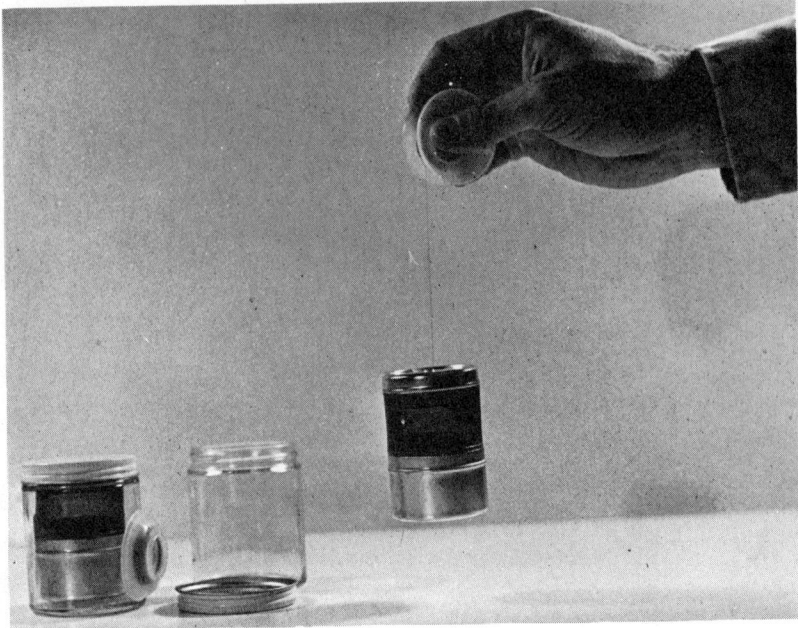

Figure 3. Lipo-pore screen sampler

Jar at left is self-contained sampler. For use in sampling a slick, inner cannister is removed, and attached monofilament line is used to lower sampler to water. Lower portion is weighted to maintain sampler in upright position. Upper portion consists of Lipo-pore screen that is permeable to oil but impermeable to water. After sampling, sampler is replaced in jar, line is cut, and sample is sealed for shipping

Table I. No. of Peaks Within Range by Log-Ratio Method

Range (%)	Wx No. 4 No WF	Wx No. 4 With WF	Wx No. 6 No WF	Wx No. 6 With WF
0–5	10	11	2	12
5–10	6	6	9	6
>10	2	1[a]	7	0

[a] 700 cm^{-1} slightly out.

Figure 4. Infrared spectra showing real-world oil spill before (lower curve) and after (upper curve) removing carbonyl-containing impurities by saponification procedure

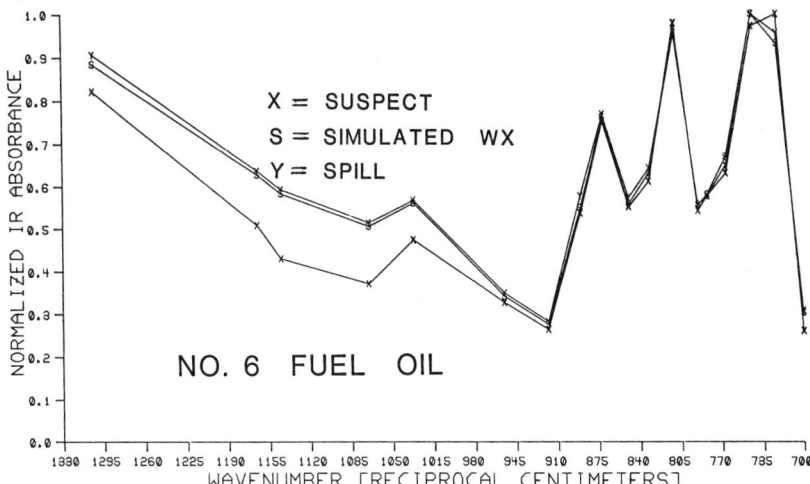

Figure 5. Computer plot of normalized absorbances for 18 data points taken in oil spectra

Plot shows differences between unweathered oil (suspect "X") and weathered oil (spill "Y"). It also shows how closely computer-simulated weathering of "X" lies relative to spill (curve "S"). Note that the simulation even shows slope reversal between 744 and 722 cm^{-1}

pled. It proved to be the same as the cargo of the *Grand Zenith*, the first concrete evidence of the location of that vessel's disappearance in the January storm.

References

(1) U.S. Coast Guard, Dept. of Transportation, "Polluting Incidents in and Around U.S. Waters, Calendar Year 1976", CG-487, 6 Dec. 1977.
(2) ASTM D19 Committee, "New Standard Practice for Preparation of Sample for Identification of Waterborne Oils", ASTM Method D3326 (Revised), Part 31, ASTM Annual Standards, 1978.
(3) G. M. Frame, G. A. Flanigan, and C. P. Chamberlain, "Cleanup Procedure for Contaminated Oils Prior to Fingerprinting by GC and IR Spectroscopy", 1978 Pittsburgh Conference on Analytical Chemistry and Applied Spectroscopy, Abstracts, Paper 554, Mar. 1978.
(4) J. W. Frankenfeld, "Weathering of Oil at Sea", Final Report, U.S. Coast Guard Contract DOT-CG-23035-A, NTIS Accession No. AD787789, Sept. 1973.
(5) C. W. Brown, P. F. Lynch, and M. Ahmadjian, "Identification of Oil Slicks by Infrared Spectroscopy", Final Report, U.S. Coast Guard Contract DOT-CG-81-74-1099, NTIS Accession No. ADA040975, Aug. 1976.
(6) G. A. Flanigan, "Ratioing Methods Applied to GC Data for Oil Identification", Proceedings of Workshop on Pattern Recognition Applied to Oil Identification, Coronado, Calif., Nov. 1976, Published by IEEE 1977, Cat No. 76CH1247-6C.
(7) C. P. Anderson, T. J. Killeen, and A. P. Bentz, "Weighting Factors for Infrared Data Points in Pattern Recognition for Oil Identification", 1978 Pittsburgh Conference on Analytical Chemistry and Applied Spectroscopy, Abstracts, Paper 397, Mar. 1978.

The opinions or assertions contained herein are the private ones of the writer and are not to be construed as official or reflecting the views of the Commandant or the Coast Guard at large.

Controversy Erupts over Mt. St. Helens Ash

Stuart A. Borman

American Chemical Society, Washington, DC 20036

Originally published in ANALYTICAL CHEMISTRY, 1980, Vol. 52, No. 11.

> "This meeting did wind up breaking down into a little bit of an argument as to the content of the material."

Photos Courtesy of U.S. Geological Survey

A full-scale analytical controversy has emerged out of the ash cloud of the Mt. St. Helens volcano. The question at issue is: Is there or is there not crystalline silica in the volcano's ash?

Health authorities in Washington state, at the National Institute of Occupational Safety and Health (NIOSH), and at the Center for Disease Control (CDC) in Atlanta, are concerned about the ash's silica content because of the potential hazard of a disease called silicosis. Analysis has focused particularly on the respirable fraction of the ash, which comprises particles roughly 10 μm in diameter and lower. But months after the first major eruption on May 18, investigators are still not in agreement about the amount of silica in the ash, or even, in some cases, about whether there is any silica there at all.

You might well wonder, in this age of half-million-dollar spectrometers, why researchers cannot come to an agreement on the silica content of an ash sample. The disagreement has certainly caused some embarrassment to health authorities, who find their sophisticated technology apparently failing them at a moment when public attention is sharply focused upon their efforts.

The controversy certainly involves substantive disagreement over the composition of the ash. But legitimate differences in the samples themselves have obscured the issue. And much of the controversy hinges on the merits of a wet chemical sample pretreatment method some of the labs have been using to eliminate interferences.

What the health authorities are worried about is the threat of silicosis, caused by inhalation of respirable-size particles that contain silica. There are two forms of the disease—simple silicosis and progressive massive fibrosis (PMF), the latter being the more serious form. In PMF, according to Daniel Banks, a Medical Officer at NIOSH, large nodules form and take the place of normal gas-exchanging lung tissue. PMF sufferers thus have difficulty breathing and shortened life spans. Simple silicosis not only may cause labored breathing in some cases, but may also lead to PMF. If volcanic ash from Mt. St. Helens contained appreciable amounts of silica, and thousands of local residents were subjected to prolonged exposure from eruptions and from resuspension of fallen ash, a silicosis epidemic could ensue.

Silica (SiO_2) is not to be confused with *silicate*, the latter being a shorthand term for the aluminosilicate minerals such as $Na(AlSi_3O_8)$ that make up perhaps 65–70% of the Mt. St. Helens ash. What is at issue is not the major component of the ash, but a relatively minor component—crystalline silica, which can cause silicosis.

In one corner of the silica controversy are the occupational health labs. The Washington state Department of Labor and Industries (DLI) and the University of Washington Department of Environmental Health (UW) are finding 5–10% crystalline silica in the ash. NIOSH is in the same ball part as DLI and UW.

On the opposing side are the "antisilica" forces. These include the Geology Department at Washington State University (WSU), Battelle Pacific Northwest Laboratories in Richland, Wash., and the U.S. Geological Survey (USGS) in Menlo Park, Calif. The scientists involved in the determinations at these three institutions are geologists. They have found little or no silica in the Mt. St. Helens ash samples.

Some of the earliest ash analyses were performed by the DLI lab. In March DLI checked the early ash emissions for fluoride, asbestos, and silica, looking for any component that might be hazardous to the people of Washington state should a big eruption occur. Their analyses suggested a possible problem with crystalline silica. So when the big eruptions did occur in May and June, a number of labs joined the search for crystalline silica in the tons of ash that fell on the Northwest.

On June 24 a meeting to discuss the potential health effects of the Mt. St. Helens ash fall was organized by CDC–NIOSH and hosted by Region X of the U.S. Environmental Protection Agency (EPA) at their offices in Seattle. According to one of the attendees, "This particular meeting did wind up breaking down into a little bit of an argument as to the content of the material."

Washington state's DLI, for one, believed the silica was there. The DLI lab uses infrared spectrometry (IR) to analyze its ash samples. Since the IR peaks from *silicates* overpower and obscure those that might arise from crystalline silica, DLI adopted the Talvitie method to remove the sili-

Mt. St. Helens in more tranquil times (1978)

cates prior to analysis. The Talvitie method, published in ANALYTICAL CHEMISTRY nearly 30 years ago (**1951**, *23*, 623–6), involves extraction of the sample with pyrophosphoric acid, which is supposed to dissolve the silicate material and leave the crystalline silica behind.

With the Talvitie pretreatment, DLI has been able to see crystalline silica peaks in the IR ash spectra. Phil Peters of DLI claims that the Mt. St. Helens ash is 5–10% crystalline silica, and that about three-quarters of this crystalline silica is in the form of cristobalite. (There are three forms of crystalline silica—cristobalite, quartz, and tridymite.) Quartz is usually at the 1–2% level, and tridymite concentration is less than 1% when it is present at all, according to DLI.

DLI is in close agreement with UW, where the Talvitie method is used prior to analysis by X-ray diffraction (XRD). According to Guy Goble and Jerry Lofgren at UW's Department of Environmental Health, a typical ash analysis is: 5–10% cristobalite, about 1% quartz, and about 1% tridymite.

NIOSH in Cincinnati is in basic agreement with DLI and UW on the silica content of the ash, though its numbers are a little lower. Respirable dust particles from a number of locations have been found to contain 3–7% free crystalline silica by NIOSH researchers. The Talvitie method was not only used by NIOSH prior to XRD analysis—the method was developed there in the early fifties.

But all is not harmony in the world of silica analysis. Other reputable laboratories disagree with these figures. One of these laboratories is Ron Sorem's at WSU.

"Ron Sorem wasn't at the [CDC–NIOSH] meeting, but his paper was passed around, and it criticizes very much anyone who says there's crystal-

line silica in the samples," says one pro-silica researcher. "I'd have to say that he and his lab are diametrically opposed to the results we are finding."

But if Sorem has not found any crystalline silica, he still presents his conclusions with less belligerence than the above quote might indicate. In his lab, the ash is analyzed with XRD outfitted with a powder camera. Sorem is the first to admit that it's very difficult to detect crystalline silica in the ash, because the silicate XRD signal swamps the part of the spectrum where silica would appear.

Sorem has not used the Talvitie method to remove the interference, and neither has any of the other antisilica investigators. In fact, one of these researchers asked ANALYTICAL CHEMISTRY for the literature reference to the method when we called to inquire about his results.

"The powder diffraction technique Sorem is using is basically a geologist's method," says one of the pro-silica investigators. "It's probably a pretty good method for obtaining the required geological information. I think, unfortunately, there's something lacking in his technique and instrumentation. I think primarily the problem is one of inexperience. It's one thing to be an expert in geological formations and rocks and things like that. It's quite another to identify exposure levels of a material of this type."

With a powder camera, such as the one Sorem is using, X-rays are shot through a powder sample. The X-rays form a diffraction pattern which is captured on a cylindrical photographic plate. At UW, on the other hand, XRD is performed with a diffractometer, which includes either a scintillation counter or a proportional counter, and a goniometer to scan all the angles. "With a diffractometer you can scan across peaks and you see the entire peak, instead of just lines on a film," explains UW's Lofgren. "We can measure an angle to about 0.05 degree. With Sorem's powder camera, I'd say 0.5 degree would be pretty difficult."

The DLI and UW investigators complain that Sorem could not possibly have seen any silica anyway, because he failed to use the Talvitie method to clean the sample up prior to XRD analysis. "In IR or XRD," says DLI's Steve Cant, "if you try to look at the ash without pretreating, you'll have a pretty difficult time identifying crystalline silica, and therein lies one of the reasons different labs have reported different results."

A number of labs are on Sorem's side, however. John Shade at Battelle NW found less than 1% crystalline silica in the ash. And Andrei Sarna–Wojcicki of USGS reports, "We didn't find much silica at all." Battelle and USGS used XRD, scanning electron microscopy (SEM), and energy-dispersive X-ray (EDAX) techniques to examine their samples. Neither lab used the Talvitie method. Interestingly, however, Sarna–Wojcicki did start to see low intensity quartz peaks arising from the background after a gravimetric and magnetic separation technique was used to remove some of the interferants.

Attention has naturally focused on the Talvitie method as the culprit in the silica controversy. Peters at DLI states, "Some people have suggested that the extraction with pyrophosphoric acid could be converting some of the silicates to silica." So Peters subjected some pure silicate minerals to the Talvitie method. After the extraction with pyrophosphoric acid, no crystalline silica was found in the residue. NIOSH ran a similar experiment, and concurs that the Talvitie method does not cause silica to form.

Another concern is that different laboratories have been analyzing completely different kinds of ash sample. Sarna–Wojcicki at USGS explains, "The health agencies are looking at material that's generally finer than about 10 μm in diameter, whereas the geologists are starting with a bulk material collected on the ground, which includes everything up to the coarsest material that's there. If you're looking for a vapor phase material, it very well might be concentrated on a phase that's less than 10 μm."

In addition, ash samples from different locations can have compositions that vary widely. As Sorem of WSU puts it, "My philosophy is that different people may be right on their own samples."

To standardize the type of sample being analyzed and to get to the bottom of the silica controversy, NIOSH has organized a so-called round robin testing program. NIOSH collected settled dust samples from a number of locations around Washington state. The samples were cycloned to isolate the respirable fraction (<10 μm particle diameter). To assure sample homogeneity, the respirable dust was combined and mixed in a tumble mill before being sent out to the laboratories participating in the program.

"Each lab will be asked to specify how they prepared the samples and how they performed the analysis," explained Donald Dollberg of NIOSH in July, the week before the samples were to be sent out. "Each lab will report back the total percent crystalline silica in the samples. They will then get a report that will show them where they stand statistically in comparison to the other labs." Dollberg expected the results to be tabulated by the end of August, after the copy deadline for this issue of ANALYTICAL CHEMISTRY, and expressed hopes the round robin program would finally resolve the dispute over silica.

Certainly, one of the problems in the dispute has been lack of interdisciplinary communication. One pro-silica investigator admitted that a good dialogue with the opposition had never been achieved, as far as suggesting that the geologists try the Talvitie method, for instance.

E. R. Crutcher of Boeing Co. in Seattle put it this way: "I know people on both sides, and I can see holes in the approach on both sides." Another researcher said, "A lot of people are doing partial work, and we're included.

"Some of these people go more for the controversy than for the actual objective, which is to find out the best method for analyzing the ash samples," says WSU's Sorem. "I just hope the round robin program finally results in a method that is truly precise and accurate. What we're all really striving for is the truth."

Scientists Still Split over Silica in Ash

Stuart A. Borman

American Chemical Society, Washington, DC 20036

Originally published in ANALYTICAL CHEMISTRY, 1980, Vol. 52, No. 12.

Controversy over the amount of crystalline silica in Mt. St. Helens ash continues. (See the September ANALYTICAL CHEMISTRY, pp 1136–40 A). On one side are those who feel there is 1–3% crystalline silica in the ash, the "low-silica" people for short—Ron Sorem of Washington State University (WSU) is one of them. On the "high-silica" front are those researchers, including Steve Cant of Washington state's Dept. of Labor and Industries (DLI), who believe the ash is 3–7%, 5–10%, or 5–12% crystalline silica, depending on which lab you speak to.

Cant explains that people involved in logging, agriculture, and other outdoor activities continue to be exposed to excessive amounts of ash. Inhalation of crystalline silica carried by the ash could cause silicosis, a disease in which nodules form in the lungs, displacing healthy tissue. The disease sometimes causes progressive deterioration and death. If the ash from Mt. St. Helens contains appreciable crystalline silica, those with long-term exposures to it could be at hazard. "So it's still important to resolve the question of free silica content," says Cant.

There are two distinctions that can be made between the protagonists in the controversy. The high-silica proponents are, for the most part, occupational health scientists, whereas the low-silica people are earth scientists. And the high-silica group has utilized the Talvitie method to pretreat the ash prior to spectrometric analysis, while the low-silica scientists have not, with one exception. In the Talvitie method, the ash is extracted with pyrophosphoric acid to dissolve and remove *silicate* interferences, leaving the crystalline *silica* behind.

Many low-silica researchers are skeptical of the Talvitie method. They believe that crystalline silica not present in the original samples may be created as an artifact of the procedure, and that this may account for the higher silica values obtained by investigators who pretreat their samples in this way. Nevertheless, one of the low-silica labs, Battelle Pacific Northwest Laboratories, has started using the Talvitie procedure. Battelle NW had found hardly any crystalline silica in the ash without Talvitie pretreatment. With pretreatment, they are finding 1–3%, but this is still a lot less than what the occupational health labs are finding. As one high-silica worker sarcastically put it, "In their latest results, they're finding 1–2% more than nothing, I guess."

Ron Sorem, Associate Dean of Sciences at WSU and a faculty member in the Department of Geology there, sums up the low-silica position: "We had a conference here the first part of July on volcanic ash, with 14 people in attendance. We discussed the whole problem and came to a consensus: There could be 1–3% free crystalline silica in the ash, but no more. Any more would be detected by some of the methods that were used, and since some of them detected none, we picked the limit of detection."

Sorem is skeptical of the high-silica results. "To me," he said, "it's still a question of identification. I'm sure people are finding something in the ash, but I'm not sure they know what it is." Sorem believes the Talvitie method has to be examined more carefully and critically.

In an attempt to resolve some of the differences over ash silica content, the National Institute for Occupational Safety and Health (NIOSH) has organized a "round-robin" testing program. They sent identical ash samples to 20 labs the first week of August, suspecting that real differences in the silica content of different samples might be obscuring the issue even further. NIOSH hopes to calculate best average silica concentrations based on the responses received. The laboratories were to report back by August 29, and Donald Dollberg of NIOSH hoped that preliminary results of the study would be ready by about the middle of September.

But prospects for resolution of the silica controversy by round-robin testing seem remote. The fact that NIOSH is itself one of the high-silica protagonists has made some of the low-silica researchers uneasy, and at least one has refused to participate. "It's like a baseball game where the umpire comes from one side," he said. "An independent group should evaluate the results. I'm not going to get involved in something the validity of which I'm not sure of."

Even if this criticism of the program is unfair, the fact remains that, while the round-robin program can generate average values and confidence limits, it cannot resolve questions about the merits or demerits of the Talvitie method. Dollberg says, "I suspect that the high-silica labs will get high values and the low-silica labs will get low values. I hope I'm wrong and they all agree, but I think it's going to be divided up again between the geologists

and the occupational health people."

The National Academy of Sciences (NAS), asked to intervene in the silica controversy, has, perhaps wisely, avoided involvement. President Glenn Terrell of WSU contacted NAS to request that an independent commission be set up to examine all the evidence in the dispute. In his response to Terrell, Philip Handler, President of NAS, said: "I have been assured that scientists of the National Institute for Occupational Safety and Health are sending samples to several different laboratories in an attempt to settle the analytical problems that have arisen. Regarding health hazards, . . .preliminary opinions indicate high risk only for persons with chronic respiratory problems and for workers with long-term exposure; I am told that such individuals are being monitored.

"Both aspects of the problem," Handler continued, "thus seem to be receiving attention. Depending on the outcome of currently ongoing studies, a decision will be made on the degree of participation by the NAS as soon as the data permit."

The dispute over silica in ash first arose at a meeting on health effects held at the Environmental Protection Agency's Seattle office in June. Another meeting on analytical procedures for ash analysis was originally planned as a follow-up, but it was never held, being replaced by the round-robin testing program. "Here it is September," said Cant of DLI last month, "and we're no farther ahead in terms of resolving the issue. I think perhaps the idea of having scientists get together to discuss measurement methods should never have been dropped."

This is surely not the first time that different analytical methods produced contradictory answers that had to be reconciled. Any analytical lab manager will tell you that it happens all too frequently. But when the samples involved are of great importance (see this month's Editorial, p 1793), and the sharp eye of publicity is focused on the scientific effort to produce a single true answer, mere disagreement may quickly degenerate into controversy and dispute.

The two quotes that illustrate this article suggest a need for closer interaction between the scientists involved in the controversy, particularly between those of different academic persuasions. The silica question is perhaps more a breakdown in scientific communication than a failure of analytical methodology. As Cant expressed it, "I think it would be reasonable to have a meeting after all, where scientists could discuss the relative merits of various techniques, perhaps after the round-robin results are in. Let's hope that before too long, we can put this thing to rest."

As George Morrison writes in this month's Editorial, "Past experience demands that the participants get together to resolve differences in methodology. Only then can controversies like the Mount St. Helens analyses be resolved satisfactorily."

The Amoco Cadiz oil spill in the waters off France, 1978. Ship at left dumps absorbent styrofoam on the slick

Ocean Pollution

Stuart A. Borman

American Chemical Society, Washington, DC 20036

Analytical chemistry aids in developing the information base to support rational decision making on utilization, conservation, and development of the nation's ocean and coastal resources

Originally published in ANALYTICAL CHEMISTRY, 1980, Vol. 52, No. 7.

The "Federal Plan for Ocean Pollution Research, Development, and Monitoring, Fiscal Years 1979–83," a comprehensive look at the national effort to keep the oceans clean, provides valuable insights into areas of study which need more attention and effort. Analytical chemistry and measurement technology are integrally involved in the endeavor to protect our ocean environment. In fact, the federal plan points out that measurement technology is frequently the limiting factor in developing information for ocean pollution problems. The further development of analytical methodology and instrumentation will thus be emphasized as the specific recommendations of this study are implemented in the next few years.

The mandate for the extensive interagency study that led to publication of the plan emerged in the National Ocean Pollution Research and Development and Monitoring Planning Act of 1978, which President Carter signed in May of that year. The plan was prepared by an interagency committee empaneled by the Executive Office of the President, with overall supervision by the Administrator of NOAA. Besides NOAA's parent agency, the Department of Commerce, six other Cabinet departments were represented on the committee, as well as the Environmental Protection Agency (EPA), the National Aeronautics and Space Administration, the National Science Foundation, the Nuclear Regulatory Commission, and the Council on Environmental Quality.

This first five-year (1979–83) plan is intended to provide overall direction and coordination for the federal activities related to ocean pollution research, development, and monitoring, an effort which entailed a federal government investment of $165 million in fiscal year 1978 alone. The plan identifies research activities that need better coordination, and determines means to ensure that critical knowledge gaps are addressed.

All these research activities relate to the very basic question of the capacity of the world's oceans to assimilate natural and manmade inputs without environmental impairment. Analytical methodology is integrally involved in determining where pollutants come from and in what quantities they enter the oceans, how pollutants are transported and transformed once they enter the oceans, and how they affect marine organisms and humans. A number of questions remain unanswered.

For example, most theories about transport and fate of toxics in the marine environment are as yet unconfirmed. Questions still exist on the extent to which pollutants change form, are absorbed by sediments, or are accumulated by marine organisms. Analytical chemistry is utilized in all phases of the research effort to answer these and other questions about the levels and dynamics of pollutants in the oceans.

At least 11 federal agencies are involved in this effort. In 1978 the Department of Interior accounted for about one-third of the funding for ocean pollution research, with EPA running a close second (21% of funding). Next in order of funding were the Department of Commerce/NOAA (11%) and the Department of Energy (10%). Chapter II of the plan describes in detail the research funding interests of the various agencies and departments involved in ocean studies.

Since financial resources from the

government agencies are always limited, it is important to assign priorities to perceived research needs, and this is exactly what the plan proceeds to do. A number of the most important priorities, as defined by the study, relate directly to analytical chemistry. These high priority areas should be prime candidates for increased appropriations, according to the plan, and should continue to be funded at present levels at least.

Among the high priority ocean pollution research needs in the domain of analytical chemistry are the following:
• determination of existing levels of critical ocean pollutants in estuaries;
• determination of processes that disperse, concentrate, and cycle municipal waste pollutants in the ocean;
• identification of sources, distribution, persistence, and bioaccumulation of synthetic organic pollutants;
• development of standard methods and criteria for evaluating pollution content of sediments;
• monitoring of industrially polluted waters to determine degree and speed of recovery; and
• development of means to detect spilled oil under ice and its dispersion into connecting waters.

This is, of course, only a partial list. Many of the other high priority projects involve analytical methodology as well.

Problem Pollutants

In the past, approximately 65% of federal funding has been pollutant specific, and of that portion of the program the major emphasis has been on petroleum (57%), metals and inorganic chemicals (11%), and synthetic organic chemicals (10%). Petroleum pollution has been in the news most often: the Campeche blowout, the Amoco Cadiz spill, and the Argo Merchant incident are among the most recent examples.

Inorganic chemicals that are considered to be particularly important marine pollutants include acids, alkalis, cyanides, and metals. Human sources of metals and inorganic chemicals include the release of sewage effluents into rivers and oceans, the combustion of fossil fuels, and the dumping of wastes into the ocean. The most toxic metals are mercury, silver, copper, cadmium, lead, zinc, chromium, tin, and selenium. In a few isolated incidents such as the Minimata, Japan, mercury disaster, metals have found their way into the human diet at highly toxic levels.

The synthetic organic category contains hundreds of different chemicals, many of them developed in the last few decades. The halogenated organic compounds, for instance, used as solvents, pesticides, and polymer intermediates, include Kepone, DDT, the PCBs, and Dieldrin. The organophosphate pesticides, such as Malathion and Parathion, have to some extent recently replaced organochlorine compounds for agricultural applications. They act on the nervous system and may be toxic to mammals and aquatic species.

Other problem pollutants are the radionuclides. At present the principal sources of marine radionuclides are fallout from above-ground nuclear tests and releases from nuclear and conventional power plants. A potential future source is leakage from ocean-dumped containers of low-level radioactive wastes, should spent nuclear fuel ever be disposed of at sea.

Recommendations

A number of specific recommendations relating to analytical chemistry are made in the plan. These include recommendations on the development of measurement technology and better quality assurance procedures.

In the domain of measurement technology, the plan states that little attention has been given to marine waters and the specialized considerations relating to their analysis: "In general, methods are not standardized and few laboratories have the capability of performing analyses with acceptable accuracy and precision. Methods are labor intensive, expensive, and not easily automated or transferred to the field."

The plan also points out that development of measurement technology is frequently underemphasized in the competition for scarce research dollars, partly because of the long lead times necessary. The following needs in the development of analytical methodology are mentioned:
• development of standardized and automated laboratory procedures for the analysis of trace metals, toxic materials, and nutrients in water, tissue samples, and sediments;
• development of reliable, accurate *in situ* portable sensors for measuring trace metals, toxics, and nutrients. Such systems are needed for unattended monitoring from buoys, monitoring compliance with pollutant discharge limits, and controlling pollutant levels in controlled ecosystem experiments;
• development of standardized automated sampling systems for water, sediments, and organisms.

For the information generated by measurement to be meaningful, quality assurance is of paramount importance. The plan thus recommends that information be provided on measurement uncertainty, including errors inherent in sampling, instrumentation, and analytical techniques. Specific recommendations include the development of requirements for periodic certification of analytical laboratories, and the listing of accredited private, academic, and federal facilities with capabilities for ocean pollution research.

This federal plan for 1979–83 is what the folks at NOAA refer to as the first plan. In fact, the plan will be updated every two years, so the second plan, for fiscal years 1981–85, will probably be rolling off the presses a little over a year from now. The first plan includes no specific framework for implementation, but NOAA staffers anticipate possible incorporation of lead agency implementation procedures in future plans.

Copies of the "Federal Plan for Ocean Pollution Research, Development, and Monitoring, Fiscal Years 1979–83" are available in limited quantities upon request to: National Marine Pollution Program Office, NOAA, Room 927, 6010 Executive Blvd., Rockville, Md. 20852.

Hope Alexander

Barrier beaches near Westerly, R.I.

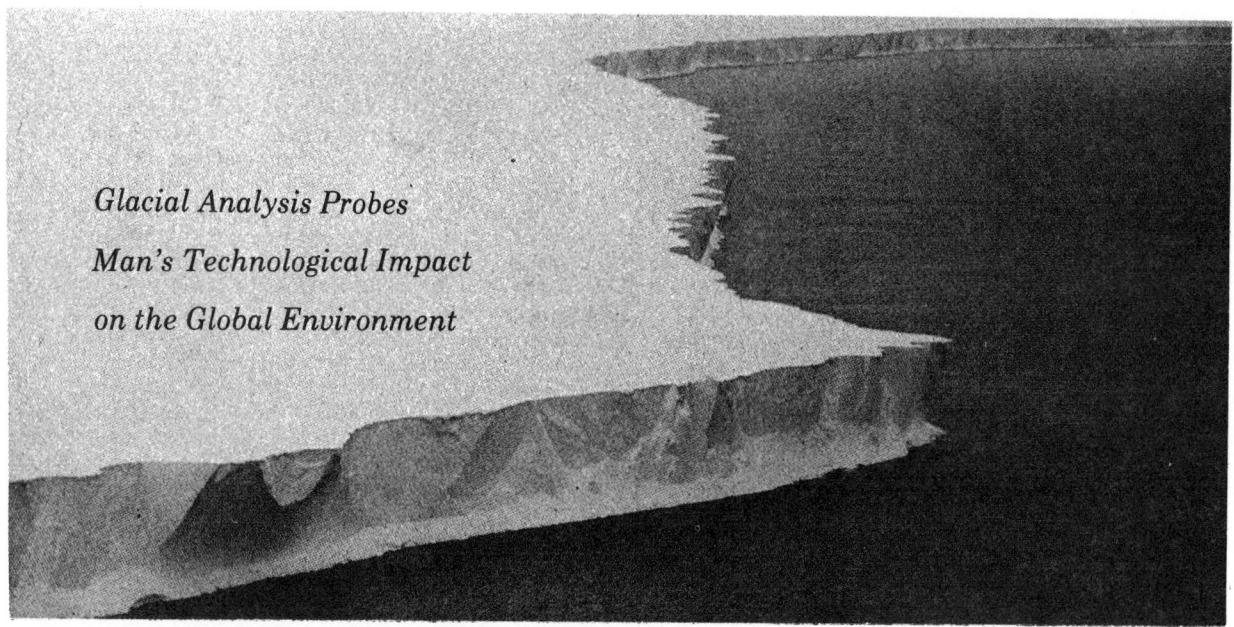

Glacial Analysis Probes Man's Technological Impact on the Global Environment

Plutonium in Glaciers

Stuart A. Borman

American Chemical Society, Washington, DC 20036

Originally published in ANALYTICAL CHEMISTRY, 1979, Vol. 51, No. 14.

Investigators from a variety of disciplines are paying quite a lot of attention to snow and ice these days. For locked inside some of the world's most forbidding glaciers is information of great potential value to modern man in his attempt to understand his environment and the effects of technology on the environment.

When DDT, transuranic radioactive nuclides, products from the combustion of fossil fuels, inorganic compounds, metals, organochlorine compounds, and hundreds of other substances are released to the atmosphere, they linger for a time before their return to the earth and the seas. And these processes have been going on for millions of years. "Where does one go to find a record of such fallout?" asks Edward D. Goldberg of the University of California at San Diego. "One of the promising places is in gla-

The ice front of the Ross Ice Shelf (see above photo by M. Herron), which meets the Ross Sea in a 50-meter-high cliff. Some of the most spectacular Antarctic icebergs, which may be several square miles in area, are formed when portions of the ice shelf break away

ciers. The reason that glaciers are terribly attractive is that they exist at every latitude on the surface of the earth, even the equator." Kilimanjaro, for example, has a permanent glacier.

"The precipitation in these regions is dominantly affected by what's in the atmosphere at the time," explains Michael M. Herron of the State University of New York at Buffalo. "Once it lands on the ice sheet or ice shelf it remains frozen, so you have a generally continuous record of the composition of the atmosphere over time spans as long as 120 000 years, and up to perhaps several million years in portions of East Antarctica."

The goal of this research effort is the elucidation of the impact of man on the global atmosphere. We need a better picture of how atmospheric pollutants are dispersed and how long various species persist in the environment. In addition, we need to establish points of reference with which to evaluate our concerns about pollutant levels. "A few years ago fish were found to contain 'dangerous' levels of mercury," explains Herron. "But nobody knew how high the levels had been a hundred years previous to that. We can now go back 100, 1000, or more years and find out what the atmosphere of Greenland was like,

which can't be done by any other means."

A number of recent investigations in Greenland and Antarctica have focused on the deposition of radionuclides in glaciers, and on the element plutonium (Pu) in particular. Manmade Pu is now a global pollutant, though it was but a short time ago that the natural level of Pu in the environment was effectively zero, except for the low-level presence of ^{239}Pu in pitchblende.

On a weight basis, Pu is one of the most biologically hazardous substances known. The more abundant isotopes are alpha-emitters with half-lives of up to 379 kiloyears. Pu enters the body primarily by inhalation and ingestion. Once it enters it deposits in the skeleton and the liver, where its highly energetic alpha particles may destroy surrounding tissue (1).

Most of the Pu in the environment today was produced in atomic bomb tests. Radioactive fallout patterns from nuclear weapons testing depend on the size of the detonation and its location. For instance, fission products in the troposphere usually reach the earth in an irregular band centered roughly at the latitude of the explosion, since prevailing winds are usually easterly or westerly. High-yield explosions propel fission products and other radioactive nuclides into the stratosphere, where material may persist for months or even years before it returns to earth. The recent deposition studies in Greenland and Antarctica have added an historical perspective to our understanding of these dispersion processes.

South Greenland

One such study is found in a 1977 *Nature* paper (2) by Minoru Koide and Edward D. Goldberg of Scripps Institution of Oceanography, and Michael M. Herron and Chester C. Langway, Jr. of the State University of New York at Buffalo. The group analyzed samples from the South Greenland ice sheet, where the snow and associated dissolved and particulate substances are essentially immobile following deposition. They dated the ice layers by ^{210}Pb geochronology, a technique developed by Goldberg in Switzerland in 1960. Radon gas (^{222}Rn) in the atmosphere from the decay of ^{238}U in the earth's crust gives a relatively constant rain of radioactivity. To date the ice layers, the activity of ^{210}Pb, a persistent daughter of ^{222}Rn with a half-life of 22 years, is determined at the surface. A layer of ice with half that much radioactivity is then 22 years old, and so on.

Drilling for the South Greenland study took place at South Dome (64 °N, 45 °W) in 1975. Plutonium isotopes were first separated from the melted snow samples by ion exchange chromatography, then determined by alpha spectrometry with a pulse height analyzer and a surface barrier detector. The detector contains a Si diode capable of distinguishing between alpha particles of different energy. The alpha spectrometer cannot distinguish between ^{239}Pu (5.16 MeV) and ^{240}Pu (5.17 MeV) because of the similarity of their alpha particle emission energies, but it can give a combined reading for these isotopes, and readily distinguishes them from ^{238}Pu (5.5 MeV).

The South Greenland group observed two distinct $^{239+240}$Pu maxima of nearly equal size for the periods 1955–60 and 1963–65. Fallout from above-ground tests of nuclear weapons detonated in the late 50's and early 60's produced the maxima. Testing ceased during a moratorium between the two series, producing a valley between the two peaks.

An additional $^{239+240}$Pu peak was detected in the strata deposited in 1946–48, a ghostly reminder of the detonations at Almagordo, N.M., Hiroshima, and Nagasaki.

Much less ^{238}Pu is produced in an atomic blast than ^{239}Pu, but the group's data indicated a significant influx of ^{238}Pu in 1966, compared to a background deposition rate that is effectively zero. The sudden appearance of ^{238}Pu in this stratum is attributed to the burnup of SNAP-9A. On 21 April 1964, a navigational satellite including a Systems for Nuclear Auxiliary Power generator, SNAP-9A, containing about one kilogram of ^{238}Pu, was launched from Vandenburg Air Force Base in California. The rocket failed to boost the satellite into orbit, and the payload came down in the Indian Ocean. Subsequent stratospheric checks indicated that the generator completely burned up during reentry and turned into small particles at an altitude of about 50 km. The Atomic Energy Commission had, in the words of John W. Finney in the *New York Times,* "lost $1 million worth of lethal plutonium somewhere in space near Africa" (3).

Dome C

Further investigations relating to Pu deposition were later conducted in Antarctica by G. A. Cutter and K. W. Bruland of the University of California, Santa Cruz, Center for Coastal Marine Studies, and R. W. Risebrough of the University of California, Berkeley, Bodega Marine Laboratory (4).

Samples for the work were obtained by Risebrough in January of 1977 at a forbidding spot known as Dome C, where the mean annual temperature is −53.5 °C. Summer temperatures remain well below freezing, preventing vertical percolation through successive layers and reducing potential losses from volatilization, and making Dome C an ideal place for deposition studies.

But while the site may be favorable, the problems of sampling in Antarctica can be monumental, a fact underscored by a series of airplane accidents in 1974 and 1975 at Dome C, which is only accessible by air. Although no one was hurt, at one time three Hercules LC-130 transports were stranded at the camp.

Eventually the disabled planes were repaired and evacuated from the site, but other formidable problems plague Antarctic research teams, including the threat of sample contamination. Risebrough writes: "The chances of sample contamination are very high, either from past human activities in the vicinity of the sampling site or from sampling and extraction operations. The area about the South Pole is probably much too contaminated to permit such studies: It is much more likely that pollutants originated from the nearby station rather than distant sources many thousands of kilometers away. The problem is universal: Wherever we go, we bring and disseminate our pollutants" (5).

The Dome C analyses, also run by alpha spectrometry, indicated heavy $^{239+240}$Pu deposition in the late 50's and early 60's, as the South Greenland work had, except that the late 50's peak far and above dominated the early 60's peak, whereas the peaks were of equal magnitude in the South Greenland study. The tests in the early 60's were conducted by the USSR at Novaya Zemlya in the Arctic. Fallout from this series was thus much more prominent in Northern latitudes.

Significant quantities of $^{239+240}$Pu first began to appear in the Antarctic snow in 1955. Cutter remarks that the late 50's peak's size and timing suggest that initial fallout was substantially contributed by the U.S. Castle test Bravo, detonated on 28 February 1954, at Bikini, a particularly dirty bomb which generated controversy and protest at the time. Ten days after the blast the Atomic Energy Commission admitted that 28 Americans and 236 natives of the Marshall Islands, 330 miles from Bikini, had been exposed to radiation. Cutter et al. also detected the ^{238}Pu fallout from SNAP-9A, which appeared as a rapid activity increase in the 1965–66 stratum.

Ross Ice Shelf

A second Antarctic study was recently conducted by Minoru Koide, Robert Michel, and Edward Goldberg of the University of California at San Diego, and Michael M. Herron and Chester C. Langway, Jr. of the State University of New York at Buffalo, with ice samples from Antarctica's Ross Ice Shelf (6). Their deposition profiles also indicated heavy accumulations of $^{239+240}$Pu in the late 50's and early 60's, light accumulations in between due to the testing moratorium, and moderate fallout due to the French and Chinese tests in the late 60's and early 70's. Fallout from the SNAP-9A burnup was evident in their ^{238}Pu profile. The researchers also analyzed for a number of fission products such as ^{90}Sr and ^{137}Cs.

Glaciological research is today proceeding on many fronts. There are ongoing and recently completed studies on lead aerosols, dusts, sea salts, trace metals, sulfate, combustion products, aliphatic and aromatic hydrocarbons, and pesticides in snow. Edward Goldberg is presently interested in soot deposition. "We can identify the source of carbon by its surface characteristics and size, the morphology and dimensions of the carbon particles," he explains. "We can distinguish between coal particles, wood particles, and oil particles. Now we want to look at the history of burning by studying carbon particle deposition in glaciers."

Much additional work remains to be done on the transuranics and fission products as well. Deposition studies in mid-latitudinal glaciers could lead to a better understanding of interhemispheric transport. Ice sheets in the Arctic could be examined for fallout from the Russian tests at Novaya Zemlya. This work and many other projects planned or already conducted are helping to give us a better understanding of the historical context of pollution. As Michael Herron puts it, "In the coming years we will stage a full-scale research assult on as many elements and organic compounds as possible, in old snow and in new snow. This will enable us to monitor changes in concentration indicating the influence of man's activities on the environment."

References

(1) "Cleaning Our Environment: A Chemical Perspective," Second Edition, American Chemical Society, Washington, D.C., 1978.
(2) Minoru Koide, Edward D. Goldberg, Michael M. Herron, Chester C. Langway, Jr., *Nature.* **269,** 137–39 (1977).
(3) John W. Finney, *New York Times,* 24 May 1964, 1.
(4) G. A. Cutter, K. W. Bruland, R. W. Risebrough, *Nature.* **279,** 628–29 (1979).
(5) Robert W. Risebrough, *Antarct. J. U.S.* **12,** 131–32 (1977).
(6) Minoru Koide, Robert Michel, Edward D. Goldberg, Michael M. Herron, Chester C. Langway, Jr., *Earth Planet. Sci. Lett.* **44,** 205 (1979).

Crystalline Silica in Mount St. Helens Ash

Sherry O. Farwell and Dennis R. Gage

University of Idaho, Chemistry Department, Moscow, ID 83843

Originally published in ANALYTICAL CHEMISTRY, 1981, Vol. 53, No. 13.

In the months since the first significant Mount St. Helens ash eruption on May 18, 1980, there has been a great deal of concern regarding the potential crystalline silica content of the ash. This concern is founded in the fact that chronic inhalation of respirable-size airborne particles containing significant amounts of crystalline silica can induce a respiratory disease known as silicosis. Respirable-size airborne particles have aerodynamic equivalent diameters of 10 μm or lower. Our laboratory's initial efforts in the assessment of the impact of Mount St. Helens volcanic ash were focused entirely on evaluating human exposure levels via the collection of continuous, meaningful inhalable particulate data using dichotomous virtual impactor samplers (1, 2). We assumed that the crystalline silica content in the inhalable portion of the volcanic ash would be quickly and conclusively determined by laboratories with greater instrumental capabilities than our facilities at the University of Idaho.

However, early analyses by other laboratories revealed conflicting results in both the types and amounts of crystalline silica in the Mount St. Helens ash. The initial conflicting data over the crystalline silica content of the ash sparked a somewhat heated controversy that was recently described in two FOCUS articles in ANALYTICAL CHEMISTRY (3). The confusion prompted the National Institute for Occupational Safety and Health (NIOSH) to organize a round-robin testing program for crystalline silica determinations in volcanic ash. Ash samples were collected from several locations in Washington state and the respirable fractions isolated. Ash samples of 10 μm or less were sent to participating laboratories for round-robin determination of their crystalline silica content. We decided as a result of the conflict reported in ANALYTICAL CHEMISTRY to investigate the feasibility of using another technique to determine the crystalline silica content of the ash.

Several different methods can be used to identify crystalline silica (4). These include polarized light microscopy, scanning electron microscopy, transmission electron microscopy, and visible spectrophotometry. Of these methods, however, none can distinguish between the three crystalline silica polymorphs commonly found in geological samples—quartz, cristobalite, and tridymite. The potential hazardous effects on the respiratory sys-

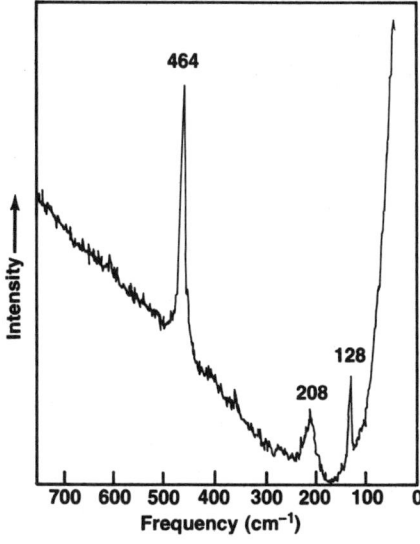

Figure 1. Raman spectrum of quartz. Laser power, 190 mW at sample; time constant, 1 s; scan rate, 1 cm^{-1} s^{-1}; slits, 3 cm^{-1}; full-scale, 20 K counts; zero suppress, 19.7 × 100

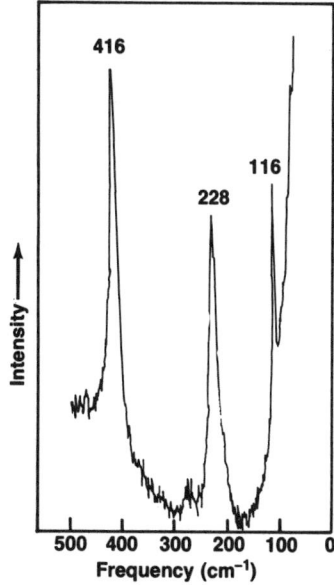

Figure 2. Raman spectrum of cristobalite. Conditions same as Figure 1 except full-scale, 12 K counts; zero suppress, 11.0 × 100

tem vary with the form of the crystalline silica, so knowing the polymorph speciation is very important. In theory, infrared (IR) spectrophotometry and X-ray diffraction (XRD) have the capability of qualitatively and semiquantitatively determining these three silica polymorphs. Earlier investigations in other laboratories revealed, however, that other matrix components in the ash produced severe interference problems in these techniques. Thus, a sample "cleanup" procedure was recommended for removing these interferences.

The sample pretreatment procedure utilized the Talvitie method (5) for digestion of the ash sample. With the Talvitie method, the sample is digested for a specified period of time (usually 16 min) in pyrophosphoric acid. This removes the silicate interferences and leaves the crystalline silica in a purified form. As reported by Talvitie and recently reconfirmed by NIOSH (4), loss of crystalline silica in the pyrophosphoric acid digestion does occur. However, if the digestion time and temperatures are accurately controlled so that they comply with the recommended procedure, maximum losses of quartz and cristobalite are in the 10–20% range. Experiments were performed in response to the concern that silicates were being converted to crystalline silica during the phosphoric acid digestion. NIOSH results showed no detectable crystalline silica artifacts in various plagioclase minerals digested via the pyrophosphoric acid procedure. Nevertheless, considerable skepticism remained regarding the Talvitie method for pretreatment of the ash prior to analysis.

An analytical approach was needed that would not require pretreatment of the sample by the controversial Talvitie method.

The Search

A review of potential analytical techniques led us to our department's relatively new laser Raman spectrometer (LRS). A subsequent search of the LRS literature revealed that like IR and XRD, LRS spectra could probably distinguish the crystalline silica polymorphs. In addition, since Raman spectra tend to be less cluttered with peaks than IR spectra or XRD patterns, one could expect component peak overlap in real-world samples to be less in Raman spectra, and the corresponding analyses to be simpler. While the results of the literature search suggested that qualitative analysis of the crystalline silica was possible, similar information was not available to suggest the potential of LRS for quantitative determinations.

Unfortunately, a reliable standard of tridymite was not available to us

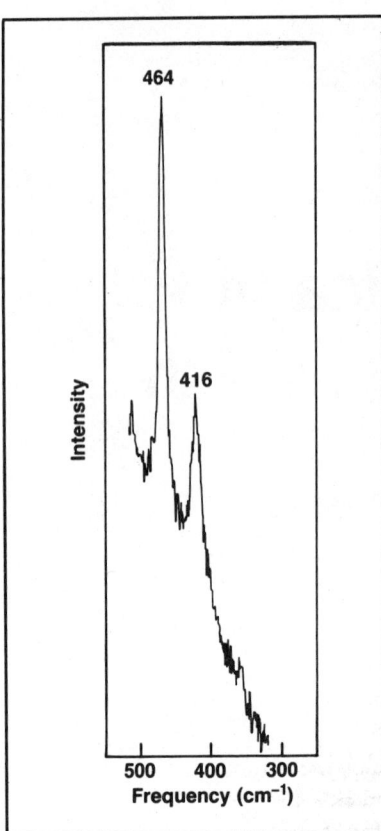

Figure 3. Raman spectrum of pellet consisting of 50% quartz–50% cristobalite

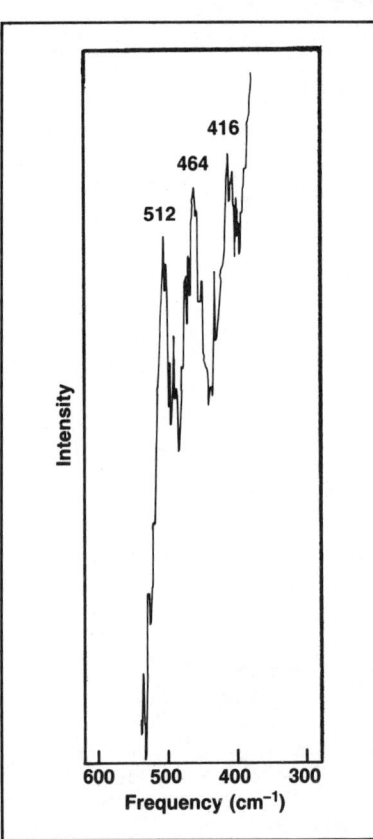

Figure 4. Raman spectrum of NIOSH volcanic ash sample

and no *definitive* Raman spectrum of tridymite was found in the LRS literature. Several LRS spectra of tridymite were located, but there was little agreement among them. To make matters worse, our standard tridymite spectrum matched none of the literature references. On the other hand, our experimental LRS spectra for quartz and cristobalite, as shown in Figures 1 and 2, were identical to the literature spectra (6). Unlike the IR spectra for these two silica polymorphs, which are characterized by relatively broad absorption bands at almost identical wavelengths (4, 7), the LRS spectra show sharp, well-resolved peaks.

To illustrate the peak resolution and relative scattering efficiencies of quartz and cristobalite, an LRS spectrum of a pellet consisting of 50% quartz and 50% cristobalite was obtained. Figure 3 shows this spectrum in the region from 300–500 cm^{-1}. It is evident the quartz peak at 464 cm^{-1} is approximately twice as intense as the characteristic cristobalite peak at 416 cm^{-1}. Also note that this quartz/cristobalite mixture spectrum reveals distinct, completely resolved peaks for the 464 cm^{-1} and 416 cm^{-1} bands; the characteristic peaks for quartz and cristobalite are separated by 48 cm^{-1}. The Raman peaks in the lower wavenumber region of the quartz and cristobalite spectra overlapped and could not be used to differentiate between these two silica polymorphs in a mixture.

As previously known and reconfirmed in our studies, the peak intensities in a Raman spectrum are directly proportional to the laser intensity and the analyte concentration. The majority of the ash matrix is volcanic glass, which has a very weak Raman signal. Figure 4 shows a typical LRS spectrum of a Mount St. Helens volcanic ash sample that was pressed into a pellet and *directly* analyzed by the LRS system. The peak at 416 cm^{-1} corresponds to cristobalite, the larger peak at 464 cm^{-1} corresponds to quartz, and the peak at 512 cm^{-1} is believed to be due to plagioclase feldspar in the ash matrix. However, the 512 cm^{-1} peak was present in all of the volcanic ash samples, and it was used for quantitation by the "direct comparison method" (8). Whereas the absolute Raman peak intensities are dependent on laser power, the ratio of intensities for two peaks in a spectrum will be independent of laser power. To quantify the amounts of quartz and cristobalite in the volcanic ash samples, the direct comparison method was used in combination with the method of standard addition (6). Details of our experimental techniques and results are the subject of a techni-

Table I. Comparative Results for Crystalline Silica (% by Weight)

Analyte		IR[a]	XRD[b,c]	LRS[d]
Total quartz + cristobalite sample	1	5.1	2.21–4.05	5.44 ± 0.25
	2	4.3	2.11–3.74	4.04 ± 0.10
	3	6.2	3.54–6.44	8.61 ± 1.23
	4	9.2	5.14–8.72	10.70 ± 0.27

[a] Analysis performed by M. Bolyard at NIOSH lab; [b] Combined results of NIOSH round-robin testing program from 14 participating labs; [c] Data represent lower and upper 95% confidence limits; [d] Analysis performed by authors at University of Idaho

cal paper appearing in this issue of ANALYTICAL CHEMISTRY (p 2123).

The Data

The four NIOSH round-robin ash samples were analyzed by LRS and the results sent to NIOSH. NIOSH, in turn, gave us the compiled analytical results of the collaborative study. Table I compares our LRS results with the IR data obtained in the NIOSH laboratory and with the summarized XRD results from the 14 different laboratories that participated in the NIOSH testing program. An article describing in detail the IR and XRD procedures and results is being prepared for publication by NIOSH scientists (9). The comparative analytical data in Table I show the LRS results for total cristobalite and quartz to be somewhat higher than the corresponding IR and XRD data. However, it should be emphasized that the IR and XRD data in this table were obtained on the ash samples after pretreatment by the hot phosphoric acid procedure to remove the silicate interferences; our LRS determinations were performed on chemically untreated volcanic ash. If one assumes that 10–20% of the crystalline silica is lost due to dissolution during the pyrophosphoric acid digestion procedure, then no significant unaccountable differences exist between the LRS and IR data and the high range of the XRD data for total quartz and cristobalite.

Our LRS results agree well with the crystalline silica concentrations obtained by two other spectroscopic techniques that employ different methods of sample preparation. Although our independent analytical approach supports the presence of crystalline silica in inhalable ash of ≤10 μm aerodynamic equivalent diameter, the ultimate impact on human health of inhalable particulates derived from Mount St. Helens volcanic ash needs to be evaluated. The more urgent tasks include the following:

- an accurate, independent measurement of tridymite in volcanic ash of inhalable size since tridymite is apparently the crystalline silica polymorph with the smallest dose to silicosis response effect (10);
- the determination of the types and relative amounts of crystalline silica in the ash of less than 2.5 μm aerodynamic equivalent diameter since these very small particles can reach and deposit in the alveolar portions of the lung (1); and
- the development of an appropriate human health risk level criterion for the evaluation of measured airborne crystalline silica levels at various durations of continuous exposure.

Acknowledgment

The contribution of the round-robin analytical data from D. D. Dollberg and associates at the NIOSH laboratory in Cincinnati, Ohio, is gratefully acknowledged.

References

(1) Farwell, S. O.; Gage, D. R.; Jernegan, M. F.; Felkey, J. R. *JAPCA* **1981**, *31*, 71.
(2) Farwell, S. O.; Gage, D. R.; Jernegan, M. F.; Felkey, J. R. "Inhalable Particulate Studies of Mount St. Helens Volcanic Ash"; paper #312 presented at the 1981 Pittsburgh Conference on Analytical Chemistry and Applied Spectroscopy, Atlantic City, N.J., March 1981.
(3) *Anal. Chem.* **1980**, *52*, 1136 A and 1272 A.
(4) Dollberg, D. D.; Bolyard, M.; Sweet, D. V.; Carter, J. W.; Stettler, L. E.; Geraci, C. L. "Mount St. Helens Volcanic Ash: Crystalline Silica Analysis"; National Institute for Occupational Safety and Health, Cincinnati, Ohio, 1980.
(5) Talvitie, N. A. *Anal. Chem.* **1951**, *23*, 623.
(6) Farwell, S. O.; Gage, D. R. *Anal. Chem.* **1981**, *53*, 2123.
(7) Taylor, D. G.; Nenadic, C. M.; Crable, J. V. *Am. Ind. Hyg. Assoc. J.* **1970**, *31*, 100.
(8) Cullity, B. D. "Elements of X-ray Diffraction"; Addison-Wesley: Reading, Mass.; pp 391–6.
(9) Dollberg, D. D. NIOSH, Cincinnati, Ohio, personal communication, April 16, 1981.
(10) Zaidi, S. H. "Experimental Pneumoconiosis"; The John Hopkins Press: Baltimore, Md., 1969.

Partial support for this work was provided by special funding from the Region X Office of the Environmental Protection Agency. Funds for the LRS instrumentation were provided by NSF grant CHE-7727395.

Identification of Organic Compounds in an Industrial Wastewater

Ronald A. Hites

Indiana University, School of Public and Environmental Affairs, Bloomington, IN 47405

Viorica Lopez-Avila

Midwest Research Institute, Kansas City, MO 64110

Originally published in ANALYTICAL CHEMISTRY, 1979, Vol. 51, No. 14.

For several years, our laboratory has been engaged in the identification of organic compounds in industrial wastewaters and in rivers (1–6). The purpose of these studies has been to identify the exact molecular structures of the compounds present so that the origin, environmental fate, and toxicity of a wide variety of compound types could be assessed. Our approach has been based on survey analyses ("What is present?") rather than specific compound analyses ("How much of compound x is present?"). Because the compounds are found at very low concentrations (ppb) and in quite complex mixtures, we have used gas chromatographic mass spectrometry (GC/MS) as our primary analytical tool. Using modern, commercially available instrumentation, GC/MS data can usually be obtained without excessive effort; the interpretation of these data, however, is not always so straightforward. In most cases, we have had to collate many different types of information in order to solve these interpretive problems. This paper presents such a problem and its solution.

One industrial wastewater which we have analyzed quite extensively is that of a specialty chemicals plant (3). This plant operates in a batch production mode, generally following a weekly schedule. A wide range of compounds including pharmaceuticals, herbicides, antioxidants, thermal stabilizers, ultraviolet light absorbers, optical brighteners, and surfactants is produced. Water is used in synthesis processes, in the recovery of solvents, in steam jets, and in vacuum pump seals. The wastewater was neutralized in either of two 1-million-gal equalization tanks, passed through a trickling filter for biological degradation, and clarified in a 150 000-gal tank with a residence time of 3 h. The water spills over from the clarifier at a rate averaging 1.3×10^6 gal/day and enters a river through an underground pipe about 100 yards away. Only about one-fourth of the total biochemical oxygen demand (BOD), which averages 12 000 lb/day, is removed by the waste treatment system.

In the course of the complete organic analysis of this wastewater, several unknown but related mass spectra were encountered. Figure 1 gives these six mass spectra. Spectra A and B were obtained by GC/MS, while spectra C–F were obtained by collecting fractions after separation by high pressure liquid chromatography. Each fraction was collected in a 5 ml pear-shaped flask and was evaporated to dryness on a rotary evaporator. The sample was redissolved in 5–10 μL of dichloromethane and transferred to a capillary tube (1.5–1.8 × 50 mm) which was introduced into the mass spectrometer via the direct probe. This technique required 10–15 min per fraction; even when the fraction contained more than one compound, clean mass spectra for each compound could usually be obtained because of differential volatility of the various components.

The use of HPLC as a supplement to GC/MS is an important feature of this study. Among other benefits, it indicates that the compounds are indigenous to the sample and are not formed by pyrolysis during the GC/MS analysis itself.

Clearly spectra A–F are related to one another. They all show isotopic clusters which indicate the presence of two or more chlorine atoms; several have intense peaks at m/e 218 and 252; and spectra A, E, and F indicate molecular weights differing by 126 amu (288, 414, and 540). None of these spectra were in an index of known mass spectra (7), nor were they retrieved by a computerized search of mass spectra of known compounds (8).

Spectrum A indicates a molecular weight (MW) of 288 and has an isotopic cluster corresponding to Cl_3. (For reference, the expected distributions of the Cl_1 to Cl_5 isotopic clusters are given in Figure 2). We consulted a list of commercially manufactured organic compounds which was ordered by exact molecular weight; the only compound listed with MW = 288 and with three chlorines was 5-chloro-2-(2,4-dichlorophenoxy)-phenol. (This list is available in microfiche form on request to the author.)

$$Cl-\bigcirc-O-\bigcirc-Cl \quad (A)$$
$$\quad\quad OH \quad Cl$$

Furthermore, this compound is manufactured by the company whose wastewater we were studying. Spectrum A agrees with this structure; it shows ions due to the loss of one Cl (m/e 253) and two Cl (m/e 218). The authentic compound A was obtained; its mass spectrum and GC and HPLC retention times agreed with the unknown. We can now consider structure A proven and use this structure as a springboard from which the structures of compounds B–F can be reached.

At this point it is helpful to our subsequent discussion if we outline the synthetic scheme which we presume is used in the commercial production of compound A (see Figure 3). The first step is the reaction of 2,5-dichloronitrobenzene with 2,4-dichlorophenol; the resulting trichloro-nitro-diphenyl-ether is then reduced with Raney nickel to the amino compound. Finally, the diazonium salt is formed and

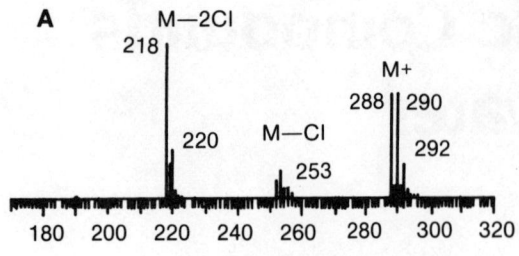
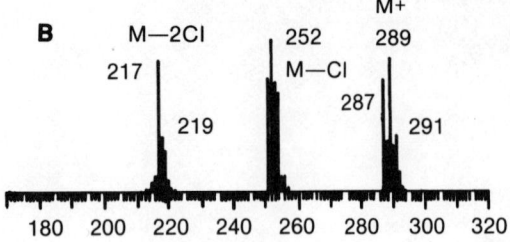
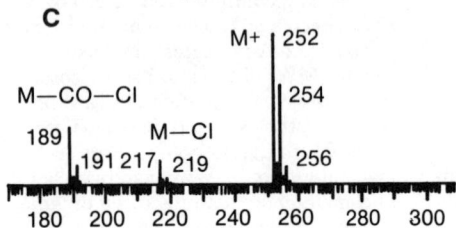
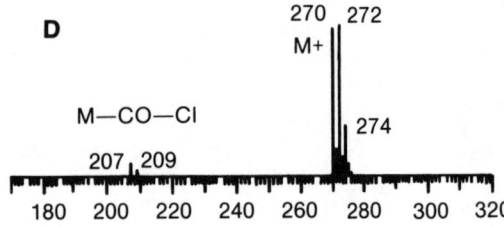
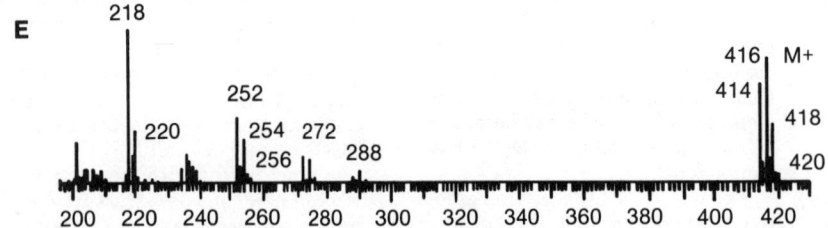
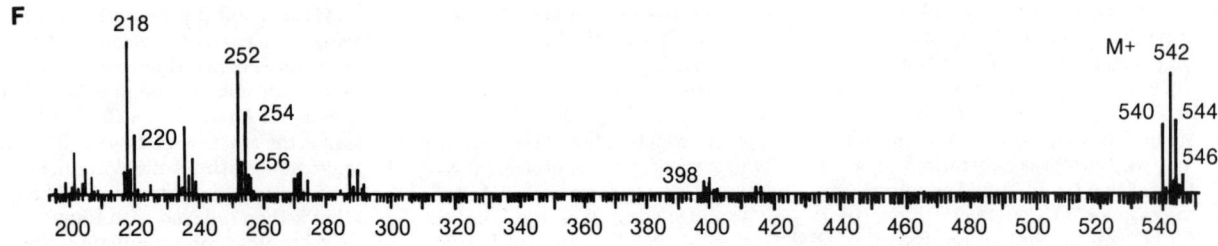

Figure 1. Mass spectra of unknown compounds A–F which were isolated from the wastewater of a chemical plant
Conditions: electron impact, 70ev, HP 5982A mass spectrometer

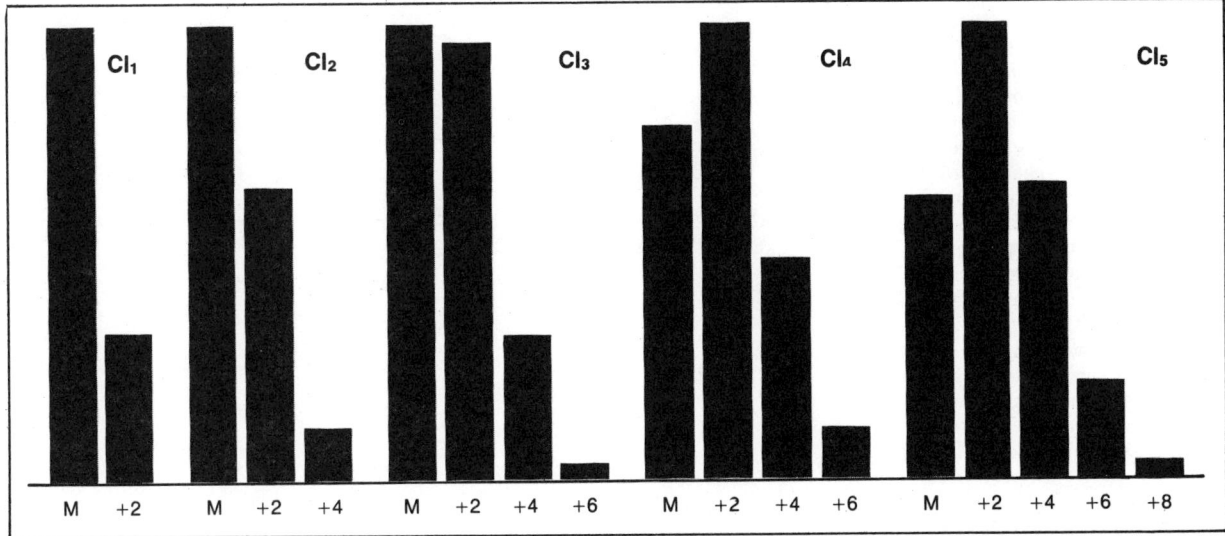

Figure 2. Characteristic multiplets of peaks caused by the presence of one to five chlorine atoms in a molecular or fragment ion; they are all spaced two mass units apart

hydrolyzed to give compound A. None of these reactions is 100% complete; thus, many of the precursors can be present in small amounts at any step of the process.

Let us now move on to spectrum B; it indicates a molecular weight of 287 and the presence of three chlorine atoms. The spectrum is very similar to spectrum A except it is shifted down by one mass unit. Replacing the OH (17 amu) in compound A by NH_2 (16 amu) could account for this shift. This compound is commercially available and is, in fact, the synthetic precursor of compound A (see Figure 3).

(B)

In spectra A and B, the group of peaks due to the loss of Cl and HCl is complex (see m/e 251 to 256). In spectrum B, for example, m/e 252, 254, and 256 are due to the loss of Cl from the intact molecule. Since there are still two chlorine atoms in this ion, the expected isotopic abundance is 252: 100%, 254: 65%, and 256: 11%. The intense ions at m/e 251, 253, and 255 in spectrum B are due to the loss of HCl from the molecule and probably have the structure

The corresponding ions in spectrum A are at 252, 254, and 256 and probably have the structure

Spectrum C shows a similar ion cluster at 252, 254, and 256 which is also the molecular weight. By analogy with the above ion structure, we formed the hypothesis that compound C is a dichlorodibenzodioxin which was formed as a by-product during the production of compound A. This hypothesis is supported by the ion at m/e 189 which is due to the loss of Cl and CO from the molecule. This is a characteristic fragmentation for chlorodibenzodioxins (9). Spectrum C, of course, cannot tell us where on the ring the chlorines are positioned. However, since we know the substituent positions of the precursor, we can predict that compound C is the 2,7-dichloro isomer

(C)

Spectrum D shows a molecular ion at m/e 270 which contains three chlorine atoms and an ion at 207 due to the loss of CO and Cl. An elemental composition of $C_{12}H_5Cl_3O$ is suggested by these data and by analogy with compound A. It takes very little imagination to put together a trichlorodibenzofuran structure from this information. Such a compound could be formed by the loss of nitrogen followed by intramolecular coupling during hydrolysis of the diazonium salt intermediate

(D)

Assuming that this is the correct reaction, we can position the chlorines as shown.

Spectra A, E, and F indicate molecular weights differing by 126 amu and show the presence of 3, 4, and 5 chlorine atoms, respectively. The 126 amu difference is most likely

We can attach this moiety to compound A in at least two ways (see Figure 4). These particular positional isomers were selected because either could be formed by the reaction of compound A with 2,4-dichlorophenol which is probably present as residual starting material (see Figure 3).

Compounds E_1 and E_2 can be distinguished from each other based on the electron impact induced cleavages indicated in Figure 4. We see that both compounds would give fragment ions at m/e 253 and 269 (with 2 chlorines) due to the loss of the dichloro substituted ring; however, only compound E_2 could give ions at 287 or 271 (with three chlorines). Spectrum E (see Figure 1) shows a small ion cluster at 272 with a Cl₃ isotopic pattern. This could correspond to the fragment expected at m/e 271 with an additional proton added by a rearrangement of the phenolic hydrogen. In addition, the methane chemical ionization mass spectrum of compound E (see Figure 5) shows an ion cluster at m/e 287 with 3 chlorines; this corresponds to the two-ring, two-oxygen fragment shown in Figure 4. This chemical ionization spectrum definitively indicates that the hydroxy group is on the terminal ring rather than the middle ring; thus, we believe structure E_2 is the correct assignment for compound E.

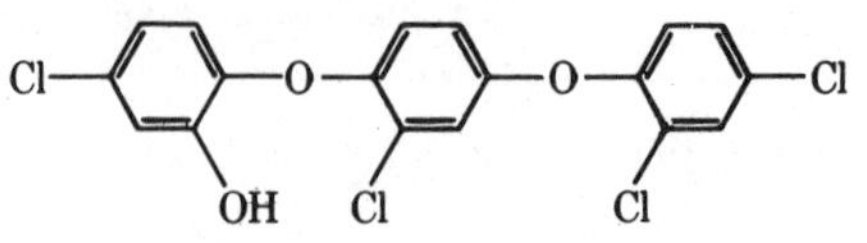

Spectrum F indicates a molecular weight of 540 and has an isotopic pattern indicative of five chlorine atoms. Since its molecular weight is 126 amu greater than compound E, we hypothesize that it was formed by the reaction of 2,4-dichlorophenol with compound E; this is analogous to the formation of compound E itself. The positions of the substituents can be determined from the assumed precursors and from the mass spectrum in a manner similar to compound E. Note that the fragment ions at m/e 272, 274, and 276 in spectrum E are shifted by 126 amu in spectrum F and now appear at m/e 398, 400, and 402 and that these ions have a four chlorine isotopic pattern. These facts indicate that the hydroxy group in compound F is also in the terminal ring. Therefore, compound F has the structure:

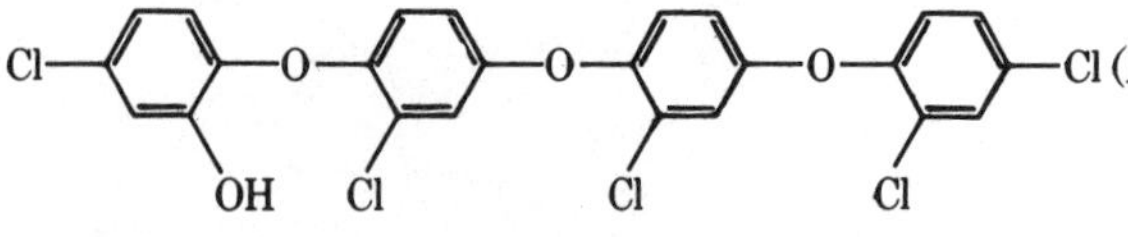

In order to verify these structures, which all seem to result from the thermolysis of compound A, 1 g of this compound was heated at 250 °C for 60 min in a 100 mL round-bottom flask fitted with a reflux condenser and a thermometer. The products were extracted with dichloromethane and analyzed by the HPLC/MS technique outlined above. The chromatogram and peak identifications are shown in Figure 6. Clearly, several reactions have occurred. Degradation of compound A to dichlorophenol (MW = 162, presumably the 2,4-isomer) has taken place; this is apparently followed by a reaction of compound A *with* dichlorophenol to produce the

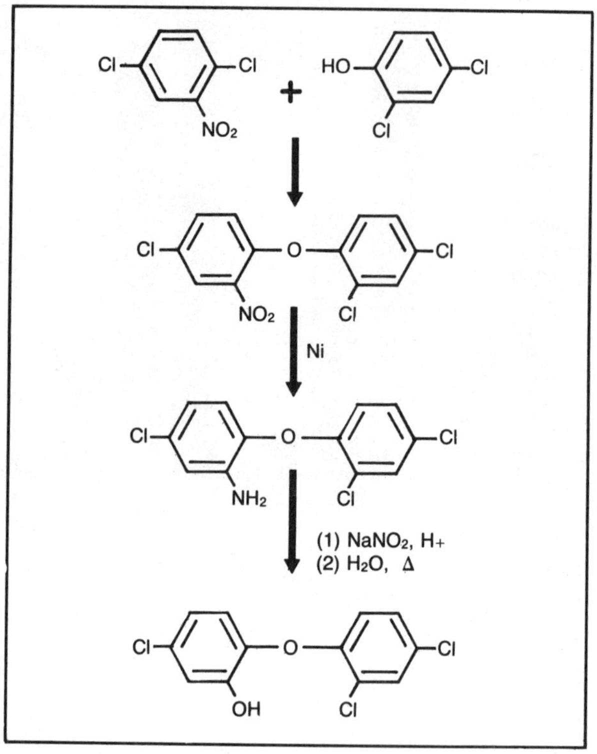

Figure 3. Synthetic scheme which presumably was used for the industrial production of 5-chloro-2-(2,4-dichlorophenoxy)-phenol [compound A]
Reference: E. Model and J. Bindler, Swiss Patents 428,758 to 428,760 (see *Chem. Abst.* **68**, 12690-2)

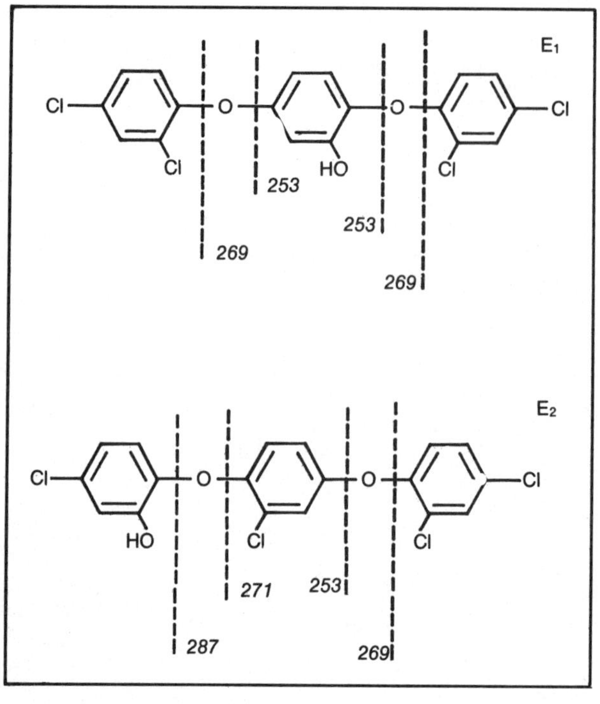

Figure 4. Two possible structures for compound E showing masses of expected fragment ions

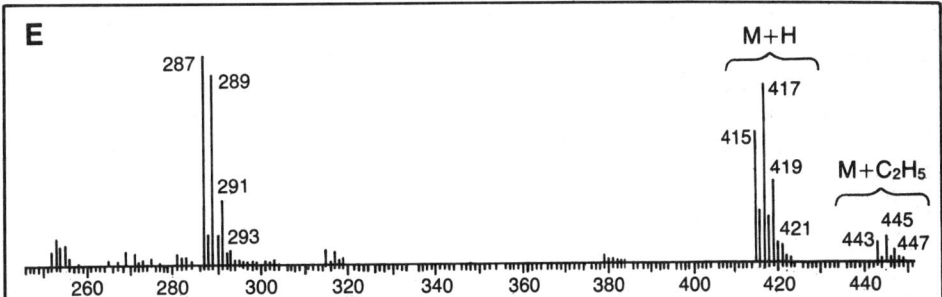

Figure 5. Methane chemical ionization mass spectrum of compound E

three ring compound E. This compound in turn reacts with dichlorophenol to yield compound F. Further reactions of this type give even larger compounds of this nature; note the compound of MW = 666 (540 + 126) in Figure 6 which has five rings. Similar compounds containing six to nine rings are probably formed as well (see arrows in Figure 6) but good mass spectra could not be obtained of these peaks. The peaks numbered 2 to 9 in Figure 6 are compounds of the form

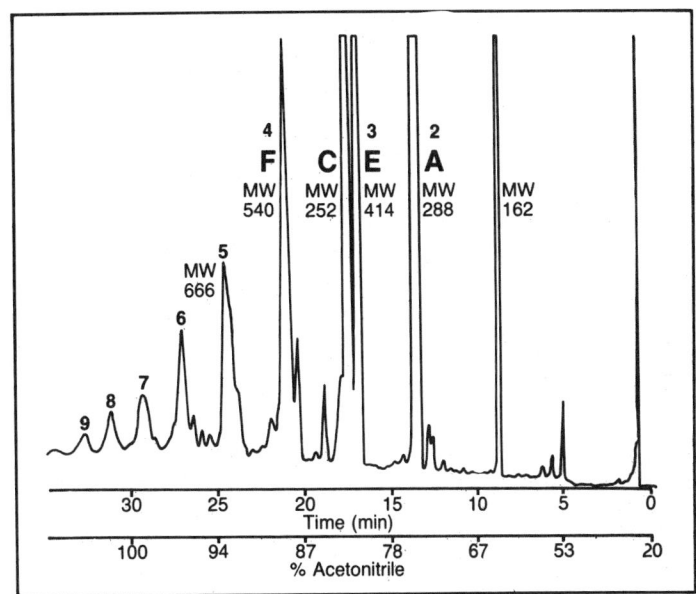

Figure 6. High pressure liquid chromatogram of the reaction products formed by heating 5-chloro-2-(2,4-dichloro-phenoxy)-phenol [compound A] at 250° for 60 min

HPLC conditions: μ Bondapak C_{18}, 20–100% CH_3CN in H_2O, 2 ml/min, 254 nm. The peaks which were identified by mass spectrometry as compounds A, C, E, and F are so indicated. The compound with MW = 162 is 2,4-dichlorophenol

where n = 2, 3, ... 9.

Compound C (the dichlorodioxin) is also formed by the thermal cyclization of compound A (see Figure 6, MW = 252). Compound B (the amino species) and compound D (the trichlorodibenzofuran) are not found in the products of this thermolysis. This is expected since compound B is actually the synthetic precursor of A, and D is formed from its diazonium salt.

This set of identifications demonstrates a number of things:

- Several by-products are formed during the commercial production of compound A and are discharged with the process wastewater.
- The interpretation of groups of mass spectra of related compounds is far more productive than treating each as an isolated case. For example, it would have been very difficult to interpret spectrum F without first having deduced structures A and E.
- The use of production information (compounds in commercial production and their industrial syntheses) is essential for the efficient interpretation of mass spectra of compounds isolated from the environment.
- Considerable information about an unknown chemical production process can be obtained by a careful analysis of the wastewater of that process.

Acknowledgment

The cooperation of the company officials and plant personnel is appreciated. This work has been supported by the Chemical Threats to Man and the Environment Program of the National Science Foundation (Grant No. ENV-75-13069) and by the U.S. Environmental Protection Agency (Grant No. R 806350).

References

(1) G. A. Jungclaus, L. M. Games, and R. A. Hites, "Identification of organic compounds in tire manufacturing plant wastewaters," *Anal. Chem.*, **48**, 1894–96 (1976).
(2) L. M. Games and R. A. Hites, "Composition, treatment efficiency, and environmental significance of dye manufacturing plant effluents," *ibid.*, **49**, 1433–40 (1977).
(3) G. A. Jungclaus, V. Lopez-Avila, and R. A. Hites, "Organic compounds in an industrial wastewater: a case study of their environmental impact," *Environ. Sci. Technol.*, **12**, 88–96 (1978).
(4) L. S. Sheldon and R. A. Hites "Organic compounds in the Delaware River," *ibid.*, **12**, 1188–94 (1978).
(5) L. S. Sheldon and R. A. Hites "Sources and movement of organic compounds in the Delaware River," *ibid.*, **13**, 574–79 (1979).
(6) L. S. Sheldon and R. A. Hites, "Environmental occurrence and mass spectral identification of ethylene glycol derivatives," *Sci. Total Environ.*, **11**, 279–86 (1979).
(7) *Eight Peak Index of Mass Spectra*, Mass Spectrometry Data Centre, Reading, United Kingdom (1974).
(8) S. R. Heller, G. W. A. Milne and R. J. Feldmann, "A computer-based chemical information system," *Science*, **195**, 253–59 (1977).
(9) N. P. Buu-Hoi, G. Saint-Ruf and M. Mangane, "The fragmentation of dibenzo-p-dioxin and its derivatives under electron impact," *J. Heterocycl. Chem.*, **9**, 691–93 (1972).

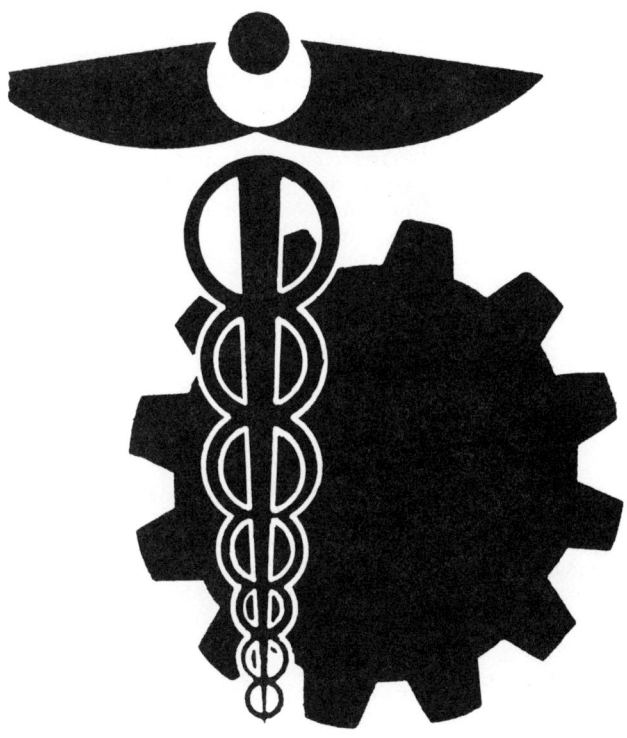

Industrial Analytical Chemists and OSHA Regulations for Vinyl Chloride

**S. P. Levine, K. G. Hebel,
J. Bolton, Jr., and R. E. Kugel**

Stauffer Chemical Co., Eastern Research Labs
Dobbs Ferry, NY 10522

Originally published in ANALYTICAL CHEMISTRY, 1974, Vol. 46, No. 14.

In 1971 the newly created Occupational Safety and Health Administration (OSHA) with the advice of its technical arm, the National Institute for Occupational Safety and Health (NIOSH), adopted a 500 parts-per-million (ppm) by volume permissible level for worker exposure to vinyl chloride monomer gas (VCM). In 1974 a review of this standard was prompted by reports of several deaths from a rare form of liver cancer called angiosarcoma among polyvinyl chloride (PVC) plant employees. This suggested a possible relationship to PVC production. Animal studies and epidemiological surveys indicated that exposure to VCM might be a causative agent involved in the development of angiosarcoma in humans (1). These facts led OSHA to issue in April 1974 a temporary emergency standard of a 50-ppm ceiling exposure. This standard also provided for regular monitoring of the work space by personnel monitoring systems able to assay 5-ppm VCM with a relative precision of ±20% (average for a 10-min air sample). This was to have a profound effect on the vinyl chloride and the polyvinyl chloride industries which employ approximately 360,000 workers in over 7,500 plants.

Following public hearings, OSHA published a final standard for VCM in October 1974 (Figure 1) (2, 3). However, the emergency temporary standard remained in effect until April 1, 1975. The final standard sets a maximum permissible level of 1 ppm for an 8-hr time-weighted average exposure. A ceiling limit of not more than 5-ppm VCM over a 15-min period has also been set. In addition, an action level of 0.5 ppm was set up; exposures above the action level require periodic monitoring, medical examinations, and training (4).

The emergency and the final standards called for vastly different approaches from an industrial hygiene point of view. Under the temporary standard, a survey was made to determine areas of emission, and only "grab" sampling was performed. It was only necessary to ensure that work areas did not exceed 50 ppm of VCM in the air. This sampling approach was changed with the advent of the permanent standard. Now areas must be regulated by both the ceiling (maximum) value and by the time-weighted average exposures of workers in those areas. These requirements call for classifying the areas and types of jobs in plants and for monitoring the actual exposures of workers over a typical workday. The combined sampling and analytical methods used have to be capable of determining VCM down to the 0.25-ppm level with a precision at the 95% confidence limit of ±50%.

Analytical Approach

To aid industry in monitoring programs designed to comply with this standard, NIOSH published a preliminary procedure for VCM sampling and analysis that was classified as "operational, but not thoroughly characterized" (5). This method calls for collection of VCM in glass adsorption tubes containing one of the specific NIOSH-approved lots of activated charcoal. Air from the breathing zone of the worker is drawn through the adsorption tube with the aid of a small low-flow battery-operated pump. After sample collection is completed, the tube is capped and sent to the laboratory for analysis. VCM is desorbed from the charcoal with CS_2, and the resulting solution is injected into a gas chromatograph (GC) for analysis. Separation of VCM from other components is performed with an SE-30 column. Since NIOSH realized that this procedure had not yet been thoroughly characterized, the final standard allowed this method or any equivalent method to be used.

Preliminary testing by our laboratory, as well as by others, indicated that

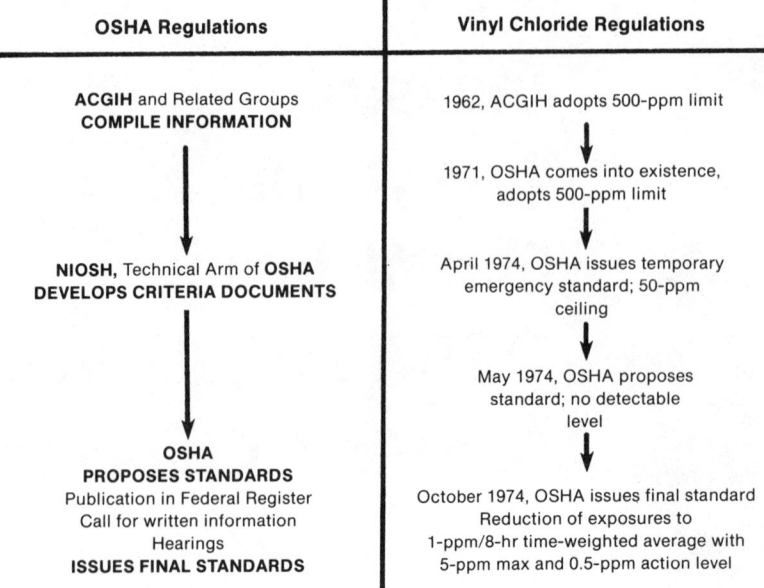

Figure 1. Genesis of OSHA regulations
ACGIH = American Conference of Government Industrial Hygienists. NIOSH = National Institute of Occupational Safety and Health. OSHA = Occupational Safety and Health Administration

there were some disadvantages with the recommended method. These were: poor storage stability of VCM on the charcoal tubes, lot-to-lot variations in charcoal, low and variable desorption efficiency of VCM with CS_2, the inadequacy of the SE-30 column to resolve VCM from other components (of the plant air) and/or CS_2 impurities, and the toxicity and flammability of CS_2. In addition, the volatility of CS_2 made it difficult to prepare stable standard solutions of VCM in CS_2.

Because of these problems, our laboratory, the Analytical Section of Stauffer Chemical Co.'s Eastern Research Center, sought to develop an improved method capable of VCM personnel monitoring for Stauffer Chemical Co.'s PVC resin and fabricating plants. In addition to the requirements set forth by OSHA, we had several other considerations to include when deciding which analytical approach to use:

• The sampling device had to be capable of storing VCM with no losses for periods of up to one week to permit shipping of samples from several plant locations to a central laboratory for analysis.

• The GC column must cleanly resolve VCM from interfering substances that might be found in plants employing VCM or VCM-containing materials used in a wide variety of synthetic and/or fabricating formulations.

• The analytical procedure should exceed in both accuracy and precision the stated OSHA requirements so that the number of personnel monitoring samples could be minimized. This requirement, plus a well-designed sampling program, was needed to ensure the validity of the resulting VCM exposure data, because the variations due to personnel, work shift, process, and even day of the week are not always controllable.

• Due to the effective date of the standard (January 1, 1975, delayed to April 1, 1975, by court order) and the time required to train personnel, strict time limits were imposed on the analytical method development stage of this project. This timing precluded the use of semiautomated VCM analysis systems that have since appeared on the market. Although many of these commercial systems are perfectly satisfactory, their precision, reliability, and delivery date were all unknown at the time that the VCM personnel monitoring surveys were started by Stauffer Chemical Co.

In an industrial environment, method development frequently involves more method adaptation than actual invention. The development and adaptation problems for this project involved two categories, the sampling system and the analysis system.

Sampling System

We have investigated the utility of two types of personnel sampling systems for organic gases. The first involves concentrating the sample in an adsorbent tube, such as that used in the NIOSH procedure. Although certain drawbacks have been noted in the NIOSH procedure, variations of adsorbent tube design, adsorbent, and/or VCM desorption techniques have been applied successfully by several groups. These variations involve the use of modified reusable charcoal tubes, heat desorption devices, head space analyzers, and desorption with CS_2 at Dry Ice temperatures. Advantages of the adsorbent tube approach are the small size of the sampling apparatus and the fact that large volumes of air can be drawn through the tube, thereby concentrating the VCM by several orders of magnitude. A drawback in the heat desorption and head space analysis procedures (which are applied to the adsorbent tubes) is that gas chromatographic analysis can be performed only once; repetitions are not possible since the sample is either totally consumed in a single determination or its concentration has been substantially changed. The use of Dry Ice baths to minimize losses of VCM and/or CS_2 during desorption from charcoal tubes was developed by Dow Chemical Co. (6). This procedure is a variation of the NIOSH-developed method and has been tested by our

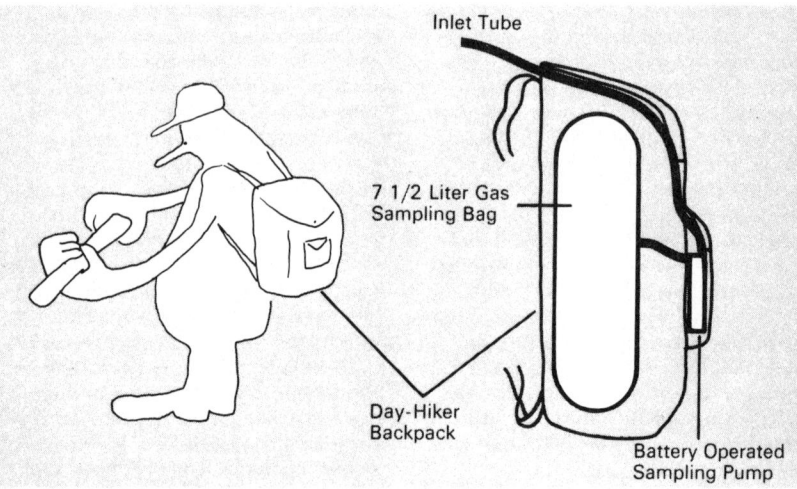

Figure 2. Vinyl chloride personnel sampling unit

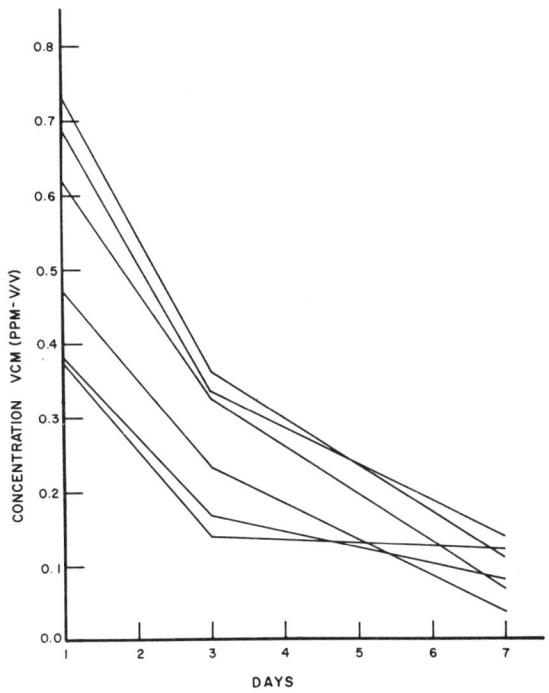

Figure 3. Loss of vinyl chloride gas from Teflon gas sampling bags

laboratory. Although somewhat time-consuming, it is more reliable than the original NIOSH procedure.

A second basic sampling procedure involves the collection and storage of the sample gases without concentration. The method of choice for personnel monitoring involves the use of gas sampling bags. A battery-operated pump is used to draw air from around the worker's breathing zone and exhaust it into the bag. The contents of the bag are then analyzed directly by gas chromatography or any other suitable analytical technique. The pump and the bag are placed in a small day-hike backpack which is then worn by the worker for a complete workshift (Figure 2).

Gas sampling bags are commercially available and are usually fitted with a metal twist-lock valve, although some are also equipped with a permanent or replaceable septum or with a filling snout. Although the storage stabilities of a wide variety of volatile materials in these bags have been summarized in the literature (7–11), none of these reports has dealt with 0.2–1.0 ppm concentrations of VCM in air. Therefore, the storage stability, memory effect (from previous samples), and losses of VCM in two commercially available gas sampling bags were studied. In addition, the precision and accuracy of the total sampling system (pump, bag, and tubing) were defined. Chosen for this study were a Teflon bag equipped with a replaceable septum and a twist-lock valve and an aluminized Scotchpak three-layer bag equipped with a valve.

Figure 3 shows the loss of VCM from Teflon bags to be in the range of 20% per day. It was not determined whether this loss resulted from the permeability of Teflon or from mechanical problems. There is really little need for a septum on a gas sampling bag since maximum GC precision can more easily be achieved by using gas sampling valve injection rather than gas syringe injection techniques.

Figure 4 illustrates the storage stability of VCM in aluminized Scotchpak bags. There is no detectable loss of VCM for a period of one week over the concentration range of 0.1–1.1 ppm VCM in air. Because of the possibility of leaks in gas sampling bags, it is recommended that they should be leak tested with clean compressed air for a period of several hours before use or reuse. In actual field use, we find about a 10% "mortality" rate for aluminized Scotchpak bags when they are used repeatedly.

All further studies were carried out on only the aluminized Scotchpak bags. Bags experimentally filled with between 1.0 and 10 ppm VCM had no detectable amount (<0.03 ppm) of VCM remaining after two repetitions of vacuum pumping of the bags and refilling with compressed air. It is, therefore, our practice to perform three pump-and-fill cleaning cycles before each reuse of a sampling bag.

The performance of the entire sampling and analytical system was checked by pumping 1.0 ppm of VCM in air from a full gas sampling bag through connecting tubing and a sampling pump into a second bag which had been evacuated prior to the experiment. The lengths of Teflon-lined neoprene tubing used and the pump were the same as would be used in the field. This was a simulation of the

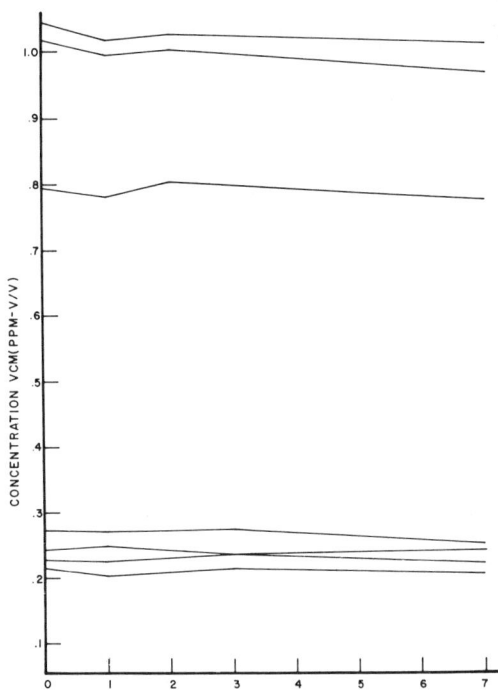

Figure 4. Stability of vinyl chloride gas in aluminized Scotchpak gas sampling bags

losses that could be expected under actual use conditions. The data showed a mean recovery of 96–97% of the VCM.

Analytical System

Because of the inadequacy of the NIOSH-recommended SE-30 gas chromatographic column, we have chosen to use a Durapak (Carbowax 400 on Porasil) packing. The use of Durapak requires a low inlet pressure which results in only minor baseline disruption during sample injection with the gas-injection valve. Sample loops of up to 5.0 ml can be used without significant loss of resolution. We have found nominal analysis times of between 3 and 5 min to yield more than adequate resolution and results with relative precisions (95% confidence level, single injection) of ±3% at the 1.0-ppm level and ±11% at the 0.25-ppm level. Since the precision increases as 1/square root of the number of injections, we routinely inject each sample two or more times. This system can detect less than 0.03-ppm concentrations of VCM in air.

Several valving configurations have been developed by other laboratories which incorporate the function of column backflush and injection, and cut-and-backflush and injection in a single automated valve. These valves minimize the time necessary to clean the GC column of constituents from the air sample having longer retention times than VCM. A cut-and-backflush system presently in use by many laboratories incorporates a precolumn of Durapak (n-octane on Porasil) with porous polymer and a short analytical column of porous polymer. An automated 10-port valve is used (12).

In conclusion, the industries which use VCM for resin production and PVC for fabrication of finished vinyl products were faced with developing, field proving, and installing a complete VCM personnel monitor system in a short time. Sampling and analysis guidelines supplied by NIOSH were not adequate for the goals of our program, thus necessitating extensive method adaptation and development. The method outlined above represents a viable analytical approach for an industrial setting.

The long-range goal for determining worker exposure to VCM must be, however, the utilization of continuous monitors. These VCM monitors must have the capability of correlating variable ambient air VCM concentrations with the results of corresponding personnel monitoring samples.

References

(1) H. Falk, J. L. Creech, Jr., C. W. Heath, Jr., M. N. Johnson, and M. M. Key, *J. Am. Med. Assoc.*, **230**, 59 (1974).
(2) "Threshold Limit Values for Chemical Substances in Workroom Air Adopted by ACGIH," American Council of Governmental Industrial Hygienists, Cincinnati, Ohio, 1973.
(3) G. Clack, *Job Saf. Health*, **3**, 4 (1975).
(4) *Fed. Regist.*, **39**, 194 (1974).
(5) "NIOSH Manual of Analytical Methods," U.S. HEW 75-121, pp 178, 1–178, 9, 1974.
(6) A. A. Allemang, L. W. Severs, and L. K. Skory, "Monitoring Personnel Exposed to Vinyl Chloride and Other Chlorinated Solvent Vapors in an Industrial Work Environment," American Industrial Hygiene Conference, Minneapolis, Minn., June 4, 1975.
(7) R. T. Maykoski and C. Jacks, "Review of Various Air Sampling Methods for Solvent Vapors," NTIS AD-752-525, 1970.
(8) F. J. Schuette, *Atmos. Environ.*, **1**, 515 (1967).
(9) "Methods of Air Sampling and Analysis," Interscience Committee, American Public Health Assoc., pp 7–8, 138, Washington, D.C., 1972.
(10) F. B. Higgins, Jr., "Sampling for Gas and Vapors," in "Source Sampling of Atmospheric Contaminants, Symp. Proc.," H. G. McAdie, Ed., Chem. Inst. Canada, Ottawa, Canada, 1971.
(11) R. E. Dilgren, Shell Chemical Co., Analytical Method HC-604-74, 1971.
(12) F. Zado, Western-Electric Co., Princeton, N.J., personal communication. This system is not marketed by Western-Electric Co.

Detecting and Defining Air Pollutants: One Laboratory's Experiences and Approaches

James L. Lindgren, Henry J. Krauss, and John S. Mgebroff
Texas Air Control Board, Austin, TX 78723

Originally published in ANALYTICAL CHEMISTRY, 1980, Vol. 52, No. 9.

The Texas Air Control Board is the agency responsible for maintaining the quality of the ambient air of Texas. An extensive network of continuous and noncontinuous monitors is used to detect air quality trends in the state. Regulations of the board limit the emission of air contaminants from stationary sources.

The agency's well-equipped laboratory analyzes network samples obtained by noncontinuous monitors and samples submitted for analysis by our regional investigators.

The samples submitted may be gaseous, liquid, or solid; they are submitted for a variety of reasons. Analyses are sometimes necessary before a permit can be issued to a new source, to check the performance of existing sources, or to answer citizens' complaints. The submitted samples are reviewed by the Laboratory Director (Figure 1) and assigned to the appropriate analyst, depending on his or her discipline. In addition, any special handling or preparation of the sample is discussed at this time.

Elemental Analysis

Early in the history of the Texas Air Control Board, the administration asked the analytical staff several questions:
- Can the general public's exposure to some of the more common and toxic elements be determined economically?
- Can the analytical results be used to identify polluting sources of which the investigative staff of the agency is unaware?
- Could the analytical technique benefit the agency in other ways?

At that time, atomic absorption (AA) was being used to perform metal analyses. Those acquainted with AA are aware of the digestion and dilution necessary to prepare the sample. If elements to be determined are in a complex matrix, as samples from polluting sources often are, it is necessary to prepare time-consuming internal standards to correct for interelemental interferences. After considering several approaches, a comparison of AA with an emerging technique called energy dispersive X-ray fluorescence spectroscopy (XRF) was undertaken (1). The manpower, cost of chemicals, response time, and capital investment were studied in analyzing for some 12 metals from 10 sampling sites (2). The cost effectiveness of the XRF technique quickly became apparent. The end result has been the employment of the XRF system, so that the agency's 100-plus particulate samples, collected every six days, are each analyzed for some 32 elements.

A computer controls the XRF during the analysis of the samples, then massages the data and provides a printout of the results in the appropriate units. Some 48 samples are analyzed for 32 elements in a 12-hour period. The quantity of data available on the public's exposure to some 32 elements in Texas is now rather extensive. The cost per determination is far

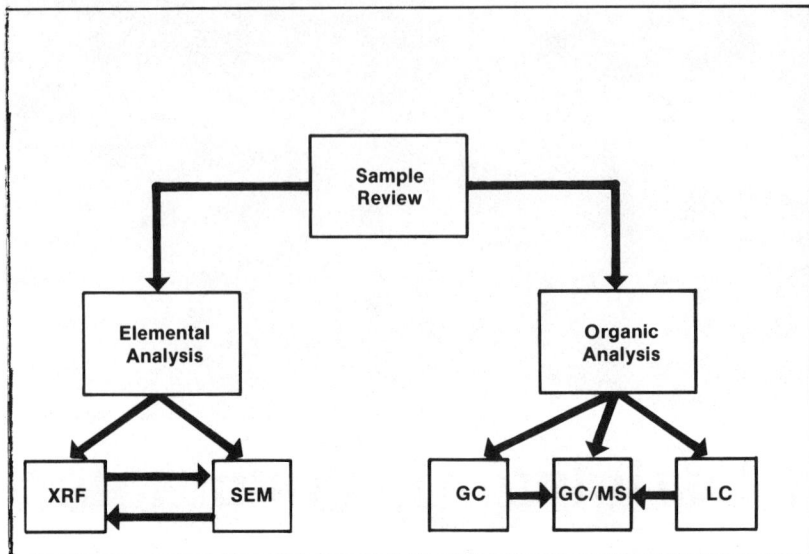

Figure 1. Sample flow and multiinstrumental options available to accomplish the analysis of complex samples

less than can be achieved by other analytical approaches.

The second of the questions was also answered early in our search for an alternative to AA. Analysis of the results produced by that first comparative study by the analytical staff, who were not familiar with the industry in the area sampled, indicated a source of Zn. By correlating the analytical results with the meteorology of the sampling day, the location of the suspected source was narrowed and later verified by authorities familiar with local industry.

In addition, several previously unsuspected sources were found to be polluting neighboring communities with undesirable levels of certain potentially toxic metals. A smelter was closed when undesirable levels of Mo, Sb, and other metals were detected in its emissions. Several other instances demonstrated that our analytical approach had satisfied the second question.

It was only after several years of experience that the third question was adequately answered. The first demonstration was in a court case against an emitter of lead. The defendant claimed that automotive emissions were the source of the lead. Several years of XRF data observation had established that a definite Pb/Cl and Pb/Br ratio existed on the particulate samples collected where automobiles were the only source of lead emissions. The XRF, which is capable of detecting Cl and Br adsorbed on the particulate matter, failed to establish this definite ratio, though a substantial number of samples near the smelter were collected and analyzed. The XRF data won our case, and since the samples were analyzed nondestructively, they were available for verification had this been necessary.

A second answer to question three surfaced when a permit engineer commented to a laboratory staff analyst that there should be an easier way to determine the emissions from a stack than the traditional, time-consuming stack sampling technique. The result of that conversation was a collaborative study with the Source Evaluation Division to determine if XRF analysis of the fuel supply could be a reliable indicator of the sulfur emissions from a stationary source, since SO_2 emissions are regulated. The results of the XRF approach consistently agreed with the stack sampling technique (3). The analysis time is only 5–10 min for a sulfur analysis of boiler fuels including the sample preparation, which requires no digestion or dilution.

Existing facilities, which propose to change their source of fuel, and new facilities, before they come on line, are routinely sampled by this fuel analysis approach for percent sulfur in the fuel. The permit engineer uses the XRF data to calculate the potential sulfur dioxide emissions from the plant and only if the results are marginal is stack sampling considered.

Routinely, the fuel samples, be they oil, coal, or lignite, are analyzed for 32 elements. Sulfur content attracts the most interest, but a 30-min analysis time by the XRF provides the engineer with the elemental concentration of the fuel so that he may decide if toxic metals emissions could be a potential health hazard as well. The XRF approach has been most beneficial to our analyses but is not the complete answer. Our XRF capability covers only an area of the periodic table, and also lacks sensitivity for some of the elements. Flameless AA is the technique of choice for Be since the XRF cannot detect it, and for As since the XRF is much less sensitive. Indeed, AA is used for a number of other elements.

There are occasions when an elemental analysis does not explain the source of a pollutant. Optical and scanning electron microscopy are then employed as complements to XRF.

Organic Analysis

Gas chromatography (GC), high pressure liquid chromatography (HPLC), and gas chromatography/mass spectrometry (GC/MS) are employed in organic analyses as complements to each other. Environmental samples submitted for detection of organic components are seldom clean or simple and often require special concentration steps.

HPLC has been effectively used to shorten and simplify the analysis of samples for herbicides. An example of this is the analysis of vegetation for the herbicide Stam (Propanil) (4). The original GC procedure calls for a 17-h hydrolysis of the vegetation in 25% NaOH followed by steam distillation into 2N HCl. The distillate is washed with benzene, the pH adjusted to 10, and again extracted with benzene, passing the benzene through sodium sulfate. The collected extract is concentrated and analyzed by GC for 3,4-dichloroaniline using a halogen-sensitive detector. The substitution of HPLC for GC does not remove the need for the hydrolysis and steam distillation, but the distillate can be made alkaline and extracted with benzene and the benzene extract concentrated at this point. It is not necessary to dry the extract since the C_{18} reversed phase column used in HPLC handles the water content of the extract with no difficulty, and the detectors normally used in HPLC are not harmed by water, as are some electron capture detectors. The gradient elution capability of the HPLC system allows one to choose solvents and gradient conditions which generally will enhance the separation of compounds of close polarity.

A dual channel UV detector is the means of monitoring the eluting sample components of an HPLC analysis.

A fluorescence spectrophotometer equipped with a 10 µl flow-through cell follows the dual channel UV. When the ratio of the two UV channels fails to confirm an eluting peak, it can be stopped in the 10 µl sample cell and a fluorescence spectrum can be obtained to further aid in the qualitative confirmation. Of course, the compound has to have fluorescing properties. This combination of HPLC and detectors provides confirmation of some polyaromatic hydrocarbons in ambient air after they have been concentrated onto commercially available C_{18} bonded liquid chromatographic column packing material (5).

For the concentration of C_2-C_{12} organic contaminants in ambient air, the laboratory has developed a technique which is effective and useful, though not necessarily unique to our organization (6). Figure 2 depicts the 6" × 6 mm pyrex tubes, which are packed with 3 cm of 60-mesh activated charcoal and 7 cm of 80/100-mesh Tenax GC. The brass caps use a soft front ferrule in order to form a diffusion-free seal.

These absorber tubes have been successfully used to concentrate contaminants from air samples submitted to the laboratory and from the atmosphere surrounding industrial sites. In a demonstration of the latter, a field investigator was provided with a quantity of tubes for the purpose of quickly sampling a specific site for vinyl chloride monomer (VCM) emissions (7). One of the more practical aspects of using these tubes at remote sites is that only a gas-tight syringe and reasonable care in handling are needed for loading. Electrical power or battery packs are not required for pumps—an obvious appeal to field investigators for spot, unannounced sample acquisitions. The tubes can be shipped or, as in this instance, conveniently carried to the laboratory.

To analyze these tubes for VCM, the tube oven (Figure 2) is mounted directly on a gas chromatograph, replacing the septum cap. The analysis is accomplished by interrupting the carrier gas flow, inserting a loaded tube into the tube oven, retightening the fittings, reconnecting the carrier gas flow, initiating the power to the tube oven, and initiating the temperature program of the column oven. The sample is desorbed onto an n-Octane on Porasil column, 1/8" × 2 m, held at 0 °C for 7 min, then increased at 6 °C/min for 14 min, and finally a 30 °C/min increase to a final temperature of 180 °C.

The detection of VCM is accomplished with a photoionization detector (PID) and flame ionization detector (FID) in series. The simultaneous trace of the PID and FID responses by a two-pen recorder allows the two responses to be ratioed, thus adding confidence to the qualitative determination. Three to five ppb by volume of VCM can confidently be detected with this analytical scheme when only 300 cc of air is sampled.

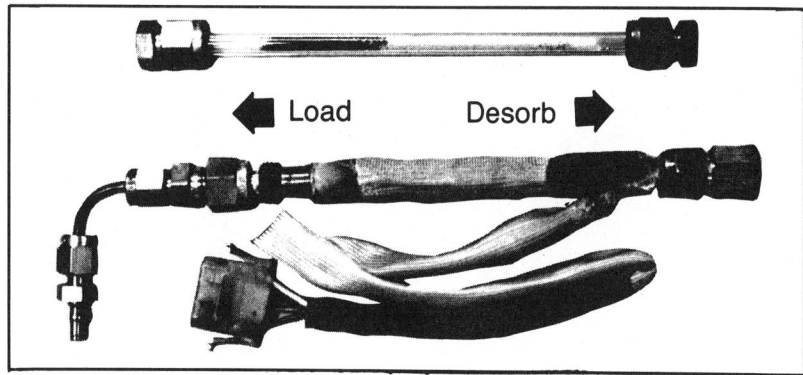

Figure 2. Charcoal/Tenax sample concentration tube (top) and tube oven used for desorbing collected sample (bottom)

The sample tubes are versatile enough to be used for the concentration of a variety of compounds. Caution should be exercised if the tubes are to be used for collection of benzene or toluene, because heat desorbing of Tenax routinely results in peaks at the retention times of those compounds. Some investigators have developed rigorous clean-up procedures which prevent this problem (8, 9).

There are many examples of a multiinstrument approach to analysis of samples. Our laboratory was involved in sampling and analyzing for nitrobenzene, which was suspected of originating from waste oils that had been applied to public roads in several communities. Originally, water samples from the ditches of the affected area were submitted. These were rather dirty and could not be directly analyzed by GC/MS without clean-up and concentration. High pressure liquid chromatography was employed to hasten the determination of nitrobenzene's presence. GC/MS was then used for confirmation of the suspected peak. After determining that nitrobenzene was indeed present in the runoff, true ambient air samples were obtained by absorption in ethylene glycol and by using the previously described concentration tubes with subsequent GC analysis. HPLC became the technique of choice for the analysis of these samples because of the simplicity of the analytical scheme. Monitoring continued until the oil had been removed from the area.

The complexity of ambient air samples often requires a multiinstrumental approach if the sample is to be qualitatively and quantitatively analyzed so that the information can be used for corrective action or to identify potential problems of air pollution. Competency in staff proficiency must be maintained, and sophisticated instrumentation must be available. The need for effective and efficient sample concentration techniques cannot be overlooked. It is an area which needs to be pursued, and advantages and limitations of a particular procedure should be identified before it is routinely used for the collection and concentration of ambient air contaminants.

Acknowledgment

The generous support of the Texas Air Control Board and the contributions and assistance provided by our staff and administration in the preparation of this presentation are gratefully acknowledged.

References

(1) J. R. Rhodes, *Am. Lab.*, (7), 57 (1973).
(2) J. R. Rhodes, A. H. Pradzynski, C. B. Hunter, J. S. Payne, and J. L. Lindgren, *Environ. Sci. and Technol.*, 6 (10), 922 (1972).
(3) J. S. Mgebroff and J. S. Payne, "The Determination of Trace Elements in Fuel Oil by Energy Dispersive X-ray Fluorescence Analysis," presented at the 29th Annual Meeting of the Pittsburgh Conference on Analytical Chemistry and Applied Spectroscopy.
(4) J. L. Lindgren, H. J. Krauss, and J. S. Payne, "Determination of Pesticides in Environmental Samples: Substituting HPLC for GC," presented at the 30th Annual Meeting of the Pittsburgh Conference on Analytical Chemistry and Applied Spectroscopy.
(5) J. L. Lindgren, H. J. Krauss, and M. A. Fox, *Journal of the Air Pollution Control Association*, 30 (2), 166 (1980).
(6) W. Bertsch, R. C. Chang, and A. Zlatkis, *J. Chromatogr. Sci.*, 12 (4) 175 (1974).
(7) J. L. Lindgren and G. Speller, "Determination of Vinyl Chloride Monomer in the Ambient Air Near Point Source Emissions," presented at the 30th Annual Meeting of the Pittsburgh Conference on Analytical Chemistry and Applied Spectroscopy.
(8) E. D. Pellizzari, private communication.
(9) L. T. Freeland, private communication.

Sampling and Analysis of Radioactive Solutions

David H. Smith, R. L. Walker, and J. A. Carter
Oak Ridge National Laboratory, Analytical Chemistry Division, Oak Ridge, TN 37830

Originally published in ANALYTICAL CHEMISTRY, 1982, Vol. 54, No. 7.

The international program on safeguards for nuclear materials is administered from Vienna by the International Atomic Energy Agency (IAEA). Scientists involved in the program are responsible for maintaining a material balance account of the radioactive materials used or generated by nuclear facilities in countries that have signed the Non-Proliferation Treaty. Radioactive materials such as uranium and plutonium are monitored, since these are the elements that can be used to manufacture nuclear weapons. The advent of the breeder reactor in many countries has necessitated the establishment of the safeguards program, since purified uranium and plutonium are recovered from the spent fuels from these reactors at special reprocessing facilities. The nuclear reactors currently operating in the U.S. are not of the breeder type, and therefore spent fuels from these reactors are not reprocessed. Thus, unauthorized use of uranium and plutonium from these solutions is not a problem. Despite this, the U.S. supports the international safeguards program both financially and by supplying the IAEA with scientists from various disciplines.

Sampling solutions of spent reactor fuels is one of the most difficult problems in the safeguards program. These solutions are highly radioactive, containing many long- and short-lived fission products in addition to excess fuel and the plutonium generated in the nuclear breeding process. The conventional techniques used in the international safeguards program for analysis of these solutions require that large samples be shipped to the IAEA in Vienna for costly and time-consuming separation of the individual elements. Because shielding of plutonium is mandatory, the weight of the packing material far exceeds that of

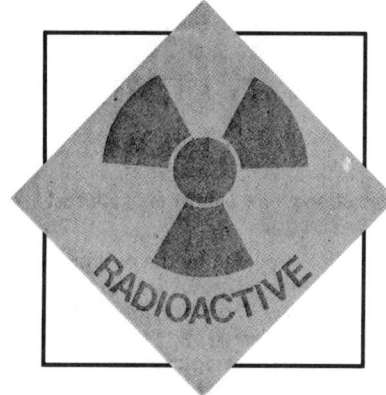

the sample. It is presently extremely difficult to ship samples of this size. This difficulty, combined with recent restrictions on the amount of material that may be shipped, has required scientists to reexamine the currently used analytical techniques with a view toward minimizing the amount of sample required.

With this goal in mind, our laboratory recently developed a new sample-loading technique that involves the use of anion resin beads (1–3). This microsampling technique offers significant advantages in three general areas. First, the quantities adsorbed are so small (1–3 ng of each element) that shipping is no longer a problem. Second, uranium and plutonium adsorb on the beads under appropriate conditions and are thus separated from fission products and most other actinides. Finally, each bead serves as a convenient vehicle for loading the sample onto a filament for ultimate analysis by mass spectrometry.

Sampling

In our technique, a known amount of spent-fuel dissolver solution is isolated and adjusted to 8 M HNO_3. This aliquot is divided into two fractions and enough of it taken so that there will be about 1 μg of uranium per resin bead. A known quantity of a spike of high isotopic purity is added to one fraction; our spikes are >98% ^{233}U and ^{242}Pu. The sample and spike are equilibrated; this is a crucial chemical step necessary for any quantitative determination by the isotopic dilution technique. Equilibration of uranium requires only thorough mixing of sample and spike solutions, but because of its multiplicity of oxidation states and tendency to form polymers and complexes, plutonium is a problem demanding special attention. Marsh et al. have compared the efficacy of various equilibration techniques (4) and have found that the most consistent results were obtained by reduction with Fe(II) and sulfamic acid followed by oxidation with $NaNO_2$. This is the method we use, and we have encountered no serious problems since adopting it. Knowledge of the isotopic compositions of the spike, the unspiked sample, and the mixture of spike and sample allows calculation of the amount of uranium or plutonium present.

One thousand resin beads are introduced into each solution; each resin bead is 100–200 μm in diameter and constitutes one sample for the mass spectrometer. Using 1000 beads allows us to sample about 1 mg of uranium, an amount comparable to that used in several other techniques. Risk of contamination is thus no greater using our technique than it is for more conventional ones. The beads are agitated on a vortex mixer in contact with the solution for 10 min, after which the solution is removed and the beads washed to remove surface contamination.

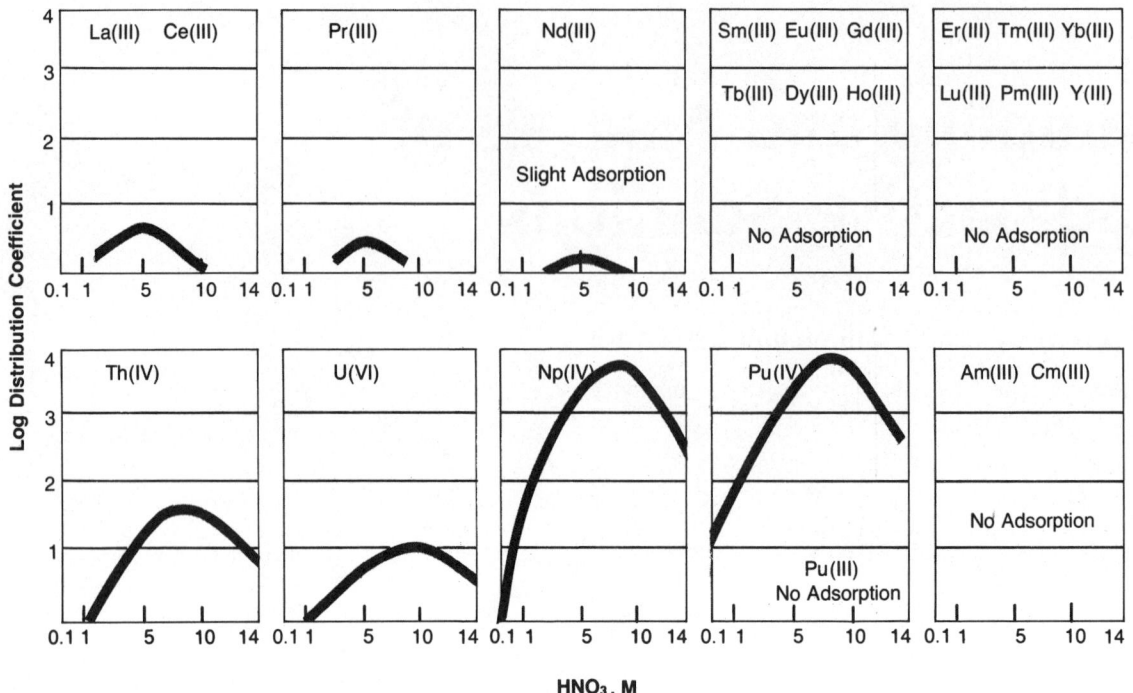

Figure 1. Plots of diffusion coefficients vs. acid strength for elements of interest in the nuclear fuel cycle

Complete separation of uranium and plutonium from fission products is achieved by this simple technique. Figure 1 presents plots of distribution coefficients vs. acid strength for several elements of concern. It is based on results originally described by Faris and Buchanan (5). Note that maximum adsorption for plutonium and uranium on strongly basic anion exchange resin (Dowex 1-X2) occurs at about 8 M HNO_3. The only elements adsorbed to any significant extent besides uranium and plutonium are thorium and neptunium. Neptunium is present only as mass 237 and causes no mass interference with either element of interest. Thorium (mass 232) also causes no interference and will be of interest if a thorium-based breeder system is ever adopted. (We have already developed a technique of sequentially analyzing Pu, U, and Th from a single bead.) In normal commercial operation, reactor fuels will be highly irradiated—on the order of 30 000 MW days/ton. Spent-fuel solutions from such materials have a U/Pu ratio of about 100, but because the adsorption coefficient of plutonium on the bead is about 100 times that of uranium, the quantities of the two elements taken up by the bead are about equal (1–3 ng uranium and plutonium per bead). This amount of plutonium corresponds to about 10^{-9} curies, which is one-tenth the daily exposure level allowed an office worker; the shielding provided by the packing material reduces exposure in handling even further. Therefore, no special shielding is necessary.

Shipment of samples on beads is easy and convenient. A number of beads may be affixed to a glass microscope slide with collodion and shipped on the slide, or the beads may be shipped directly in the special container we have designed for this purpose. In either case, the cost of shipping one sample is about $1.00 (in contrast to the $700 per sample required by older techniques).

Mass Spectrometry

The use of anion resin beads to load samples for mass spectrometry was first suggested by Freeman et al. (6). Their primary concern was the development of microstandards, but we have adapted the technique for more general application. We use the anion

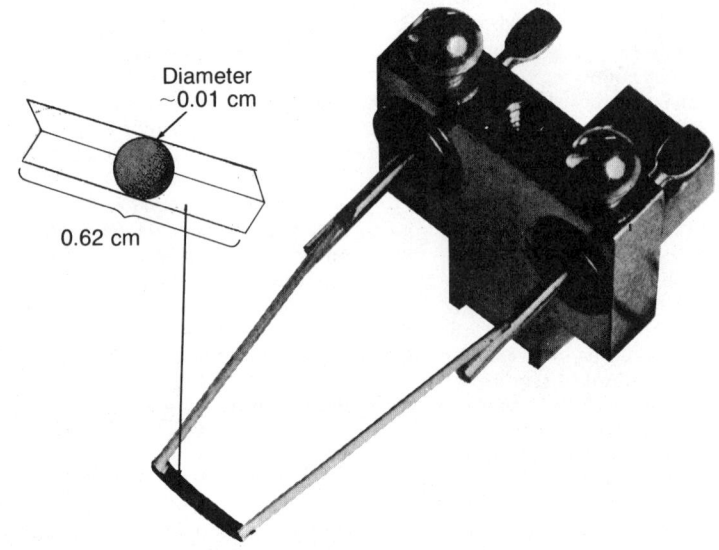

Figure 2. Photograph of a rhenium filament and its support with a sketch showing a bead emplaced in it. (The bead is not drawn to scale.)

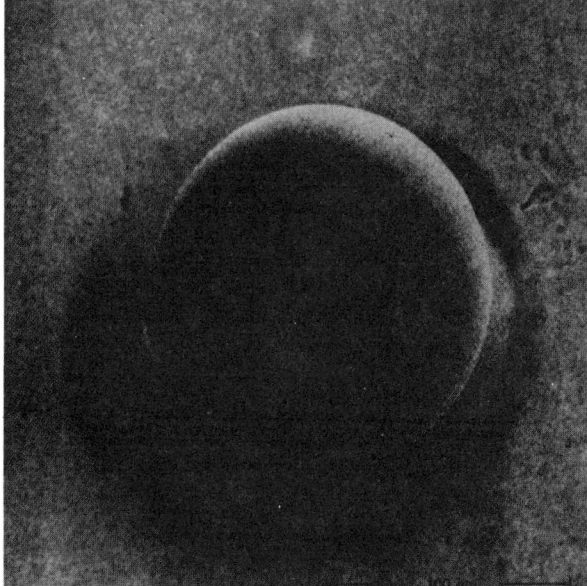
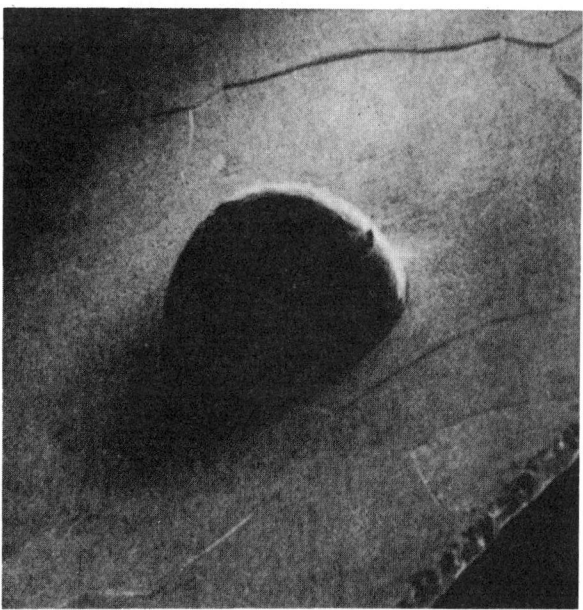

Figure 3. Scanning electron micrograph of an unheated bead. The dark material at the interface of bead and filament is the collodion used to hold the bead in place

Figure 4. Scanning electron micrograph of a bead heated for 30 min at 1700 °C

bead as a chemical-separating device as well as for introducing the sample onto the mass spectrometer filament.

Mass spectrometric analysis of such small samples requires instruments equipped with pulse-counting detection systems that count the ions individually as they arrive at the collector; we have had several of them in operation for a number of years (7–8). Analysis of the two elements proceeds sequentially from a single rhenium filament. Plutonium, which has a lower ionization potential than uranium, is analyzed first (at about 1450 °C). Excess plutonium is then burned off, and uranium is analyzed (at about 1700 °C). Computer programs automatically reduce the data to isotopic compositions, correcting for the interference between the two elements at mass 238.

Figure 2 is a photograph showing our canoe-shaped filament; a sketch showing a bead in the filament is also included. (The bead is not drawn to scale, being much smaller in proportion to the filament than shown.)

We have analyzed several hundred samples by this technique and have observed no degradation of results; indeed, precision and accuracy are in general better. Figure 3 is a scanning electron micrograph of an unheated bead showing its nearly perfect sphericity. Figure 4 is a scanning electron micrograph of a bead that has been heated for 30 min at 1700 °C. Grain boundaries in the rhenium filament are visible. Approximately half the bead has dissolved in the rhenium substrate. (See Reference 9 for a full description of this work.) The resin bead serves as a good approximation to a point source for the ion optics of the mass spectrometer and also as an efficient reducing agent to prevent loss of sample as oxide species, thus increasing our overall collection efficiency. In addition, the bead seems to serve as a reservoir of sample, feeding it to the ionization region in a more controlled manner than solution loadings (9). Average collection efficiencies for several beads gave one ion collected for each 700 atoms of uranium loaded on the filament; for plutonium, the figure was one in 80. These represent about an order of magnitude improvement over typical results from solution loadings.

Results

Results from repetitive analyses of National Bureau of Standards certified isotopic standards are given in Table I. These were analyzed over a two-month period; all results represent data taken from sequential analyses of plutonium and uranium from resin beads. The good results on the minor isotopes exemplify the ability of multistage mass spectrometers to perform precise measurements on low-abundance isotopes. Standard deviations are listed at the confidence level of 1 σ.

Experiments were performed to establish the range of U/Pu ratios over which the technique could be applied. Ratios below 10 led to saturation of the bead with plutonium and made uranium analysis extremely unreliable. Ratios greater than 1000 led to saturation with uranium. It is unlikely that reactor fuel will be burned long enough to produce U/Pu below 10. Another experiment defined the minimum quantity of plutonium required for a reliable isotopic analysis. This was determined to be about 0.1 ng if all isotopes were measured and somewhat lower (<0.05 ng) if ^{238}Pu was omitted.

During the course of these studies, we observed a consistently lower value for ^{241}Pu/^{239}Pu than those values obtained by conventional techniques involving chemical separation of the elements. This is almost certainly due to the presence of ^{241}Am in the latter, which is present in spent fuel and also accumulates from the decay of ^{241}Pu. The ability of the bead to separate plutonium and americium represents a significant advantage of the new technique.

At the end of these and other studies, our confidence in the resin bead technique was such that we began to use it routinely in preference to our previous method of loading samples as solutions.

Comparison with Conventional Techniques

The crucial test of any new analytical technique lies in comparing results obtained by it with those obtained from the older technique it is seeking to supplant. Conventional isotopic and quantitative analysis involves chemical separation of uranium and plutonium, both from other reactor products and from each other, prior to mass spectrometric analysis. Such separations are performed on samples shipped as residues of evaporated solutions to the laboratory involved.

Table I. Atom Ratios from Analyses of NBS Standards

	NBS-947 Pu isotopic standard			
	238/239	240/239	241/239	242/239
Measured value (nine samples)	0.00371	0.24156	0.04281	0.01559
NBS certified values*	0.00370	0.24147	0.04309	0.01559
Standard deviations	0.00002	0.00057	0.00025	0.00008

* Corrected to Aug. 13, 1978.

	NBS-500 U isotopic standard		
	234/235	235/238	236/235
Measured value (nine samples)	0.01034	0.99851	0.001522
NBS certified values	0.01042	0.99970	0.001519
Standard deviations	0.00005	0.00209	0.000005

Table II. Isotopic Analyses of Spent Fuel Solutions

| Laboratory | Technique | Uranium, weight percent | | | | Conc (g/L) |
		234	235	236	238	
PNC	Conventional	0.0191	1.0942	0.3746	98.512	164.70
IAEA	Dried	0.0211	1.0876	0.3718	98.519	164.83
IAEA	Beads	0.0206	1.0905	0.3699	98.519	164.68
ORNL	Beads	0.0201	1.0965	0.3722	98.511	163.95
Average		0.0202	1.0922	0.3721	98.515	164.54
Standard deviation		0.0009	0.0039	0.0019	0.004	0.40

| Laboratory | Technique | Plutonium, weight percent | | | | | Conc (g/L) |
		238	239	240	241	242	
PNC	Conventional	1.458	60.178	22.623	11.515	4.227	1.507
IAEA	Dried	1.369	60.301	22.642	11.483	4.206	1.492
IAEA	Beads	1.375	60.171	22.599	11.624	4.231	1.498
ORNL	Beads	1.366	60.232	22.635	11.529	4.239	1.492
Average		1.392	60.221	22.625	11.538	4.226	1.497
Standard deviation		0.044	0.060	0.019	0.061	0.014	0.007

To effect such a comparison, Japanese personnel trained in the resin bead technique at Oak Ridge National Laboratory (ORNL) sampled the reprocessing facility operated by the Power Reactor and Nuclear Fuel Development Corporation (PNC) at Tokai. Samples on resin beads were shipped both to ORNL and the IAEA for analysis. The same samples were analyzed at PNC with their normal technique; dried residues were shipped to the IAEA to be analyzed for comparison purposes. A detailed description of this experiment has been published (10), and a summary of the results is presented in Table II. The good agreement reflected in this table shows that the technique is a viable one for areas difficult to sample by other techniques. In particular, it should be noted that the two sets of analyses from resin beads do not differ statistically from the two obtained by conventional methods.

Adoption of this technique by international safeguards programs will result in reduced transportation costs and significant reduction of the health hazards involved in exposure to large samples, with no sacrifice in quality and amount of information obtained.

References

(1) Walker, R. L.; Eby, R. E.; Pritchard, C. A.; Carter, J. A. Anal. Lett. 1974, 7, 563.
(2) Carter, J. A.; Walker, R. L.; Eby, R. E.; Pritchard, C. A. "Safeguarding Nuclear Materials"; IAEA-SM-201/9: Vienna, 1976; Vol. 2, p 461.
(3) Walker, R. L.; Pritchard, C. A.; Carter, J. A.; Smith, D. H. USDOE Report ORNL/TM-5505, July 1976.
(4) Marsh, S. F.; Abernathey, R. M.; Rein, J. E. Anal. Lett. 1980, 13 (A17), 1487.
(5) Faris, J. P.; Buchanan, R. F. Anal. Chem. 1964, 36, 1158.
(6) Freeman, D. H.; Currie, L. A.; Huehner, E. D.; Dixon, H. D. Anal. Chem. 1970, 42, 203.
(7) Smith, D. H.; Christie, W. H.; McKown, H. S.; Hertel, G. R. Int. J. Mass Spectrom. Ion Phys. 1972, 10, 343.
(8) Smith, D. H., Ed. USDOE Report ORNL/TM-6485, November 1978.
(9) Smith, D. H.; Christie, W. H.; Eby, R. E. Int. J. Mass Spectrom. Ion Phys. 1980, 36, 301.
(10) Walker, R. L.; Smith, D. H.; Carter, J. A.; Donohue, D. L.; Deron, S.; Askaura, Y.; Kagami, K.; Irinouchi, S.; Masui, J. J. Inst. Nuc. Materials Management 1981, 10(3), 43.

Research sponsored by the U.S. Department of Energy, Division of Basic Energy Sciences and Office of Safeguards and Security, under Contract W-7405-eng-26 with the Union Carbide Corporation.

Toxicity

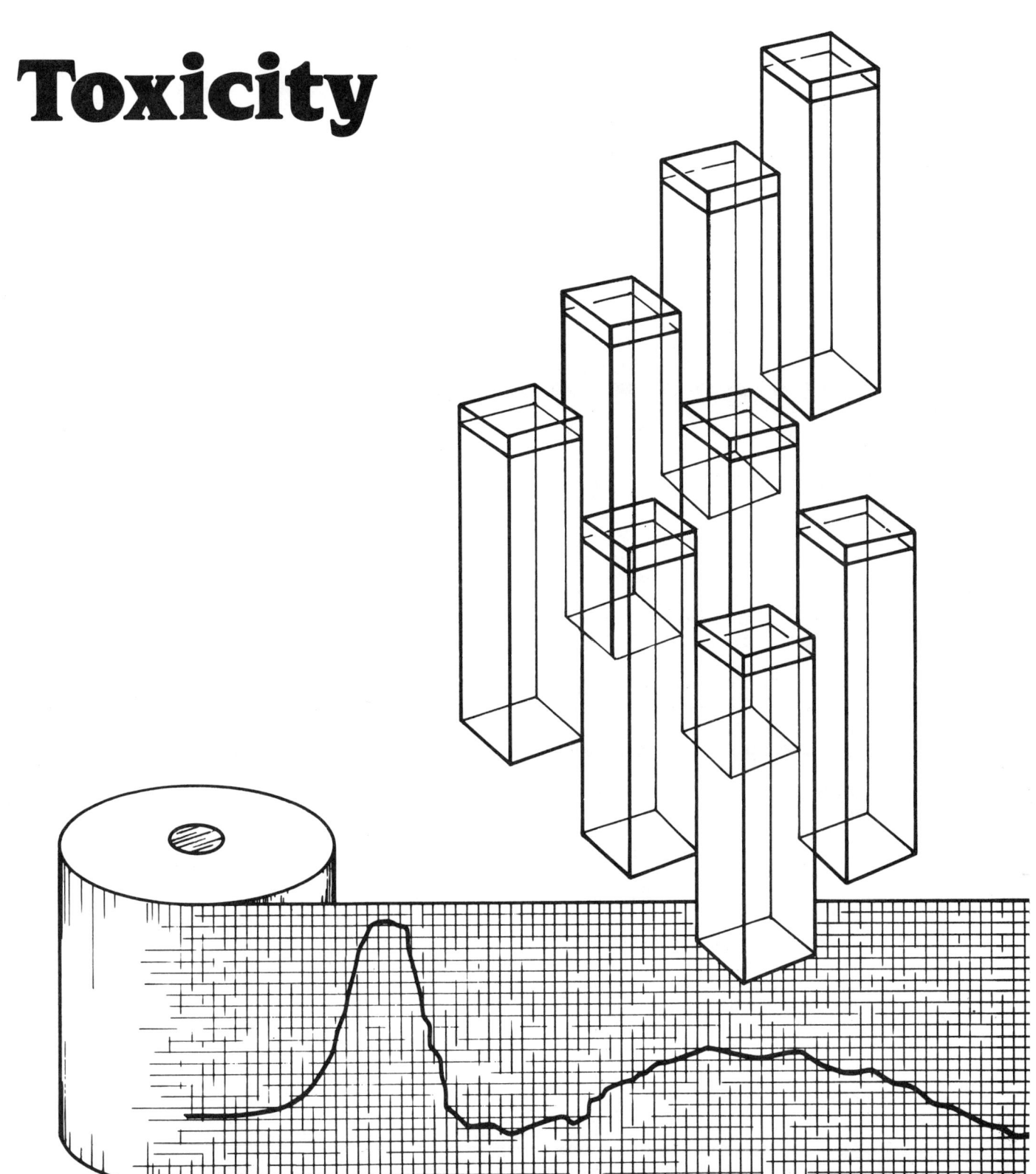

Quite clearly, the identification and analysis of toxins is an important field to all of mankind. It is complicated by the tendency of toxic materials to metabolize or alter chemically in a rapid and significant manner. The analytical plan must take this into account. Here, as much as anywhere, the use of combined techniques is necessary for positive identification of such complex materials. Good separations are a key to these analyses. In addition, it has been said that we can find vanishingly small concentrations of potentially toxic ingredients everywhere in everything. The analytical chemist has developed trace analyses to a fine art.

The approval of a new food packaging material is the important subject of "**Analytical Chemists Vital in Commercialization of New Food Packaging Material.**" This paper describes a multi-disciplinary team of an attorney, a polymer chemist, and an analytical chemist who interacted with business groups within the company and scientists from the FDA to conduct extraction tests on food containers made from a novel barrier resin. An elaborate analytical plan, relying largely on IR and UV spectroscopy, and polarography was successfully followed to attain an FDA clearance for the material.

The OSHA bill passed in 1970 marked a new direction for analytical chemistry. "**Industrial Hygiene**" traces the evolution of this science from the time of Hippocrates and assesses its status today. Although only 10% of occupational accidents and disease can be related to chemicals, this is an important field. Monitoring programs must be based on a sequence of planning, calibrating, sampling, and analyzing. An effective plan must be cost and time effective. The author stresses the need for "answers," which come from analysts, rather than "numbers," which come from determinators.

Testing of synthetic chemicals for their carcinogenic potential is an important weapon in evaluating long-term exposure effects. "**Carcinogenesis Testing and Analytical Chemistry**" discusses the role of analytical chemistry in ensuring that animal tests are sound. Analytical work is required before, during, and following testing. Much of the work involves determining the identity, purity, and stability of test materials. An important task proving that the feeding mechanism does not alter the nature of the feed material. Through this work, the value of the testing program is greatly enhanced.

In "**Search for the Cause of Legionnaires' Disease**," describes one phase of the search for this elusive killer. Transmission of disease occurs through a vector, such as an insect, water, air, or food. This paper discusses the search for a water carrier. It was first necessary to define disease victims to distinguish them from ordinary pneumonia cases. A total of 182 suspected cases were found. Their interaction with the hotel water and air-conditioning system was studied. Questions addressed where to sample, what to look for, how to compare sample profiles, and how to identify components (a GC/MS method was used). While the study was in progress, the culprit was found by the Center for Disease Control to be a virus, probably airborne or soilborne.

Nitrosamines are well known carcinogens in many animals and have been implicated as potential human health hazards. They are efficiently formed from secondary amines and nitrite. Because secondary amines are natural constituents of many foods and nitrite is present in saliva, these investigators studied the possible formation of nitrosamines in the digestive system. In "**An Unknown Salivary Morpholine Metabolite**," the authors used GC/MS with total ion/single ion detection to ascertain the structure of a material formed from morpholine in saliva. Difficulties arose from the low concentrations (μg/mL), and the impracticality of using large quantities of fresh human saliva. The key element in the study was the use of a new GC detector (the thermal energy analyzer (TEA)), which is highly sensitive and specific for the N-nitroso functionality. Analysis and subsequent synthesis showed the metabolite to be a non-mutagenic cyanamide.

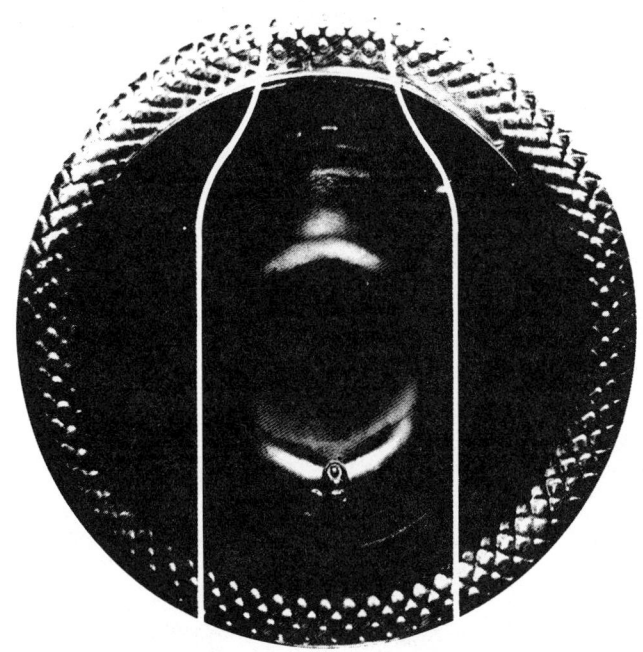

Analytical Chemists Vital in Commercialization of New Food Packaging Material

V. F. Gaylor

Standard Oil Co., Cleveland, OH 44128

Originally published in ANALYTICAL CHEMISTRY, 1974, Vol. 46, No. 11.

Discovery and development of a new family of barrier resins several years ago plummeted Sohio analytical chemists into an unusual problem-solving area. The thermoplastic, impact-resistant resins developed by the polymer research chemists were very effective barriers against transmission of oxygen, carbon dioxide, and most other vapors. Food packaging was thus a logical marketing goal for Sohio's first commercialized resin. Concurrent with this marketing decision, management recognized the need for obtaining FDA approval for the new resin, trade named Barex 210, and the importance of analytical chemistry in obtaining it. Analytical research on this problem was therefore initiated early in the development program and became a vital part of the whole commercialization process as shown in Figure 1.

The extent of the analytical work is indicated by the following figures. At least 1,632 Barex 210 bottles were ex-

Figure 1. Role of analytical research in commercialization of new food packaging material

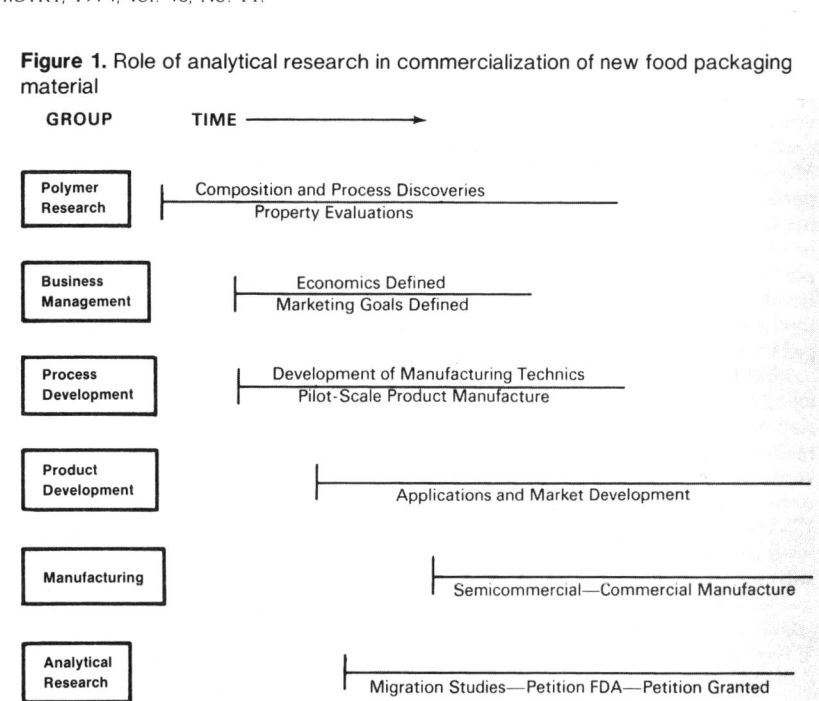

142   The Analytical Approach

tracted, 302 liters (80 gal) of ultrapure water was used, 168 liters (44 gal) of extract was slowly evaporated from 100-ml evaporating dishes, and 1,053 analytical determinations were made.

Organization

Responsibility for obtaining an FDA regulation for Barex 210 was delegated to a team representing three different disciplines. The team and its interaction with FDA and with the appropriate parts of the company are shown in Figure 2. The three team members represented a spectrum of expertise in administrative law; resin composition, properties, and processing characteristics; and instrumental and chemical analysis technics. Each member of this multidiscipline team had access to the total scientific resources of the R&D organization; thus, good two-way communication with all the various scientific and business groups involved in the resin development system was insured. Additionally, the team took advantage of advice and help available from FDA officials in the Petitions Control Branch of the Bureau of Foods. Invaluable advice on the required analysis program and on the supporting documentation requirements was received. The information developed in these joint meetings also helped the team guide process development pertaining to specific ingredients of the resin, i.e., potential migrants, and associated limitations.

Requirements for Food Packaging Regulation

Before regulating a new food packaging material, the FDA must be convinced that no harmful materials migrate from the container to the food. Migration levels are determined experimentally by contacting or extracting the packaging material with food or food simulating solvents. The exposed foods or solvents are then analyzed for any migrants, i.e., indirect food additives, extracted from the packaging material.

Migration studies on our food packaging candidate, Barex 210, were carried out in bottles made from the new resin and with the food simulating solvents listed in Table I. The solvents were "cooked" in the resin bottles at 125° or 150°F to equilibrium, i.e., until migrant levels measured in the solvents showed no increase with time.

The complete program consisted of the sequential steps outlined in Figure 3. Exploratory extraction experiments defined temperatures and approximate equilibrium times for each food simulating solvent. Nonvolatile mi-

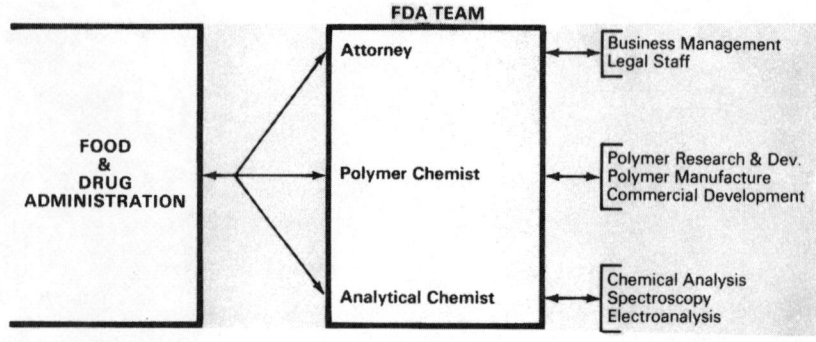

Figure 2. Organization approach for obtaining FDA approval

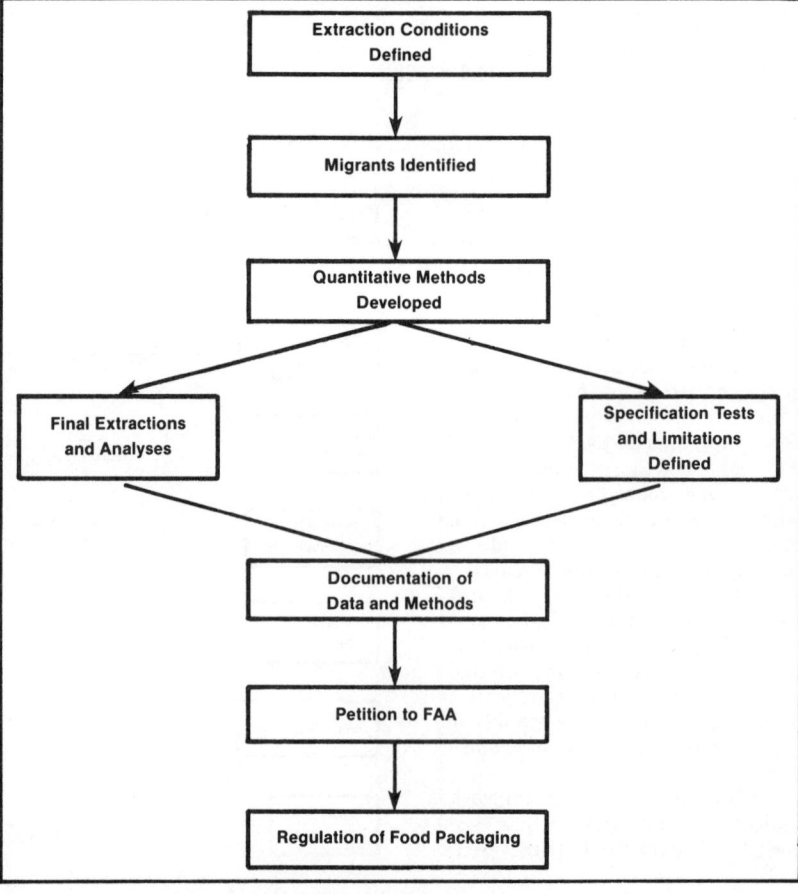

Figure 3. Sequence of FDA approval project

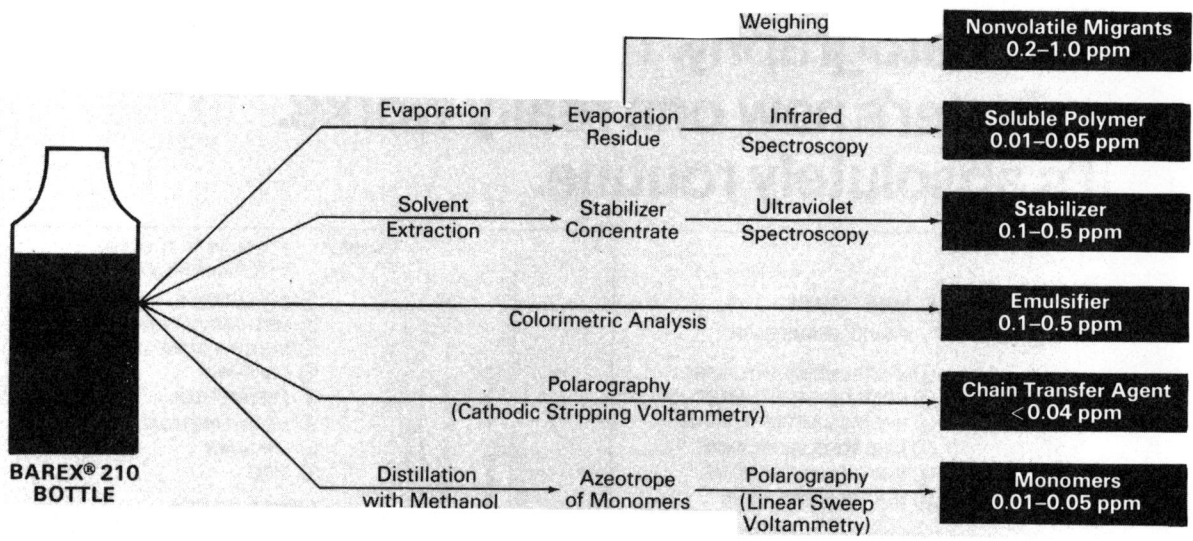

Figure 4. Analytical methods and results

grants were identified qualitatively by IR and UV spectrometry inspection of evaporation residues. Methods for quantitative measurements of both total and single migrants were then developed, in preparation for equilibrium extraction studies on several different resin batches. Quality control tests and specifications for a food-grade resin product were defined concurrently by relating bulk properties to results of the migration studies. The formal petition for the food packaging regulation contained the complete results of all the analytical studies, along with written procedures and copies of original records.

Analytical Methods and Results

Requirements for analyzing extracts of a new food packaging material partly depend on composition. As a minimum, the FDA requires measurement and identification of total nonvolatile extractables. Nonvolatile extractables of Barex® 210 were primarily emulsifier and stabilizer, both FDA-regulated food additives. Each of these was measured quantitatively by spectrophotometric procedures. We were also required to analyze the extracts for monomers, polymer, and chain transfer agent.

The complete analysis system is outlined in Figure 4. Total nonvolatile migrants were measured by weighing evaporation residues, as required by the FDA. Nonvolatile migrants from Barex 210 totaled less than 1.0 ppm in most cases. Gravimetric measurement of these low levels required the highest standards of solvent purity, clean room handling technics, and a controlled humidity atmosphere for tare and final weight measurements.

The amount of polymer in the evaporation residue was measured by infrared spectrometry. A considerable amount of technic development was needed to develop a quantitative IR method. The evaporation residues were often invisible to the eye and, at best, looked like stains in the platinum evaporating dishes. Quantitative transfer for IR analysis was achieved by redissolving the "stains" and evaporating the solutions on KBr. Analysis of the resulting KBr pellet for the low levels (<0.05 ppm) of migrating polymer required 10× scale expansion on a high-resolution, grating spectrophotometer.

The higher levels (0.1–0.5 ppm) of stabilizer and emulsifier were comparatively easy to measure. Some method development was required to obtain adequate sensitivity for stabilizer measurement by UV spectrometry. This problem was solved by a solvent extraction step in which the stabilizer was concentrated by a factor of ten.

Polarography was employed to measure both chain transfer agent and monomers. The chain transfer agent is a mercaptan, and we used the polarographic anodic depolarization wave of the sulfhydryl group at a mercury electrode. No (<0.04 ppm) chain transfer agent was detected in any of the bottle extracts. Migration levels of the two monomers, acrylonitrile and methyl acrylate, were less than 0.05 ppm in every case. Measurement of these low levels by polarography required a concentration step. Both monomers formed azeotropes with methanol and were concentrated in the first few milliliters of distillate. In the distillation step the monomers also are isolated from possible traces of interfering alkali metals.

The migration studies showed that Barex 210 is a safe food packaging material. Regulation 121.2614, entitled "Nitrile Rubber Modified Acrylonitrile–Methyl Acrylate Polymers," was published in the Federal Register on June 11, 1970, and is now part of Title 21 of the Code of Federal Regulations. Publication of the regulation cleared the way for commercial production and large-scale market testing of the new resin, and both events followed shortly thereafter.

Acknowledgment

Direct contributions as part of the "FDA team" were made by J. F. Jones, attorney, and B. F. Vincent, Jr., polymer chemist. Members of the larger analytical team who made equally important contributions include J. G. Grasselli, M. C. Helms, I. P. Horner, N. J. Meyer, C. Paxton, and M. K. Snavely.

Industrial Hygiene: Safer Working Through Analytical Chemistry

Ronald E. Hemingway

E. I. duPont de Nemours and Co., Haskell Laboratory for Toxicology and Industrial Medicine, Newark, DE 19711

Originally published in ANALYTICAL CHEMISTRY, 1980, Vol. 52, No. 8.

The Occupational Safety and Health Act of 1970 has been fondly referred to as the "Analytical Chemist's Employment Act of 1970." This facetious name has the ring of truth because the analytical chemist plays an important role on the industrial hygiene team. However, before we get too involved in the role of the analytical chemist, I will try to give a brief overview of industrial hygiene.

Industrial hygiene is the science and art of recognizing, evaluating, and controlling workplace environmental factors that may affect the health, comfort, or efficiency of the worker. These environmental factors can be divided into four stress types: chemical, physical, biological, and ergonomic. Some examples of these are shown in Table I. As you can see, the industrial hygiene program encompasses quite a number of fields and thus necessitates the team approach. The key figures on this team are the industrial hygienist, the physician and nurse, the health physicist, the engineer, the analytical chemist, the work supervisor, and the employee. The first six develop the overall plan and the latter two make the plan viable by carrying it out. The industrial hygienist is the proverbial "jack of all trades" and is responsible for the direction and maintenance of the overall program. The physician and nurse handle the medical care and surveillance program; they must be able to recognize the first signs of occupationally related disease. Radiation protection is provided by the health physicist who

Table I. Workplace Stresses—Examples

Chemical stress

Corrosives—can destroy living tissue by chemical action (e.g., sulfuric acid)

Toxic chemicals—gases, liquids, or solids that present dangers when they are swallowed, breathed or in contact with the skin

Flammable liquids—liquids with a flash point of 38 °C or less

Oxidizing materials—will decompose readily to yield oxygen, thus increasing the hazard of fire or violent reaction

Physical stress

Noise—"unwanted sound" that can annoy, cause hearing damage, or interfere with speech communication

Heat and cold stress—can cause discomfort and serious injury (e.g., heatstroke, heat exhaustion, frost bite)

Ionizing Radiation—α, β, γ, and x radiation can cause direct damage to tissue

Nonionizing Radiation
    Microwaves—100–100,000 MHz, heat damage of tissue
    IR—thermal burn
    Visible—good lighting
    UV—sunburn
    Lasers

Ergonomic stress—this arises from the improper match of an employee's physical and mental abilities to his job

Mechanical vibration, if excessive, can impair circulation

Lifting—overexertion, improper lifting

Biological agents

Occupational disease
    Farmer's Lung—fungi from grain dust inhaled
    Q Fever—livestock handlers
    Anthrax—bacterial infection
    Tuberculosis—medical field

must monitor and control the hazards presented by exposure to ionizing (X-ray, β-ray, etc.) and nonionizing (UV, IR, microwave, etc.) radiation. There are many areas that involve the engineer, but his main function on the team is in the design of control features, such as adequate fume ventilation. The analytical chemist becomes involved in all three of the previously mentioned steps by analyzing the workplace for the presence of chemical hazards. Supervisors and employees take the recommendations for control and safe work practices and implement them on the job. This team, working together, can then provide a safe workplace.

For most of us, industrial hygiene is a relatively new discipline that came into prominence with the advent of the Occupational Safety and Health Act in 1970. However, it probably began with Hippocrates in the fourth century B.C., with his recognition of lead poisoning as an occupational disease associated with the smelting of metals, and has evolved into its present form over the centuries. Table II highlights this evolutionary process. From this table, one gets the idea that industrial hygiene is not new, but, as with many scientific fields, its growth has been greatly accelerated in the last 40 years. Its growth in the last 10 years has been so great that the number of potential problems has far outpaced the number of quality solutions. Particularly in the area of workplace monitoring, there has been a real dearth of complete methods development. Only well-trained analysts and time can accomplish such methods development.

The recent increased interest in chemical stresses can be traced through the legislation (OSHA, TOSCA, etc.) and the public awareness of potential chemical hazards (e.g., carcinogens, mutagens, embryotoxins, etc.). Although it is important that industry develop responsive programs for chemical stresses, it still must be recognized that chemicals account for only about 10% of the occupational accidents and disease in the U.S. However, it is this 10% that involves analytical chemists; therefore, it will be the portion on which this paper will focus.

The analytical chemist is involved in the recognition, evaluation, and control of chemical hazards. Chemical hazard assessment is the combination of the extent of chemical toxicity and the potential for personnel exposure. Chemical toxicity is measured by toxicological testing that assesses the ability of the chemical to produce harm to the body as it is introduced by the routes of absorption through the skin (dermal), ingestion via the mouth (oral), and breathing into the respiratory tract (inhalation). The extent of damage produced by a chemical usually is dependent on both the amount in contact with the body and the length of exposure time. Thus, in assessing the potential for personnel exposure, the analytical chemist must determine the concentration of material and the duration of exposure. This type of personnel exposure assessment is usually called monitoring. The analyst is involved in the initial recognition of chemical hazards by providing survey data that yield clues to the types of chemicals, their levels, and origins of the potential exposure. If there are indications of exposure potential after the initial survey, then a more elaborate monitoring program is set up and a more thorough evaluation is made. After the problems have been evaluated, engineering controls are installed, and the monitoring program determines the effectiveness of these controls. It is in this area, monitoring, that the analytical chemist helps to generate data by analysis and the development of sampling and analytical methods. It is, therefore, very important for the analyst to understand all the aspects of the monitoring program.

Monitoring Program

A monitoring program is made up of four basic steps: (1) planning, (2) calibration, (3) sampling, and (4) analysis. Typically, in an ongoing program, the industrial hygienist will carry out the first three steps and the analytical chemist is only involved in the last step; however, he must be familiar with all the steps to better assess the quality of his analytical results and to enable him to develop new sampling and analytical methods that are accurate, precise, and useful.

As in any program, planning is the most important step. Here the monitoring program is made to yield the most useful information for the least amount of time and money. Figure 1 shows the suggested NIOSH employee monitoring strategy. This figure emphasizes not only that planning is an important initial step in defining the

Table II. Historical Evolution of Industrial Hygiene

Fourth century B.C.	Hippocrates recognized the problems of lead toxicity in mining industry
First century A.D.	Pliny the Elder designed a dust mask to protect miners
Second century A.D.	Galen recognized hazard of acid mists to copper miners
1473 (actually published 1524)	First book printed on specific industrial poisons by Ulrich Ellenbog
1531–34 (published 1567)	Paracelsus—book on the diseases of miners
1556	Agricola's "De Re Metallica" reported poisoning attributable to metals
17th century	Ramazzini—"father of industrial hygiene" wrote "De Morbis Artifican Diatriba" ("On Diseases of Tradesmen")
18th & 19th centuries	Increased interest in Europe and first effective legislation to protect worker health
1910	Bureau of Mines—to protect miners
1912	Esh Law—placed a prohibitive tax on matches made using white phosphorous. This substance caused necrosis of the jaw in match workers
1913	Department of Labor—one objective was the collection and distribution of information on industrial hygiene
1915	Public Health Service
1916	American Association of Industrial Physicians and Surgeons
1938	American Conference of Governmental Industrial Hygienists started (ACGIH)
1939	American Industrial Hygiene Association (AIHA)
1946	American Academy of Occupational Medicine
1966	Metal and Nonmetallic Mine Safety Act of 1966
1969	Federal Coal Mine Health Safety Act
1970	OSHA

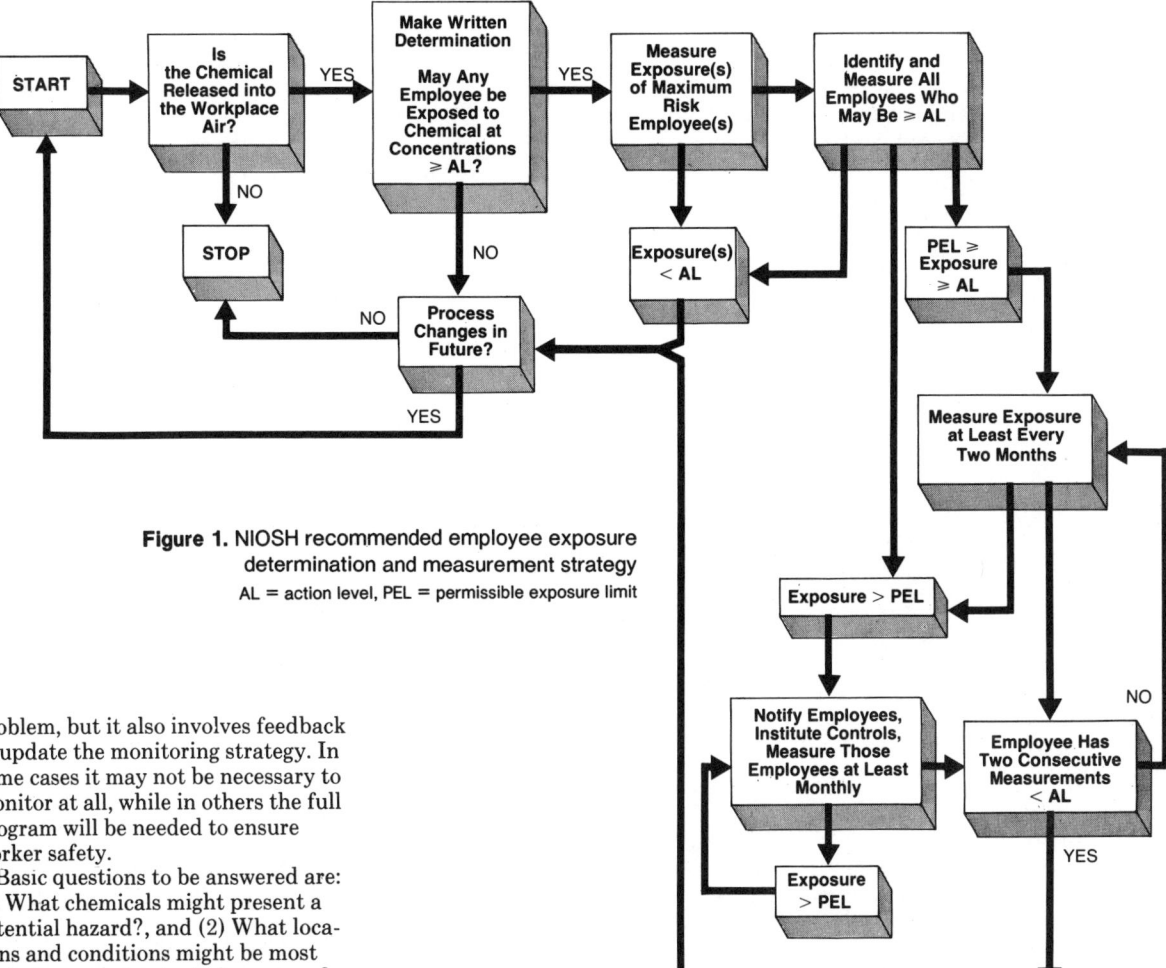

Figure 1. NIOSH recommended employee exposure determination and measurement strategy
AL = action level, PEL = permissible exposure limit

problem, but it also involves feedback to update the monitoring strategy. In some cases it may not be necessary to monitor at all, while in others the full program will be needed to ensure worker safety.

Basic questions to be answered are: (1) What chemicals might present a potential hazard?, and (2) What locations and conditions might be most likely to present potential exposure? After these questions are answered with available information, the appropriate type or types of sampling strategy must be chosen.

A key point in the planning step is for the industrial hygienist to consult the analytical chemist early. If method development is required, it will take time, and preparation for even a standard method is always important.

If the likely area of toxic hazard is not well defined, then samples taken in the vicinity of the worker's breathing zone would be most appropriate. Breathing zone samples can be taken in two ways: (1) With a sampling train, the industrial hygienist follows the worker through his routine trying to sample the worker's breathing zone; (2) A small, portable, self-powered device is attached to the worker and thus samples his breathing zone continuously. This latter technique is called personnel monitoring and it better characterizes the worker's actual exposure. However, personnel monitoring requires special equipment that can be expensive and can impede the mobility of the worker.

When a specific location is likely to present a hazard, such as an open reaction vessel, then a process sample at that specific area would be appropriate. Although the worker may not spend all of his time at this location, this would help define his maximum likely exposure.

General air samples are again fixed place samples, but are perhaps chosen to indicate the worker's potential exposure when he does not approach the area of highest exposure. By knowing the timing of a worker's routine, then, approximate worker exposure can be ascertained by combining general air sampling data with the time the worker is present in the direct exposure area (see Table III).

The last sample type is derived from biological monitoring. This technique is very powerful but takes special development and is expensive to use. Instead of monitoring the air, one measures metabolites, biochemical parameters (e.g., enzyme levels), or other physiological responses in the workers. The main advantage of this technique is that it assesses the actual amount of chemical that has entered the body, regardless of the route of exposure. Typically, the other monitoring methods only detect those chemicals that would enter the body via inhalation. Biological monitoring would also assess exposure by the oral and dermal routes. This is important since the dermal and inhalation routes are the most important routes of exposure in industry. Again though, biological monitoring is difficult to develop and

---

Table III. Calculation of Worker's Exposure from Area Samples

Worker's routine

$T_1 \equiv$ time at first work station

$T_2 \equiv$ time at lunch

$T_3 \equiv$ time at second work station

$T_4 \equiv$ time at third work station

Concentrations at the work stations

$C_1 \equiv$ air concentration at first work station

$C_2 \equiv$ air concentration at lunch ≈ 0

$C_3 \equiv$ air concentration at second work station

$C_4 \equiv$ air concentration at third work station

Time weighted average (TWA) concentration =

$$\frac{C_1 T_1 + 0 T_2 + C_3 T_3 + C_4 T_4}{T_1 + T_2 + T_3 + T_4}$$

is most applicable when there are larger numbers of workers with similar exposure potential.

While deciding on the appropriate sampling strategy, the appropriate time period for sampling must be chosen (Figure 2). Worker exposure is usually compared by the 8-h time-weighted average of exposure concentrations. The most desirable sample length is full period. However, circumstances such as the sampling method, lab protocol, or work routine may prevent full period sampling, and one of the other three approaches can be chosen. The grab sample is normally the least desirable, but it can be very useful in planning other samples and is relatively easy to develop as a sampling and analytical technique.

Calibration is something with which analytical chemists are quite familiar. However, in a monitoring program, not only does the analytical instrumentation need calibration, but the elements of the sampling train need it also. Let's take the calibration of a small personnel sampling pump as a typical example. Here the pump, connected through its sampling train to simulate the normal pressure drops that may affect the pump's performance, is attached to the stopcock end of an inverted 500 mL buret. The pump is turned on and the open end of the buret is dipped into a soap solution to form a bubble that will freely travel as the pump pulls it along. A stopwatch is then used to time the movement of the bubble between the volume graduations. The flow rate is equal to the volume displaced by the bubble divided by the time, mL/minute. The total volume sampled is the product of this flow rate and the total time sampled and is used along with the analytically determined amount of contaminant to provide meaningful air concentrations. Usually these concentrations are reported as the amount of contaminant found per volume of air sampled, mg/m^3 and used as comparison to those air concentrations recommended safe by OSHA. Thus, to ensure a safe workplace an active calibration program is a necessity.

The last two steps in monitoring will be discussed together since they are always coupled in practice. The analytical and sampling methods can be classified as either direct or indirect depending on their ability to provide a real time readout in the workplace.

The direct methods usually accomplish sampling and analysis in one instrument and give the results in real time. Table IV lists some direct devices. These instrumental methods can be used to take grab samples for survey work or, when coupled with a recorder and/or integrator, to measure TWAs. However, the instruments are moderately expensive and usually can only be justified for larger monitoring programs. Detector tubes offer a relatively inexpensive way to get a quick assessment of hazards, but their accuracy is seldom better than ±25%.

By far, most sampling and analysis schemes are two-step procedures where the sample is collected in the field and brought back to the lab for analysis. The analytical techniques used here are as broad as analytical chemistry itself. A typical sampling train is shown in Figure 3 and a number of collection devices are listed in Table V. Of these collection devices, impingers, bubblers, bags, filter paper, and solid adsorbent tubes need pumps for pulling the sample through the collection device. There are many kinds of pumps available but the most useful are small, portable, battery-operated pumps that cost $300–$500 each and have a typical precision of ±8%. The design of these pumps varies from the simplicity of a piston pump that counts the number of strokes made by a constant volume piston to sophisticated pumps with a complex electronic flow sensing device. Of the sampling techniques using pumps, probably solid adsorbent tubes are used most extensively. Figure 4 shows the NIOSH charcoal tube. There are two sections of charcoal separated by urethane foam. The front section with 100 mg of charcoal can hold approximately 10 mg vapor and the back-up section, with 50 mg charcoal, is used to detect breakthrough from the front section. To use, the glass tips are broken and one end is connected to a calibrated pump via a piece of flexible tubing. Usually the tube is located on the worker's collar within 6 in. of his mouth and nose, and the pump is attached to the worker's belt. After sampling, the ends of the tube are capped and submitted for analysis, usually by gas chromatography. The sample is prepared for analysis by breaking the tube and pouring the front and back sections into separate vials. Then the appropriate eluant (usually CS_2) is added to each vial and allowed to desorb the adsorbed chemical from the charcoal for 30 min to 1 h. The resulting solution is then analyzed by gas chromatography.

Another interesting device used for personnel sampling is the passive dosimeter. It too uses the principle of charcoal adsorption, but it requires no pump to pull through the sample. It is based on the development of a diffusion gradient between the charcoal and the outside of the dosimeter. This device can easily be worn by the worker since the whole device is typically smaller than a box of matches. The diffusion process in most of these devices produces a sampling rate equivalent to 50–100 mL/min. Analysis is essentially the same as that for the solid adsorbent tube.

These are just examples of the types of sampling and analysis methods available. The reading list below suggests some useful texts in industrial hygiene monitoring and, of course, one is free to develop one's own method.

As when developing any new analytical scheme there are some basic questions to be answered: (1) Do we have a representative sample? (2) Are there

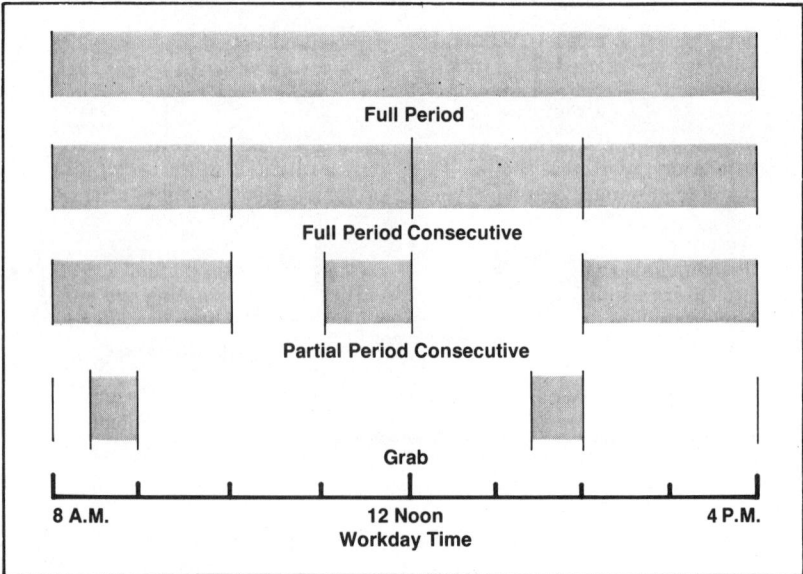

Figure 2. Types of exposure measurement strategy

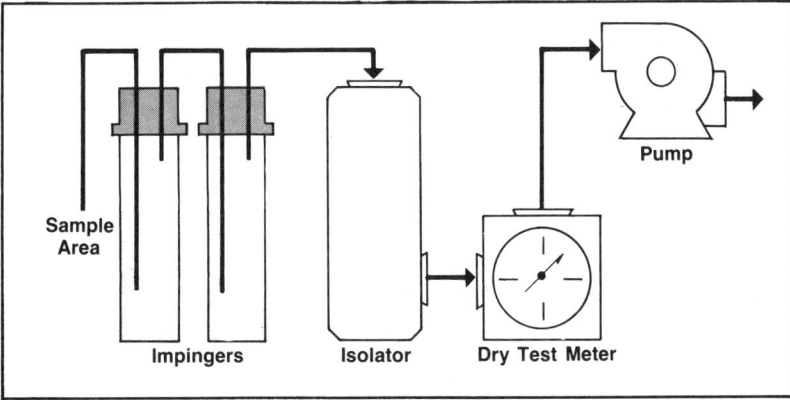

Figure 3. Typical sampling train
Impinger—collection device using solutions to trap contaminants; isolator—filters for removing contaminants from air to prevent contamination of pumps or test meter; dry test meter for measurement of air volume sampled; pump for movement of air through collection device

Table IV. Direct Sampling Devices

IR	Portable instruments for gases and vapors, having 20-m pathlengths and ppm (v/v) sensitivity
UV	Primarily used for the detection of mercury vapors
Photoionization	Absorption of UV by a molecule leads to ionization. Photons are energetic enough to ionize many species (particularly organics), but do not ionize major components of air such as O_2, N_2, CO, CO_2, or H_2O. The ions are collected on an electrode and the current is measured
GC	Portable gas chromatographs capable of doing separations and quantitation. One instrument type bypasses the column to give a continuous hydrocarbon detector based on FID
Detector tubes	Air is drawn through a glass tube packed with solid chemicals that react to indicate the presence of a contaminant by color change or length of color strain

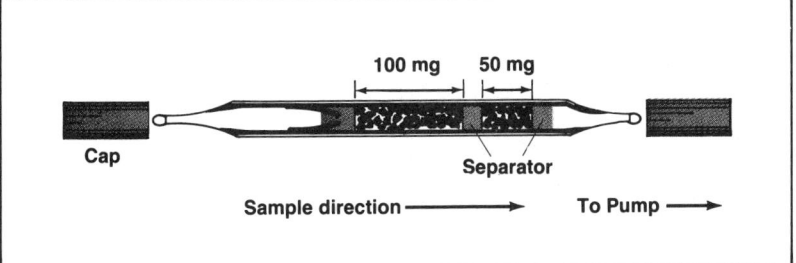

Figure 4. NIOSH charcoal tube

Table V. Collection Devices

Impingers and bubblers
Bags
Evacuated samplers
Passive devices
Solid adsorbent

any interferences? (3) Is the method precise and accurate? (4) Is it easy and inexpensive to use? All of the questions are just as hard to answer as ever. Matrix and interferences are never the same; they are at the mercy of time and space. Therefore, when developing a method, rigorous attention must be given to the environmental factors, such as temperature, relative humidity, and wind velocity, as well as the chemical interferences. Thus, field testing must be part of any method development program, with the goal to look at as many different environments as possible. At Haskell, we try to simulate varying temperatures, relative humidities, wind velocities, and interferences in an environmental control chamber designed for method development, and the methods are tested at the many sites within Du Pont that are interested in the monitor. Many times the field tests do not live up to the extensive lab trials.

Conclusions

"The Chemical Analysis of Things as They Are" was the title of a very interesting paper presented by G.E.F. Lundell at the 85th Meeting of the ACS in 1933. He discussed the differences between analysts and determinators in the analytical field. The industrial hygiene analyst produces answers; the determinator just generates numbers. "The determinators, who are by far the more numerous, may in turn be divided into two general classes: first, the common determinators who follow a method explicitly, without knowledge or concern as to the reactions involved; and second, the educated determinators who can handle systems containing one or perhaps two variables... As for the analyst, he is a comparatively rare bird... who does the best he can, guided by theory and experience, of which the most comforting is experience."

There is a place for the determinators in workplace sampling and analysis, but only after the analyst has thoroughly evaluated "... (1) the sample on which the analysis is made, (2) the method of analysis that is used, (3) the accuracy of the result, and (4) the cost of analysis."

Presently there is a real need for more analysts to be involved. The monitoring of workplace environments is in a very rapid state of growth, and, as a result, the number of methods that are truly well tested is few. More analysts are needed to provide well-conceived monitoring methods that have been tested in Lundell's four steps.

Suggested Readings

(1) Patty, F. A., "Industrial Hygiene and Toxicology," Vol. II, Interscience Publishers, 1963.
(2) "Air Sampling Instruments for the Evaluation of Atmospheric Contaminants," 5th ed., American Conference of Governmental Industrial Hygienists, 1978.
(3) "NIOSH Manual of Analytical Methods," 5 Vols., DHEW Publications, Pub. #77-157-A,B,C,78-175,79-141.
(4) "NIOSH Manual of Sampling Data Sheets," DHEW Publications, Pub. #79-159.
(5) Cheremisinoff, P. N. and Morres, A. C., "Air Pollution Sampling and Analysis Deskbook," Ann Arbor Science Publications, 1978.

Carcinogenesis Testing and Analytical Chemistry

E. A. Murill and E. J. Woodhouse
Midwest Research Institute
Kansas City, MO 64141

S. S. Olin
Tracor Jitco, Inc.
Rockville, MD 20850

C. W. Jameson
National Cancer Institute
Bethesda, MD 20815

Originally published in ANALYTICAL CHEMISTRY, 1980, Vol. 52, No. 11.

Man's environment contains a large number of natural products as well as an ever-increasing number of synthetic chemicals that, although not acutely toxic at daily exposure levels, may have harmful long-term effects, especially with respect to the induction of malignant tumors. Testing of the carcinogenic potential of chemicals to which humans are exposed both voluntarily and involuntarily (e.g., pesticides, air pollutants, industrial chemicals, food additives, drugs, etc.) is receiving increased attention from the federal government through the National Cancer Institute and the National Toxicology Program. Through the efforts of the NCI/NTP Carcinogenesis Testing Program, cancer-producing chemicals are being identified with the use of appropriate carcinogenesis bioassay tests in laboratory animals.

The analysis of these test chemicals is the first of several essential steps which lead to the successful completion of a bioassay test. Without a scientifically sound analytical chemistry program, the results of the animal studies would not be valid.

The objective of the analytical program is to serve as a resource for the analysis of the identity, purity and stability of all test materials studied by the testing program. Midwest Research Institute currently serves as the analytical chemistry resource for the Carcinogenesis Testing Program through two contracts, one with Tracor Jitco, Inc. (the prime contractor for the testing program), and one directly with the National Cancer Institute. In addition to the identity and purity profiles generated for all test chemicals, a thermally accelerated sta-

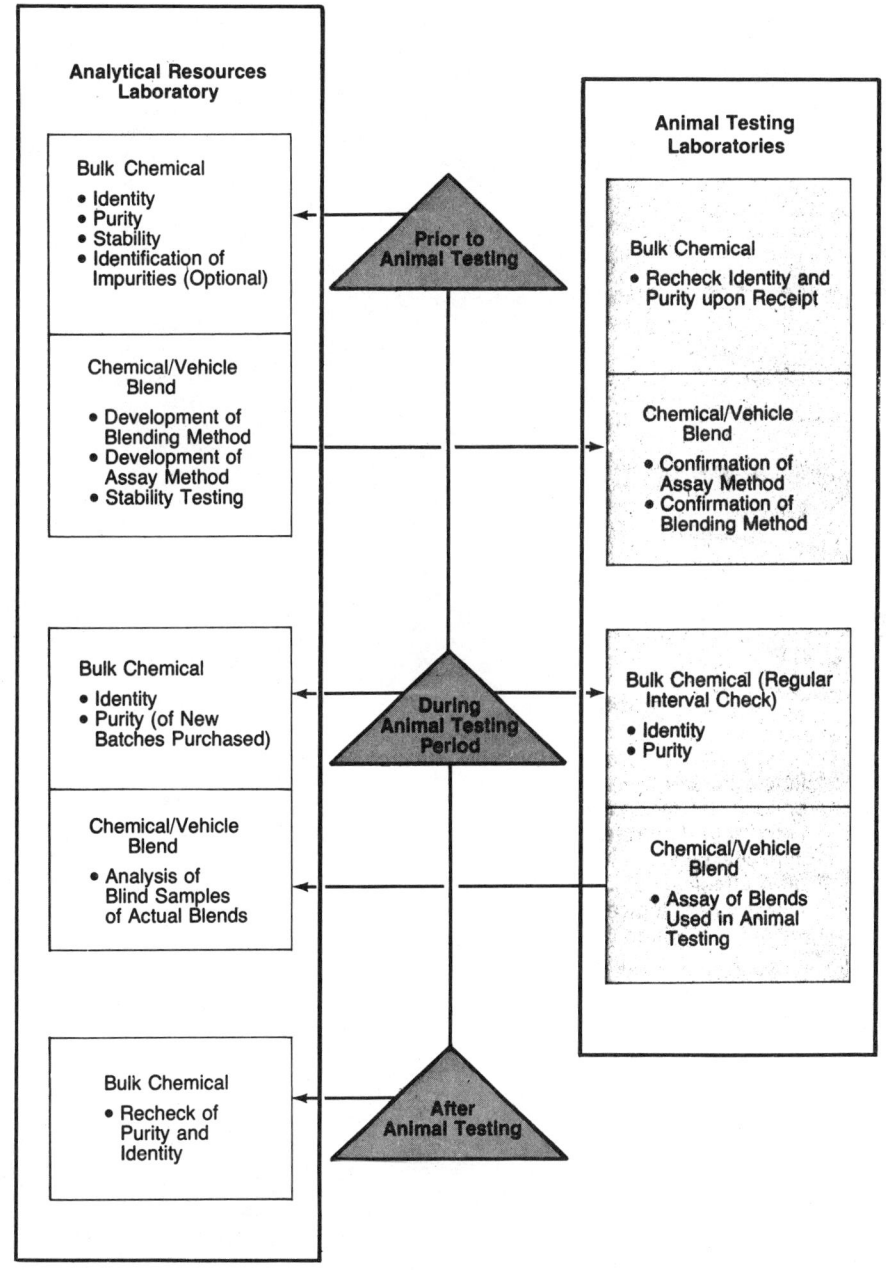

Figure 1. Analytical chemistry in the Carcinogenesis Testing Program

83/4463-0151 $07.00 © 1983 American Chemical Society

bility study is also performed to determine the stability and most appropriate storage conditions for the bulk test materials while in use at the bioassay laboratories (which may be up to four years). The analytical resource also performs methods development and protocol design for the adequate mixing of test materials in administration vehicles (rodent feed, water, corn oil, etc.), analysis of the test material in the vehicle (to ensure homogeneity and stability) and development of analytical methods suitable for use by a bioassay facility for the reanalyses of bulk test material and chemical/vehicle mixtures.

The job of the analytical chemistry resource is not completed when the bulk test material is shipped to the animal testing facility. Quality assurance programs have been established to have the analytical resource serve as a referee laboratory to spot-check the adequacy of the mixing procedure at the bioassay laboratory by analyzing randomly selected samples of chemical/vehicle mixtures. In addition, any surplus test material remaining after completion of the bioassay tests is returned to the analytical resource laboratory where additional analysis may be performed if needed. Thus, the analytical loop is completed (see Figure 1).

An interesting problem was encountered when a sample of methyl carbamate was assayed for purity. The methyl carbamate received was specified to be 99% minimum by Kjeldahl analysis. In-house nitrogen analysis indicated 99.8% purity and therefore agreed with the specifications. However, gas chromatography of the compound showed a small (0.2%) impurity before the major peak and a larger (1.2%) impurity after the major peak. Concern was raised that the methyl carbamate might contain the known carcinogen ethyl carbamate (urethane), whose presence could confuse the interpretation of the bioassay results. Fortunately, neither impurity matched an authentic sample of urethane in retention time. To identify the large impurity in the methyl carbamate sample, gas chromatography/mass spectrometry of a solution was performed using a capillary carbowax column for component separation. The reconstructed ion chromatogram (Figure 2) detected the impurity at scan number 451 and the major peak at 388. The fragment plot obtained for 451 was, however, not definitive. The base peak was 15 and the highest mass peak, 88. The closest literature search match was 1,1- or 1,3-dimethylurea, neither of which had the correct GC retention time.

Preparative gas chromatography was employed to obtain sufficient compound for identification. However, the impurity peak was difficult to locate on the preparative chromatogram because of its low concentration and tailing peak shape. The fraction containing the impurity was identified by spiking the methyl carbamate sample with ethyl urea, which had been shown on the analytical gas chromatogram to have a longer retention time than either the methyl carbamate or the impurity.

The reconstructed ion chromatogram obtained by GC/MS on the collected impurity fraction (Figure 3) showed a small amount of methyl car-

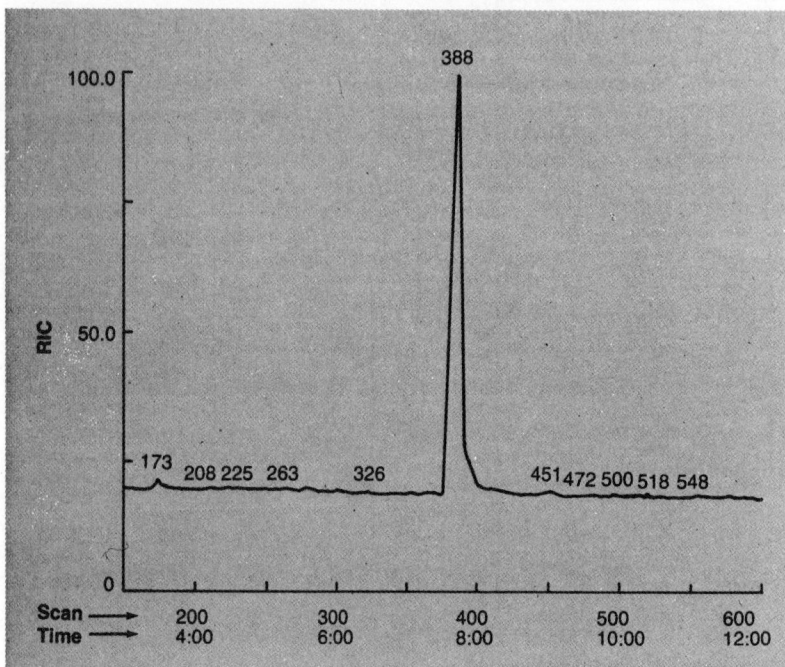

Figure 2. Reconstructed ion chromatogram (RIC) of methyl carbamate solution. 5 μL injection

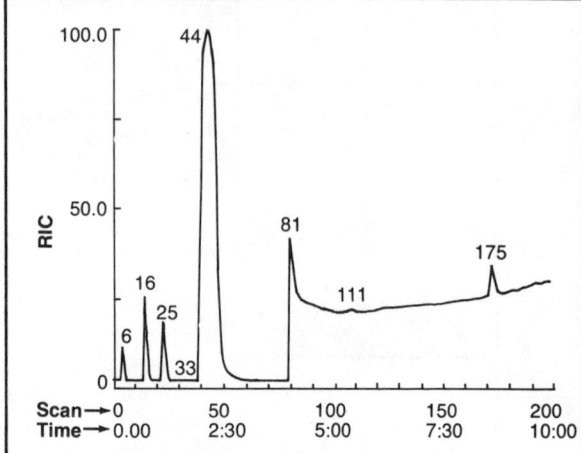

Figure 3. Reconstructed ion chromatogram of collected methyl carbamate impurity fraction. 1 μL injection

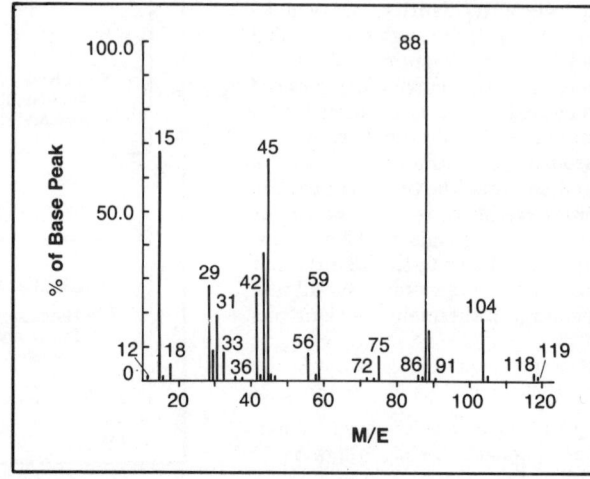

Figure 4. Mass spectrum of methyl carbamate impurity. 1 μL injection

bamate (scan 111) and the impurity (scan 175) in addition to solvent peaks at scans 44, 25, 16, and 6.

The enhanced spectrum (Figure 4) of impurity (scan 175) showed a small peak at 118 and an even smaller fragment peak at mass 119. Because of the relatively larger peak at 104 (18% of base peak, $M^+ - CH_3$), the 119 mass was considered the molecular ion. The structure of the impurity was assigned to be methyl (N-methoxymethyl)carbamate (1).

$$\underset{a}{H_3C}-O-\underset{\underset{H}{|}}{\overset{\overset{c}{H}}{\underset{|}{C}}}-\underset{\underset{H}{|}}{N}-\underset{b}{\overset{\overset{O}{\|}}{C}}-OCH_3 \quad (1)$$

The predicted fragmentation pattern of this mixed acetal of formaldehyde with methanol and methyl carbamate can account for all the peaks in the mass spectrum (Figure 5).

To further confirm the structure a Fourier transform nuclear magnetic resonance spectrum was obtained on the impurity fraction. The spectrum had resonances (Table I) which corresponded to those expected for compound 1.

Dyes and pigments frequently present interesting analytical challenges. They are typically poorly characterized mixtures defined by color matching and by physical properties. Disperse Blue 1, used in semi-permanent hair dye formulations, is identified in the literature simply as 1,4,5,8-tetraaminoanthraquinone. A partially dried factory-strength batch used in the program did indeed contain this anthraquinone derivative but it only comprised 50% of the total; in addition to 20% water, there were at least six other organic compounds present. The manufacturer verified that this was a representative batch of Disperse Blue 1.

C.I. Red 14 (Color Index No. 14720) is an azo dye used in the U.S. on wool and leather and in wood stains. Before 1966 it was approved for use in drugs and cosmetics in this country as Extract D & C Red No. 10. The textile-grade product (Carmoisine B.A.) selected for testing in the NCI/NTP program was found to contain only 67.3 ± 0.5% by weight of the azo dye II (4-hydroxy-3[(4-sulfo-1-naphthalenyl)-azo]-1-naphthalene-sulfonic acid, disodium salt) by titanous chloride titration (Figure 6). The ultraviolet/visible spectrum was consistent with the structure, and neither these spectra nor thin-layer chromatography or high pressure liquid chromatography revealed any significant organic impurities. Elemental analysis turned out to be our most incisive analytical tool in this case (see Figure 6), for C, H, N, and S were all low and Na was high. Commercial dyes frequently carry inorganic salts, either entrained in the presscake or added as filler to reduce the color intensity or modify physical properties. A follow-up analysis gave 4.76% chloride, which corresponds to 7.85% NaCl. Combining all of the data and assuming that NaCl and Na_2SO_4 were the main salts present gave a very satisfactory fit with the actual elemental analysis. (If a little Na_2CO_3 were also present, the fit would be even better.)

Thus, the bioassay of C.I. Acid Red 14 was begun with the knowledge that the test material was a typical textile-grade batch containing about two-thirds azo dye and the remainder inorganic salts and water, a considerably simpler product from a toxicologist's viewpoint than Disperse Blue 1 and other dyes with significant organic impurities.

Another major task that must be accomplished for each chemical before it can be tested in animals is the selection of the route of administration by the National Cancer Institute. While some chemicals are tested by inhalation exposure or application to the skin, most are administered by mixing with the animals' feed or by oral gavage (direct introduction into the stomach through a tube inserted carefully down the esophagus). Whatever the route, the analytical resource develops procedures for mixing the chemical with the vehicle (e.g., feed, water, corn oil) that should ensure homogeneity and checks the stability of

Table I. Fourier transform NMR spectrum of methyl carbamate impurity fraction

Proton Assignment (Compound I)	δ Value	Multiplicity	Integration Ratio
a	3.34	Singlet	3.0
b	3.72	Singlet	3.1
c	4.62	Doublet	2.2

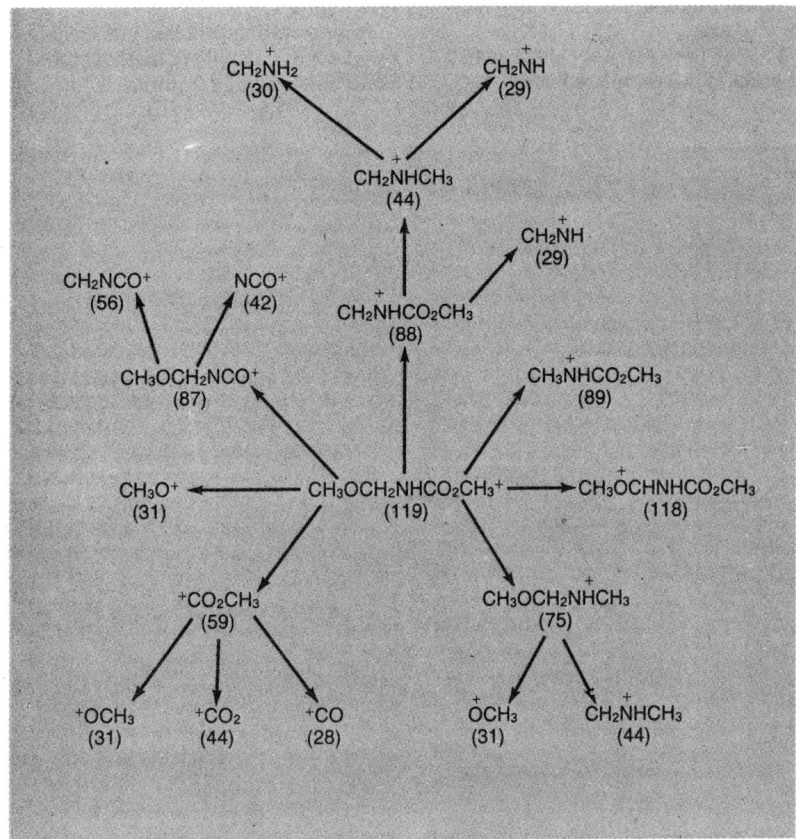

Figure 5. Predicted fragmentation pattern

Figure 6. Compound II and percent elemental compositions

	C	H	N	S	Na	Cl
II: $C_{20}H_{12}N_2Na_2O_7S_2$	47.81	2.41	5.58	12.76	9.15	0.0
67.3% II + 7.48% H_2O + 7.85% NaCl + 12.21% Na_2SO_4	32.17	2.45	3.75	11.35	13.20	4.76
Determined	33.00	2.50	3.97	11.29	14.63	4.59
	32.91	2.45	3.84	11.38	14.39	4.93

Figure 7. Chlorendic acid stability in feed

Time (at 25 °C)	Recovery
2 days	100.8 ± 0.5%
4 days	94.5 ± 1.1%
7 days	97.3 ± 1.5%

the chemical in the vehicle. The stability study typically is performed at temperatures from −20 °C to +45 °C to investigate the effects of low-temperature storage and of the ambient temperatures during preparation and use of the dose mixtures; higher temperatures help to confirm temperature-dependent instability when present.

Confirming the stability of a chemical in feed often offers an opportunity for some creative analytical work, since the test material must be recovered from the dosed feed mixture before analysis. The tremendous range of chemical substances presented for evaluation, invariably on a tight schedule, is a major aspect of the challenge. Target concentrations at which the stability must be validated may be as high as 5% (50 000 ppm) for, say, a GRAS food additive or as low as 5 ppm for a highly toxic or active substance.

When the recovery of a test chemical from feed is found to decline with increasing time or temperature, it is usually assumed that the compound is unstable in feed, and an alternate route of administration is chosen. However, the possibility also exists that the chemical is slowly binding to components in the feed and upon ingestion by the animals will still be available for absorption. Because of the complexity of the feed matrix and time/cost constraints, it is seldom practical to attempt to look for degra-

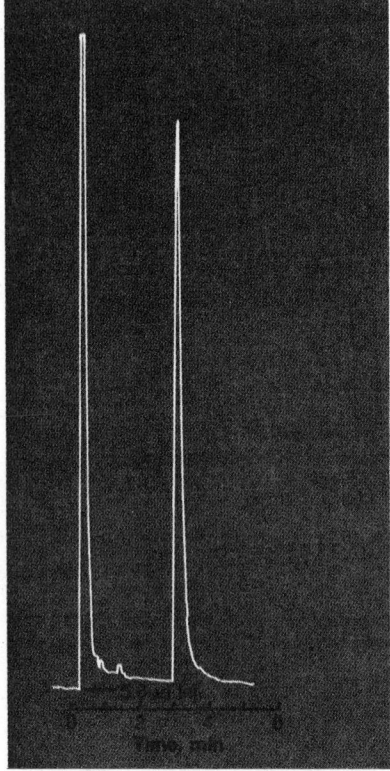

Figure 8. β-naphthylamine standard (1.08 μg/mL in toluene). Column: 3% SP2100-DB on 100/120 Supelcoport; 1.8 m × 2 mm i.d.; glass. Detector: thermionic specific, 250 °C. Carrier gas: N_2, 30 cc/min. Inlet: 200 °C. Column oven: 145 °C. Sensitivity: 4 × 10^{-12} AFS.

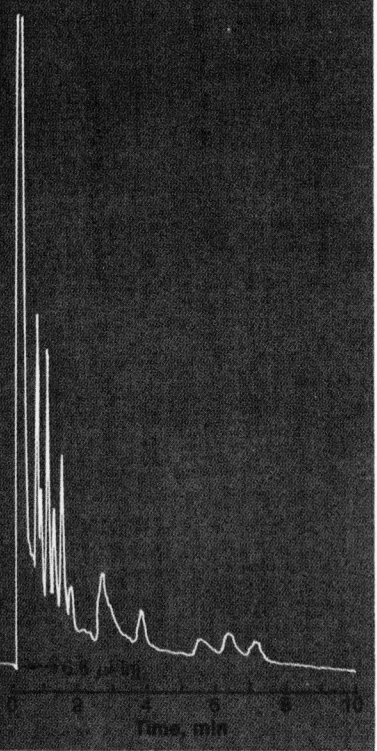

Figure 9. Blank rat urine extract. Conditions as in Figure 8

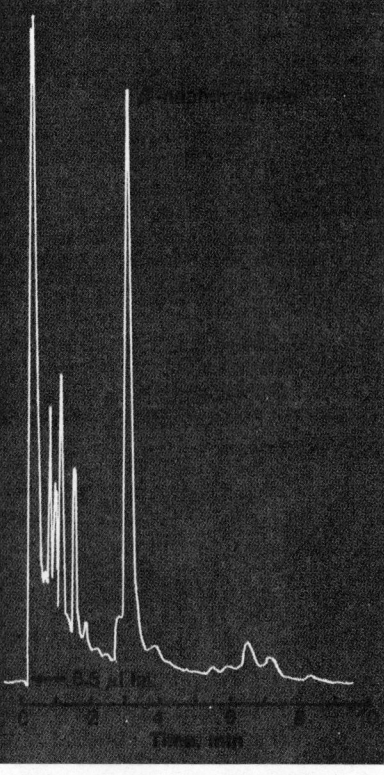

Figure 10. Spiked rat urine extract (1 μg β-napthylamine/5 mL rat urine). Conditions as in Figure 8

dation products or chemical bound to feed, but the problem sometimes can be solved more simply.

Chlorendic acid (Figure 7) is a highly chlorinated dibasic acid used in the manufacture of flame-retardant monomers, resins, and plasticizers, as well as herbicides and insecticides. It can be extracted quantitatively from feed (0.1% concentration) with 1% HCl in acetonitrile. Extracts are converted to dimethyl chlorendate with BF_3/CH_3OH (48 hr, 70 °C) or diazomethane (5 min, 25 °C) and analyzed by GC. At 25 °C the recovery declines with time over a two-week period although there is some variability among samples. From a structural point of view, chlorendic acid almost certainly is stable under these mild conditions and apparently is gradually binding to metals or amines or other reactive functions in the feed.

Another example was pentachlorophenol, the important wood preservative and herbicide. At 20 ppm in rodent feed, extraction with toluene gave quantitative recoveries immediately after mixing or after two weeks at −20 °C but only 86% recovery at 25 °C and 55% at 45 °C after two weeks. However, extraction with 1% HCl in methanol gave a 93% recovery at 25 °C and a 91% recovery at 45 °C after two weeks, suggesting that a weak complex of pentachlorophenol with feed components does form slowly and can be broken by extraction with dilute acid.

A trace analysis development problem was encountered when the analytical group was asked to develop a method for the analysis of β-naphthylamine in rat urine. N-Phenyl-β-naphthylamine was to be tested for carcinogenicity, and there were literature reports of workers exposed to this antioxidant having β-naphthylamine detected in their urine (1). β-Naphthylamine is a known human carcinogen, regulated by OSHA.

The objective in this case was to develop a simple method, preferably with minimal cleanup and without derivatization, for quantitation of β-naphthylamine in rat urine at levels of 100 ppb or above in the presence of possible N-phenyl-β-naphthylamine precursor.

Since β-naphthylamine is very soluble in toluene and only slightly soluble in water, a small column containing octadecylsilane bonded to silica particles (C_{18} Sep-Pak, Waters Associates) was used for cleanup. The rat urine was adjusted to pH 9.0 with sodium carbonate (10%); an aliquot of neutralized urine was forced through the column and the eluent discarded. Distilled water wash was then forced through the cartridge and also discarded.

An aliquot of toluene was forced through after the water wash and the eluent were collected. The toluene layer easily separated from the water layer and was concentrated for chromatographic analysis.

The concentrate was injected directly onto a 3% SP 2100-DB column and detected using a thermionic specific detector. The chromatogram of the standard in toluene, an unspiked rat urine, and an aliquot (5 mL) of rat urine spiked with 1 μg of β-naphthylamine are shown in Figures 8, 9, and 10, respectively. The recovery on spiked samples was 92 ± 4%. The detection limit was significantly less than the 100 ppb specified level.

Conclusion

The goal of the analytical chemistry program is to better define the chemicals being tested by the Carcinogenesis Testing Program. This includes not only the original test chemical but also the chemical/vehicle mixtures administered to the test animals. It is hoped that by better defining the actual test materials, the interpretation of the final results of the animal studies will be enhanced.

Acknowledgment

We would like to acknowledge the work of the members of the Bioanalytical Section at Midwest Research Institute.

References

(1) NIOSH Current Intelligence Bulletin, "Metabolic Precursors of a Known Human Carcinogen, Beta-Naphthylamine," Dec. 17, 1976.

In Search of the Cause of Legionnaires' Disease

Irwin H. Suffet

Drexel University
Environmental Studies Institute
Chemistry Department
Philadelphia, PA 19104

Patrick R. Cairo

Philadelphia Water Department
Research and Development
Philadelphia, PA 19107

Originally published in
ANALYTICAL CHEMISTRY, 1978,
Vol. 50, No. 9.

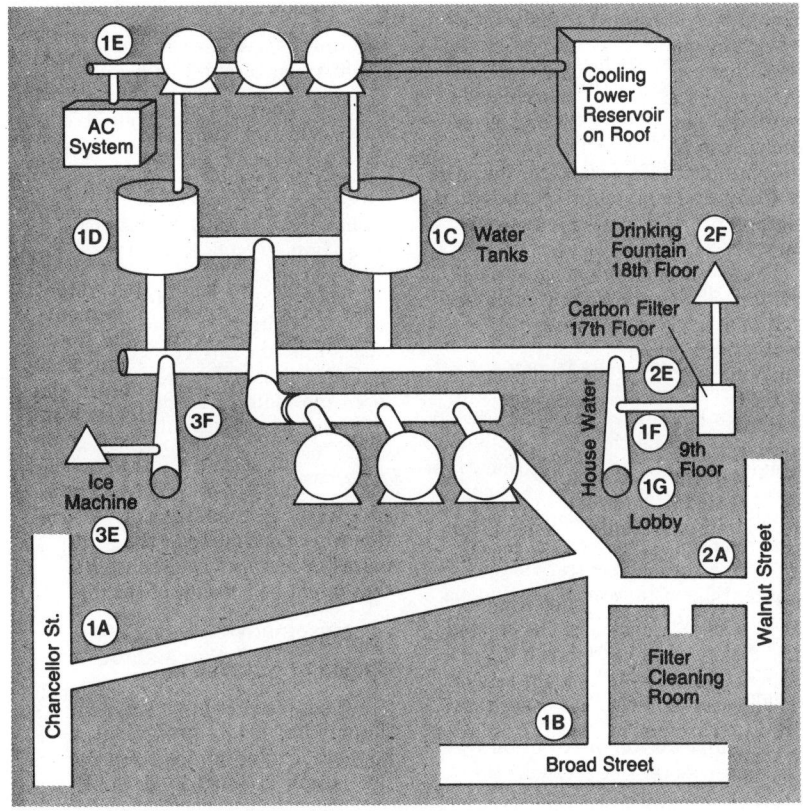

Figure 1. Schematic diagram of water system of convention center hotel

Transmission of disease occurs through a vector, e.g., insect, water, soil, air, or food. The cause (etiologic agent) and the vector of Legionnaires' disease, which occurred in Philadelphia, July 21–24, 1976, were being sought by the Center for Disease Control (CDC) (Atlanta, Ga.) and the Philadelphia Health Department. Yet, three weeks after the outbreak of the disease, no agent or vector had been found. Theories abounded!

Symptoms of Legionnaires' disease closely resembled those of pneumonia. Thus, to separate victims from background cases of pneumonia that would normally develop in a city the size of Philadelphia, CDC selected both clinical and epidemiologic characteristics. Persons were classified as victims of the disease if they possessed fever and chest x-ray evidence of pneumonia or a temperature greater than 102 °F and a cough. To further narrow the definition, they had to have attended the American Legion Convention or have visited the convention center hotel between July 1 and the onset of illness.

Statistical analysis of 182 suspected cases ultimately revealed that a typical victim became ill seven days after arriving in Philadelphia and had a temperature rise to 102–105 °F. Since these symptoms could be produced by numerous types of microorganisms and toxins, the investigation was broadened to include all possible environmental factors.

A cross-connection survey of the convention center's water system, conducted on August 8, 1976, showed many violations. Either the wastewater or the air conditioning system could have contaminated the hotel drinking water system. A cross-connection could bring toxic substances into the water supply or perhaps bring organic compounds that would react in a synergistic manner with compounds commonly found in part-per-billion quantities in drinking water. Also, possibly, an agent from a cross-connection, such as dichromate (an antifouling chemical in the air conditioning system), could change a set of trace organics in some unknown fashion into toxic substances. In addition to helping victims of Legionnaires' disease, protection of the people remaining in the hotel was also a major concern. Would there be a subsequent outbreak? Were cross-connections or toxic compounds involved? The Philadelphia Water Department, asked to investigate the environmental systems of the hotel, reviewed with health agencies their findings and learned that a thorough evaluation of the trace organics in the hotel drinking water system had not been undertaken. Thus, an effort in cooperation with Drexel University was initiated immediately to study the drinking water and the air conditioning systems.

How to study water entering and circulating in a 50-year-old hotel was a major problem. Drinking water entering the hotel could come from two treatment plants, each obtaining its water from different surface water supplies, the Schuylkill and Delaware Rivers. The two water sources then intermix within the distribution system and enter the hotel through two water mains. Where should samples be taken? How could representative samples be obtained? What isolation methods were readily available, and which should be used? How could the source of water entering the hotel be determined? How could a comparison of analyses performed at different locations be made? It should be emphasized that a comprehensive trace organic survey of any site in a complex water distribution system had never been accomplished. No protocol or standard methods existed for this type of study, and adaptation of laboratory methods to field locations was necessary.

The effect of chlorination and detention time on the trace organic content of a drinking water has only recently been studied and still is not fully understood. Data from many water treatment plants indicate that each drinking water may be unique in composition and chemical reactivity. Studies of collected samples show that even treated water is of ever-changing quality. Recent studies at one of Philadelphia's Water Treatment Plants showed a variation in trace organic content during a week of continuous composite sampling. The chlorinated material also has been shown to change with time. In fact, the rate of change may be different for each sample. Furthermore, at a specific location in a complex water distribution system, such as the one servicing the hotel, the water may be comprised of a changing mixture of river sources since water flow is controlled by consumption.

Sample Site Selection

With limited time available, the choice of sampling sites was extremely important. To ensure that no possible contamination source would be missed, sampling locations were chosen to follow the path of the water through the hotel distribution system (Figure 1) from one water main (1B) to the tanks on the 19th floor (1C, 1D) and then down through the other floors (9th and lobby, 1F and 1G). All these samples were collected simultaneously on 8/25/76. In addition, a cross-connection with the air conditioning system (1E) on the 19th floor was sampled on 8/25/76. Two special sites at the input and output water from a drinking water fountain outside of a ballroom on the 18th floor (2E, 2F) and at an ice machine on the 2nd floor (3E, 3F) were sampled along with the cooling and condenser water from the air conditioning system on 8/27/76. The second water main (1A) also was sampled on 8/27/76. These sites were representative of the location where attendees at the Legionnaires' conference could obtain water that might be a vector of the disease.

Representative Samples and Choice of Isolation Method

No sample sites are ideal, no isolation method is totally efficient, and no method of obtaining a representative sample is time tested. Consequently, one must make choices based upon the state-of-the-art at that time and the pragmatic considerations of manpower and equipment available.

The investigation reported here has enabled the development of an analytical protocol for the study of future emergency water problems.

The decision was made to utilize a composite sampling method to obtain representative samples throughout the hotel's distribution system. Eight-hour composite samples were collected during midweek working days. Macroreticular resin (MRR) XAD-2 and continuous liquid-liquid extraction (CLLE) were the isolation methods used. They have been used successfully in identification of trace organics in drinking water in Philadelphia since 1975 (1). The MRR and CLLE equipment was operational, and the manpower was experienced in its use. These isolation methods are useful for analysis of the "neutral"-type organics, boiling between 40 and 230 °C. Figure 2 shows the sampling apparatus at a location in the hotel's subbasement where a water main sample (2A) was collected.

Comparison of Samples Collected at Different Locations

A gas chromatographic "profile" method was used to compare samples (2, 3), and selected samples were run by GC/MS to identify the components of these samples. The analytical approach used to study the trace organics present in the hotel's water system is shown in Figure 3. Of prime importance is the GC profile evaluation in which the question of possible contamination may be considered. A GC profile is a fingerprint of a complex group of trace organics present in the extracted water and constitutes an information pattern of the sample. By plotting peak area or height proportional to a standard on a relative retention time scale, many GC profiles can be compared to obtain an understanding of the spatial changes in the hotel's water system. A unique GC profile computer program was used to study the difference in the trace organics between samples. It must be emphasized that differences between GC profiles can be caused by different compounds or different relative concentrations of a mixture of compounds in a sample. GC profiles are used to observe peak changes and to aid in the selection of samples to best utilize severely limited GC/MS analysis time.

A cursory evaluation of the GC profiles of the organics present through the distribution system showed it to be extremely useful in demonstrating differences between samples. When GC/MS was used to identify a compound, its identity was considered confirmed if it matched the mass spectrum of a pure reference compound and if its relative retention time on a 20% SE-30 packed column was the

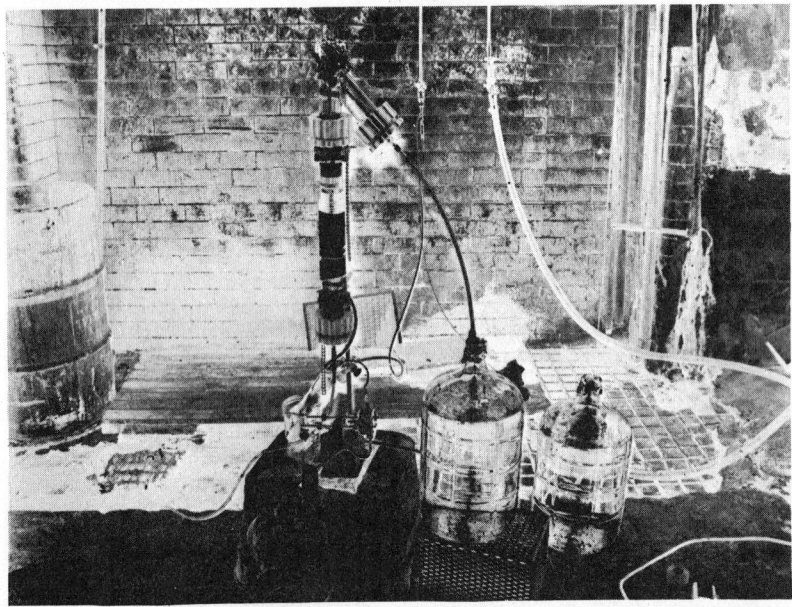

Figure 2. On-line sample collection and isolation system as operated in screen washing room located in subbasement of hotel

Walnut Street water main (2A) sample collected at this point. Continuous composite sample after dechlorination and pH adjustment fed continuously to on-line MRR-XAD-2 macroreticular resin sampler; at overflow, sample collected for subsequent continuous liquid-liquid extraction

Toxicity 159

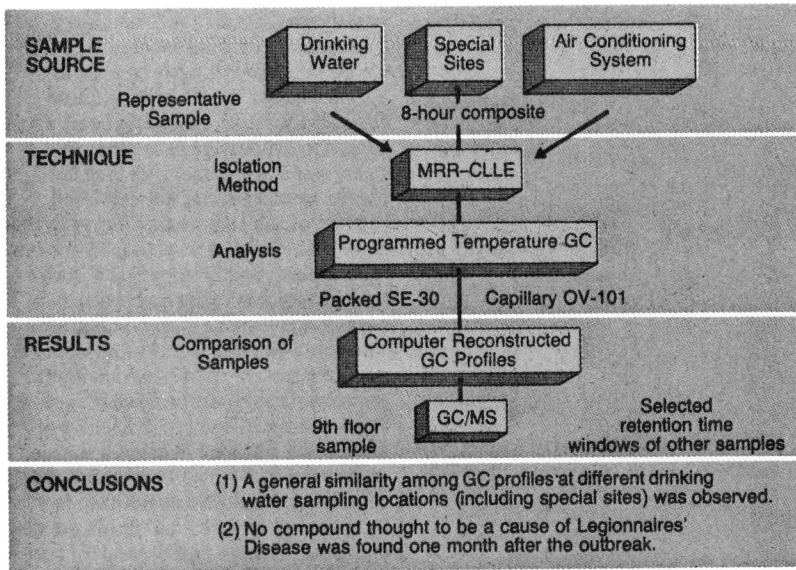

Figure 3. Analytical approach for identification of trace organics

same as the reference compound. In addition, tentative identifications were made where reference compounds were not commercially available but the mass spectrum of a peak matched uniquely characteristic literature mass spectra. When needed, computer-assisted interpretation was provided by Cornell University's Mass Spectral Identification System PBM and STIRS.

Was There a Difference Between Samples?

A detailed GC profile analysis was performed on the organics in the drinking water samples and air conditioning system. Figure 4 shows "spiked" GC profiles of samples obtained at various locations by the MRR sampling method. Each spike represents a GC peak, with the height of the peak representative of the size of the original GC peak—either small, medium, or large (relative heights on the GC peak profiles are respectively, 0.2, 0.5, and 1.0 units). A reproducibility of ±0.015 RRT units was obtained during this study.

Two phenomena were observed in drinking water samples with these spiked profiles:

• A general similarity among GC profiles of MRR samples at different drinking water sampling locations (including special sites) is obvious. The CLLE samples were very similar to the MRR samples except for concentration differences (CLLE collection volumes were around 30 L at a 10:1 water:solvent ratio, whereas MRR volumes were around 100 L).

• Minute changes could be seen less distinctly. Comparison with the original chromatograms showed that sometimes shoulders on peaks could be seen, and at other times these were obscured by large peaks. This is demonstrated more clearly by simulated 3-D profiles, with peak heights and widths of the profile actually measured (Figure 5).

The CLLE samples from the air conditioning system were the only samples to appear quite different from the drinking water system. With an equivalent sample volume, more and higher GC peaks of different RTT's were observed.

GC/MS Identification of Trace Organics in the Drinking Water and Air Conditioning System

The 9th floor samples were chosen as reference GC profiles for the hotel drinking water system since they contained the most GC peaks of the largest concentration. Thus, it was reasoned that if a compound that could have caused Legionnaires' disease was in the drinking water, it would be found in this sample or in a sample where a GC peak was missing from the profile of the 9th floor (● Figure 4).

A detailed GC/MS analysis was completed with a 6% SE-30 column on the 9th floor MRR and CLLE samples and on selected samples at RRT's where GC peaks were missing from the 9th floor MRR sample. Also, a complete GC/MS analysis was run on the MRR samples for the drinking water fountain with the carbon filter. The carbon filter had been installed on this fountain for an *unknown period of time*. Table I shows GC/MS results for the 9th floor MRR and CLLE samples. GC/MS analysis of the selected samples where GC peaks were missing from the 9th floor sample (Figure 4) showed that RRT shifts due to changes in relative concentrations

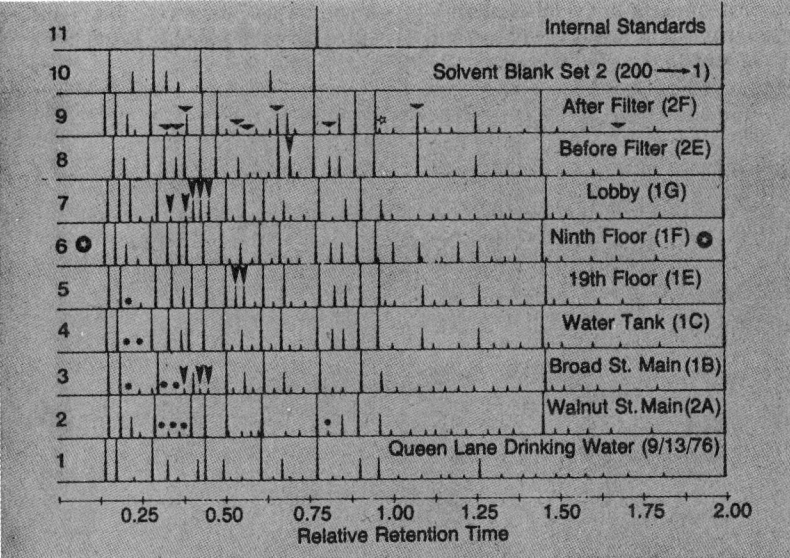

Figure 4. GC profiles on SE-30 of drinking water samples from convention center hotel

Water samples collected by MRR-XAD-2 macroreticular resin column over 8-h period on 8/23/76 and 8/27/76. Sample 1 from one of the water treatment plants that feeds water to the hotel. Sample 2 collected on 8/27/76 and listed as 2A to show it was not collected on 8/23/76. ▼ Missing in 9th floor, present in sample. ★ Missing in sample, present in 9th floor. ☆ Minor peak in carbon filter effluent, not in influent.
  Peak lowered in intensity in effluent of carbon filter as compared to influent to carbon filter. Internal standards were 2-ethyl-1-hexanol (RRT: 0.78) dibutyl phthalate (2.00). All samples eluted with 200 mL ether evaporated to 1 mL

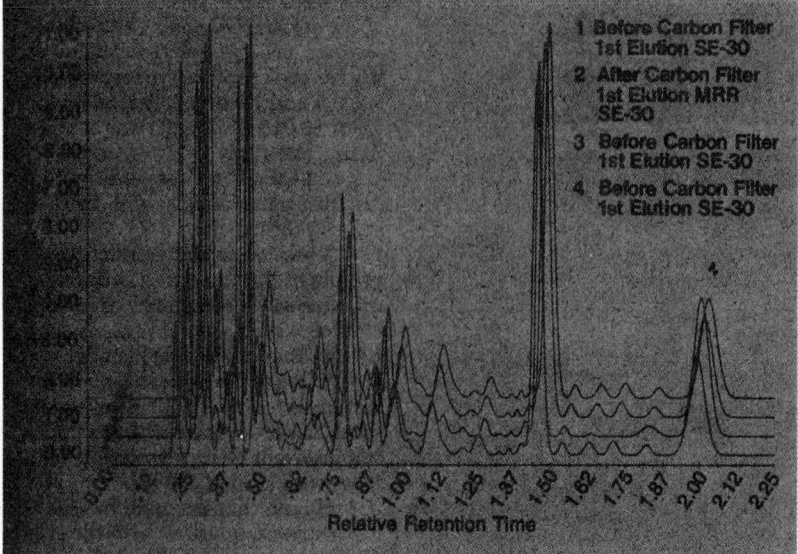

Figure 5. GC profiles of drinking water samples taken before and after a carbon filter on a drinking water fountain

Chromatograms same as sample numbers 8 and 9 from Figure 4. *Before* sample plotted as 1st, 3rd, and 4th chromatograms, respectively

of mixtures of compounds eluting as single peaks was the primary reason for differences. Benzene (19th floor) and a branched C8 alkene (19th and lobby) were the only compounds in these samples that were not found in the 9th floor sample. Two possibilities are that they were obscured by a larger peak or were added in water distribution. GC/MS of these CLLE samples added further confirmation of these components. Other CLLE samples were too dilute to allow GC/MS analysis.

GC/MS analysis of the trace organics present in the carbon filter influent showed a set of compounds which was similar to those found in the drinking water system (Table I). Two new compounds were observed: phenylacetic acid and 1,2:3,5-di-O-isopropylidene-D-xylofuranose. This may be accounted for by the fact that the influent sample to the carbon filter was collected two days after the drinking water samples. GC/MS analysis of the trace organics present in the carbon filter effluent showed the same compounds that were present in the influent except for the addition of trichloroethylene, diethyl phthalate, 2-dichlorobenzene isomers, dichloroacetonitrile, toluene, a dichloropropene isomer, and a C9 branched hydrocarbon. These compounds have previously been observed in Philadelphia's drinking water and are apparently being displaced by others that are adsorbed.

Most of the compounds found in the hotel drinking water system have been identified in drinking waters throughout the United States and are also part of the variable complement of trace organics found in Philadelphia's drinking water. Thus, it was evident that these could not be the cause of Legionnaires' disease. 1,1-Dichloroethane, a C5 alkene, methyl bromide, and phenanthrene or anthracene are compounds that had never been isolated in any other Philadelphia drinking water sample, but a search of the toxicological significance of these compounds indicated that these should not have led to the disease.

Table I shows the GC/MS analysis of the air conditioning system. Diethyl sulfate is an additional compound to the drinking water sample mixture of compounds. Since this compound was not found in any of the drinking water samples, contamination caused by a cross-connection at the time of sampling is unlikely. Once again, no obvious solution to Legionnaires' disease was found.

Capillary Column GC

The decision was made to select several samples and run capillary column GC. Its high resolving power would be used to detect minute differences between samples. Portions of

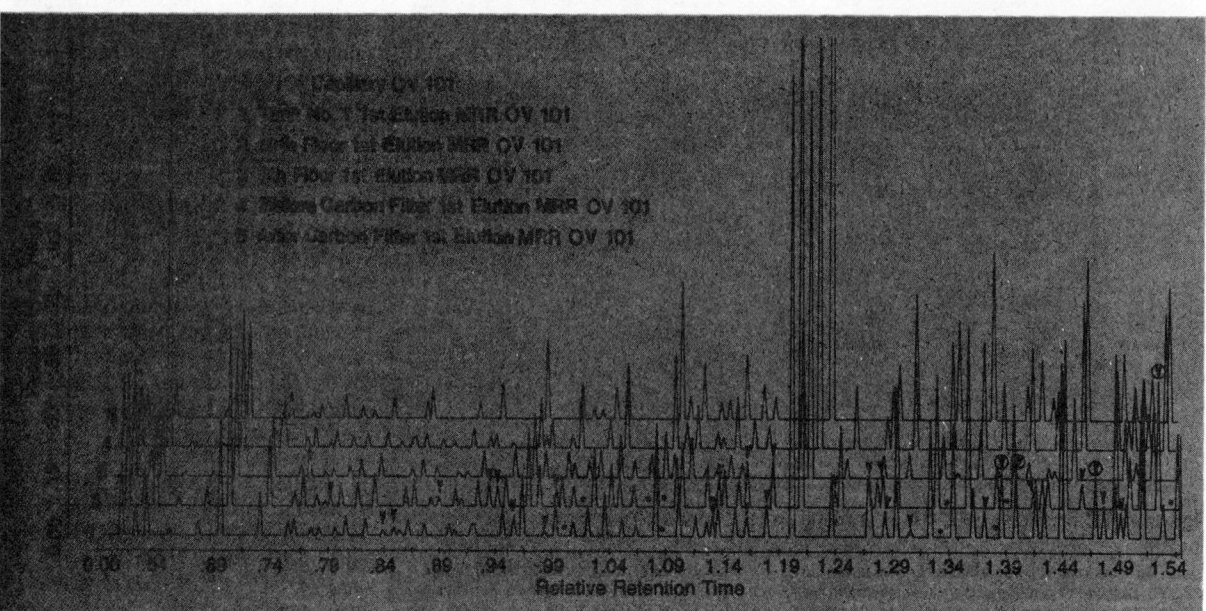

Figure 6. Portions of capillary GC profiles on OV-101 of drinking water samples from convention center hotel

These chromatograms from same samples as numbers 4, 5, 6, 8, and 9 from Figure 4. ▼ Missing in 9th floor, present in sample. ★ Missing in sample, present in 9th floor. Ⓥ If peaks are very large, they are circled

Table I. Compounds Identified by GC/MS in Bellevue-Stratford Samples Taken by Macroreticular Resin (MRR) and by Continuous Liquid-Liquid Extraction (CLLE)[a]

	drinking water 9th floor (MRR)	air conditioning system chill (CLLE)	air conditioning system condenser (CLLE)
HALOGENATED			
bromoform	X[b]	X	
carbon tetrachloride	X		
chloroform	X		
dibromochloromethane	X	X	
dichlorobromomethane	X		
dichlorobenzene isomer	X		
tetrachloroethylene	X	X	X
trichloroacetone	X[c]		
dichloropropene isomers	2[d]		
dichloroacetonitrile	X		
chloro-methyl-butene isomers	4		
1,1,1-trichloroethane		X	
1,1-dichloroethane		X	
1,2-dichloroethane	X		
methyl bromide	X		
AROMATIC			
acetophenone	X		
benzaldehyde	X[c]		X
ethyl benzene or xylene isomers	4	1	2
toluene	X		
m- or p-tolunitrile	X[c]		
dimethyl or ethyl phenol isomer	X		
dimethyl or ethyl benzaldehyde isomers[e]	2		
C5-benzene isomer	X		
phenanthrene or anthracene	X		
MISCELLANEOUS			
2,3:4,6-di-O-isopropylidene-L-sorbofuranose	X[c]		
dibutyl phthalate	X[c]		
tributyl phosphate	X[c]		
dihydroactinidiolide	X[c]		
C5 alkene	X		
C6 branched hydrocarbon	X		
C7-9 branched hydrocarbons		X	
ethanol		X	
acetone		X	
diethyl sulfate		X	
triethyl phosphate			X
TENTATIVES			
ethyl sorbate	X		
1 (2 or 4-chlorophenyl) ethanol	X		
1,1,2-trichloroethane			X

[a] All impurities found in the blank are not listed. [b] X = compound identified. [c] Confirmed by GC/MS in 9th floor CLLE. [d] Any number = number of isomers identified. [e] May also be mixed with a C4-benzene isomer.

GC profiles with true relative heights are shown in Figure 6. Approximately 125 peaks per sample were isolated by using capillary columns compared with 35 peaks on packed columns.

After a detailed examination of the capillary GC runs, it became clear that there was very little difference in the type and amount of trace organics throughout the hotel's drinking water system! Less than 10% of the approximately 125 peaks showed any noticeable variation compared with the ninth floor. Only 4 peaks on capillary GC, at RRT 1.41, 1.43, 1.49, and 1.55, were considered to be significant. Thus, our preliminary analysis with an SE-30 packed column was confirmed by capillary GC. GC profile evaluation was an effective method for examination of the trace organics throughout the hotel's drinking water system. It enabled determining similarities of samples and thus provided a screening method for subsequent GC/MS work.

The Disease Agent

In January 1977, six months after the outbreak of Legionnaires' disease, CDC reported that it had isolated a previously unidentified microorganism that is believed to be the cause of the disease. As detailed in the *New England Journal of Medicine* (4), after months of fruitless efforts, a gram-negative, nonacid-fast bacillus was isolated from lung tissues of victims by inoculating guinea pigs and then transferring their spleen suspensions to embryonated eggs. Cultivating the organism on artificial media required several more months of effort until finally a procedure was established which permitted testing of a large number of disease victims. Once confident that this bacterium was indeed the agent, CDC set out to find whether similar outbreaks of pneumonia in the past had been caused by this organism. Using serum specimens stored from previous outbreaks in 1966 at a hospital in Washington, D.C., and in 1968 in Pontiac, Mich., CDC concluded that the sera contained significant numbers of antibodies to implicate the bacterium. By November 1977 CDC reported (5) that confirmed sporadic cases of Legionnaires' disease had been found in 24 states of the United States as well as England and Spain.

The Disease Vector

Although medical authorities are confident that they have identified the agent, the mode of transmission of the disease remains a mystery. Person-to-person spread was quickly ruled out when secondary infections to families of the Philadelphia victims did not occur. Similarly, food, tobacco, and alcohol were dismissed as possible vectors. Transmission through water is unlikely, for although the epidemiologic investigation has shown that water consumption was higher than normal for victims, one-third of the cases denied drinking any water; in fact, a group never entered the hotel. The vector now believed to be the best possibility is air transmission or a soil- or dirt-borne organism. This hypothesis is predicated on the fact that many victims spent considerable time in the lobby of the hotel. A low grade or intermittent exposure of the organism may have resulted from some type of construction-related activity that would have disturbed the niche of the bacterium, causing its entry into the air environment of the hotel and its surroundings. Further work is still needed before the mode of transmission is conclusively demonstrated.

Discussion

Although the organic profile of the hotel water system did not uncover the vector for Legionnaires' disease, significant experience was gained in the development and testing of an analytical protocol to be used in responding to emergency water problems. Cross-connection incidences have been the cause of numerous gastrointestinal diseases and, in some cases, even deaths. Though in most instances microorganisms are the agents of the disease, toxic chemicals have been responsible for some occurrences. By responding promptly with thorough analytical surveys, the cause of the problem may be quickly found and corrective steps taken before serious health problems develop.

A second benefit of the investigation of the hotel water system has been the demonstration of the versatility of the isolation steps. Although designed for use in a laboratory environment, the apparatus was easily adapted to very hostile environments. These steps can be used in the field to conduct trace organic surveys of complex water distribution networks, thus obviating the need for bringing large volumes of water back to the laboratory. Finally, the role of the analytical chemist in a team of disease investigators was once again demonstrated.

Acknowledgment

The authors acknowledge the continued support of the Philadelphia Water Department for this research under the leadership of Water Commissioner Carmen F. Guarino and J. V. Radziul, chief of research. The authors thank James Coyle, Philadelphia Water Department, who coordinated the field sampling program; Lewis Brenner, Philadelphia Police Department Laboratory; and Bruce Schultz and Areta Wowk of Drexel University for GC profile work and GC/MS.

References

(1) I. H. Suffet, L. Brenner, and J. V. Radziul, "GC/MS Identification of Trace Organic Compounds in Philadelphia Waters", in "Identification and Analysis of Organic Pollutants in Water", L. H. Keith, Ed., Chap. 23, pp 375–97, Ann Arbor Science, Ann Arbor, Mich., 1976.
(2) E. R. Glaser, B. Silver, and I. H. Suffet, *J. Chromatogr. Sci.*, **15,** 22 (1977).
(3) I. H. Suffet and E. R. Glaser, *ibid.*, **16,** 12 (1978).
(4) J. E. McDade et al., *New England J. Med.*, **297,** No. 22 (Dec. 1, 1977).
(5) Center for Disease Control, "Morbidity and Mortality Report", "Sporatic Cases of Legionnaires' Disease—United States", Vol 26, No. 45, Nov. 11, 1977.

An Unknown Salivary Morpholine Metabolite

John S. Wishnok and Steven R. Tannenbaum

Massachusetts Institute of Technology, Department of Nutrition and Food Science, Cambridge, MA 02139

Originally published in ANALYTICAL CHEMISTRY, 1977, Vol. 49, No. 8.

N-nitrosodialkylamines (nitrosamines) are well-known chemical carcinogens in a large variety of animal species (1, 2) and have been implicated as potential human health hazards as well (3, 4). Nitrosamines are formed efficiently from secondary amines and nitrite via the active species, N_2O_3:

$$\underset{R}{\overset{R}{>}}NH \xrightarrow{[N_2O_3]} \underset{R}{\overset{R}{>}}N-NO$$

+ other products

This reaction occurs optimally at low pH (5) but will take place under a wide variety of conditions when amines are allowed to come into contact with nitrite (6, 7).

Secondary amines are natural constituents of many foods, and low levels of nitrite are generally present in human saliva via reduction of nitrate ion by oral microorganisms (8). Nitrite is a well-known food additive, and nitrate is found in various foods, particularly roots and leafy vegetables (9, 10). Consequently, we became interested in knowing whether or not nitrosamines could be formed under physiological conditions in the digestive system.

Preliminary experiments were carried out with morpholine, which reacted readily with saliva to form N-nitrosomorpholine:

morpholine →(Saliva) N-nitrosomorpholine

A useful analytical system for these experiments is gas chromatography/ mass spectrometry by simultaneous total-ion/single-ion detection with the single-ion monitor set at $m/e = 30$. Simple nitrosamines frequently produce a characteristic, prominent fragment (NO) at mass 30; thus, this method indicates which of the many peaks in the often complex biological extracts mass spectrum are potentially significant.

In addition to N-nitrosomorpholine, the extracts from the saliva experiments contained a number of other compounds, including one that gave an intense signal on the $m/e = 30$ mass chromatogram. Therefore, we thought it might be a new nitrosamine and placed fairly high priority on the identification of this unknown substance.

The most straightforward approach to an identification, of course, is simply to get as much pure material as possible and then obtain NMR, UV, IR, and mass spectra, which usually prove sufficient. Two major problems prevented this approach with the above saliva extract: The concentrations were low (μg/mL range), and it was impractical to run large-scale reactions since this required large amounts of fresh human saliva.

At the outset, we knew from the low-resolution mass spectrum that the molecular weight of the compound was 112, which formally corresponds to the loss of four hydrogen atoms from N-nitrosomorpholine (MW = 116). We were intrigued by the possibility that the compound might be the N-nitrosooxazine since the parent ring structure has never been synthesized.

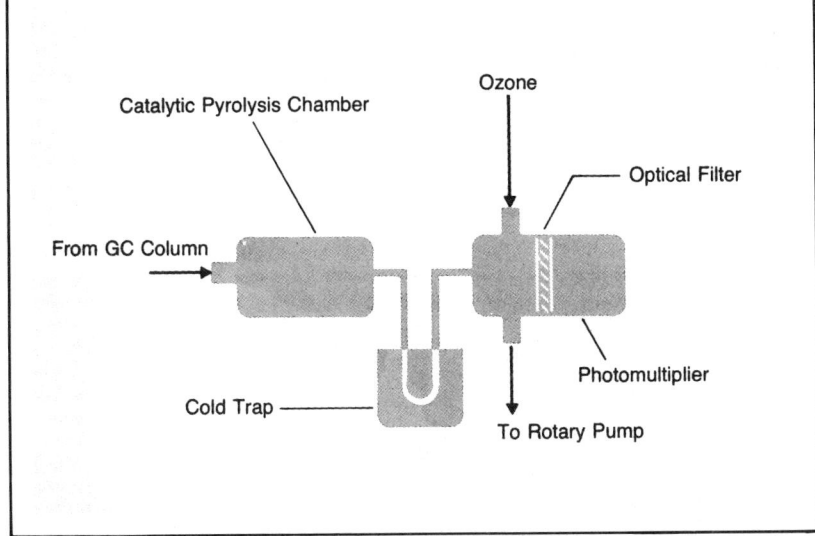

Figure 1. Schematic diagram of nitrosamine detector (TEA)

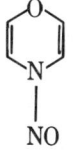

N-nitrosooxazine

This also meant, however, that direct synthesis of an authentic reference sample might be prohibitively difficult. Since an attempt to isolate a small amount of the compound by microscale preparative gas chromatography was unsuccessful, we appeared limited to information that could be obtained in conjunction with gas chromatography.

The most powerful gas chromatographic detection technique available to us at that time was high-resolution mass spectrometry, with an instrument equipped with a movable photoplate detector, at the National Cancer Institute. A concentrated saliva extract was examined at NCI by Peter Roller; the exact molecular weight of the unknown (112.0631) ruled out the nitrosooxazine structure (calculated MW = 112.0273) and indicated a molecular formula of $C_5H_8N_2O$ rather than $C_4H_4N_2O_2$. Unfortunately, a large number of structures could be drawn for this molecular formula, including simple substitution of CN at any of the three types of hydrogen on the morpholine molecule or substitution of —CH=NH at various positions on an unsaturated morpholine ring. It was not even clear that the morpholine structure had not been disrupted or that the new molecule was in fact a morpholine derivative. (Most of the new substances, e.g., phenol, found in saliva following addition of morpholine were not formed from morpholine). In addition, because all of the various possible structures seemed equally implausible from a chemical or biochemical viewpoint, there appeared to be no rational way to select a given structure for further study, e.g., independent synthesis.

Thermal Energy Analyzer

During this period we obtained a new GC detector—a thermal energy analyzer (TEA)—which is highly sensitive and specific for the N-nitroso functionality (11). This detector is based on the catalytic disruption of the N—NO bond to release molecular NO. The NO is then reacted with ozone to yield O_2 and excited NO_2^*, which decays to the ground state and emits a photon of near-infrared radiation:

$$\begin{array}{c} R_1 \\ R_2 \end{array}\!\!\!\!>\!\!NNO \xrightarrow[\Delta]{Cat.} NO$$

$$+ \text{ other fragments}$$

$$NO + O_3 \longrightarrow NO_2^* + O_2$$
$$NO_2^* \longrightarrow NO_2 + h\nu \;(0.6-2.8\,\mu)$$

The radiation is then amplified by a photomultiplier (Figure 1). Although simple in operation, the complexity of the reaction sequence assures that there will be few false positives. The TEA showed no response for the unknown compound, thus indicating that it was not a nitrosamine. Under these circumstances, we did not feel that a major effort to identify the compound could be justified, and active experiments were stopped. However, we remained interested and discussed possible structures from time to time.

The original project was gradually expanded to include the reactions of a variety of other amines in saliva, many of which yielded products with structures apparently analogous to the unknown morpholine derivative. We began to realize that structural restraints in some of these new amines could limit the possible structures of the unknown derivatives. About a year after the original observation, we carried out an experiment with diphenylamine and again observed an apparently analogous derivative:

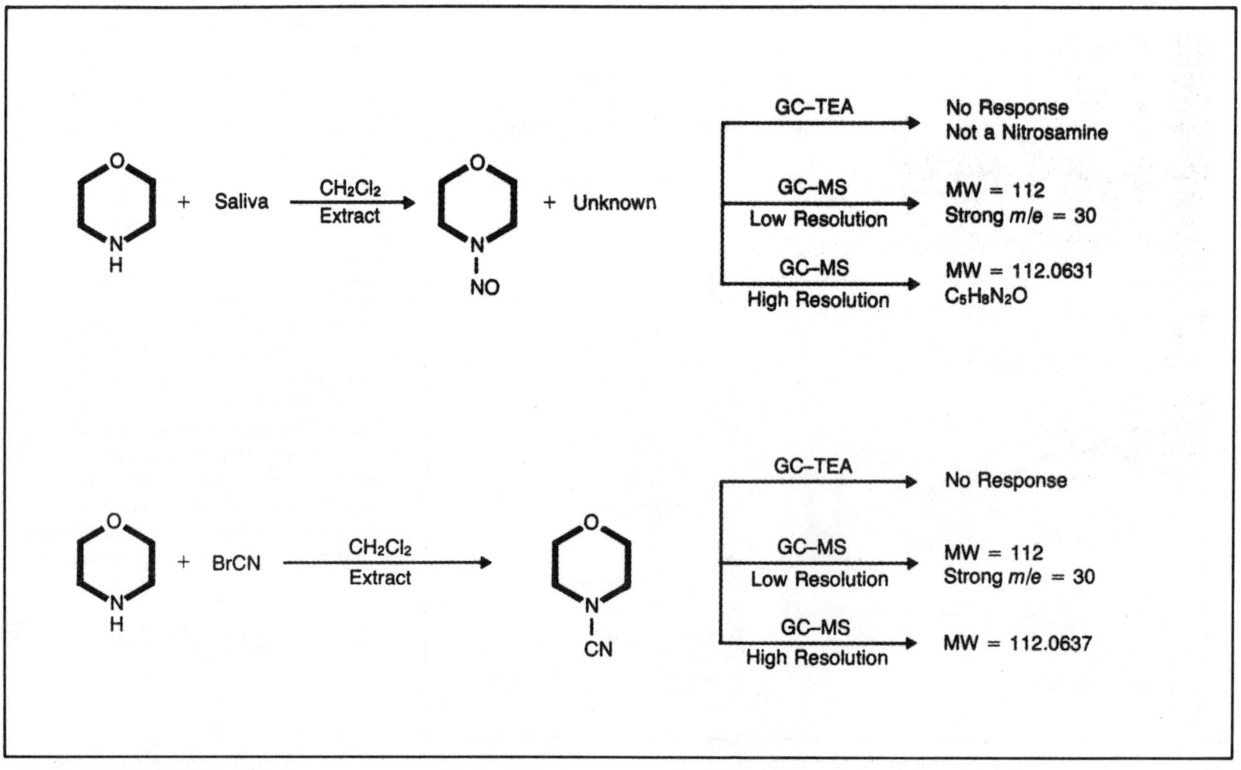

Figure 2. Analytical approach for identification of unknown morpholine metabolite

[Diphenylamine, m/e = 169] →(Saliva) [N-nitrosodiphenylamine, m/e = 198] + unknown m/e = 194

Since the only active position on diphenylamine is the amine hydrogen, the reaction could be considered formally as loss of H followed by addition of a functional group with mass = 26 (169 − 1 = 168; 194 − 168 = 26).

The most obvious identity for a moiety with $m = 26$ is CN, suggesting the cyanamide structure (**I**)

[Diphenylcyanamide structure]
I

for the diphenylamine derivative and (**II**)

[Morpholine with CN on N]
II

for the morpholine derivative.

Synthesis of cyanamides from secondary amines by reaction with cyanogen bromide is well known, and an authentic reference sample of Compound **II** was prepared and found to be indistinguishable (identical low-resolution and high-resolution mass spectrum and GC retention times) from the saliva metabolite:

[Morpholine] →(BrCN) [Morpholinocyanamide] ←(Saliva) [Morpholine]

The approach taken for the identification is shown in Figure 2.

The formation of cyanamides from secondary amines in saliva is apparently fairly general, and the identification of this compound thus effectively constituted the discovery of a new metabolic pathway for secondary amines.

The morpholinocyanamide was evaluated for biological activity and found to be nonmutagenic and moderately toxic. Although the use of mutagenicity as an indicator of carcinogenicity is controversial, it does appear to be generally true that nonmutagenic compounds are probably noncarcinogenic; therefore, we are no longer actively investigating the formation of cyanamides from secondary amines. We are, however, investigating the possibility that analogous transformations of primary amines may lead to the production of carcinogenic substances in human saliva.

References

(1) H. Druckrey, R. Preussmann, S. Ivankovic, and D. Schmähl, Z. Krebsforsch., **69**, 103 (1967).
(2) P. N. Magee and J. M. Barnes, Adv. Cancer Res., **10**, 163 (1967).
(3) P. N. Magee, Food Cosmet. Toxicol., **9**, 207 (1971).
(4) R. A. Scanlan, Crit. Rev. Food Technol., **5**, 357 (1975).
(5) S. S. Mirvish, Toxicol. Appl. Pharmacol., **31**, 325 (1975).
(6) T. Y. Fan and S. R. Tannenbaum, J. Agric. Food Chem., **21**, 967 (1973).
(7) S. R. Tannenbaum and T. Y. Fan, Proc. Meat Industry Res. Conf., Chicago, Ill., 1973.
(8) S. R. Tannenbaum, A. J. Sinskey, M. Weisman, and W. Bishop, J. Nat. Cancer Inst., **53**, 79 (1974).
(9) W. E. Phillips, J. Agric. Food Chem., **16**, 88 (1968).
(10) E. C. Hersler, J. Siciliano, S. Krulick, W. L. Porter, and J. W. White, Jr., ibid., **21**, 970 (1973).
(11) D. H. Fine, H. Rufeh, and B. Gunther, Anal. Lett., **6**, 731 (1973).

Forensic

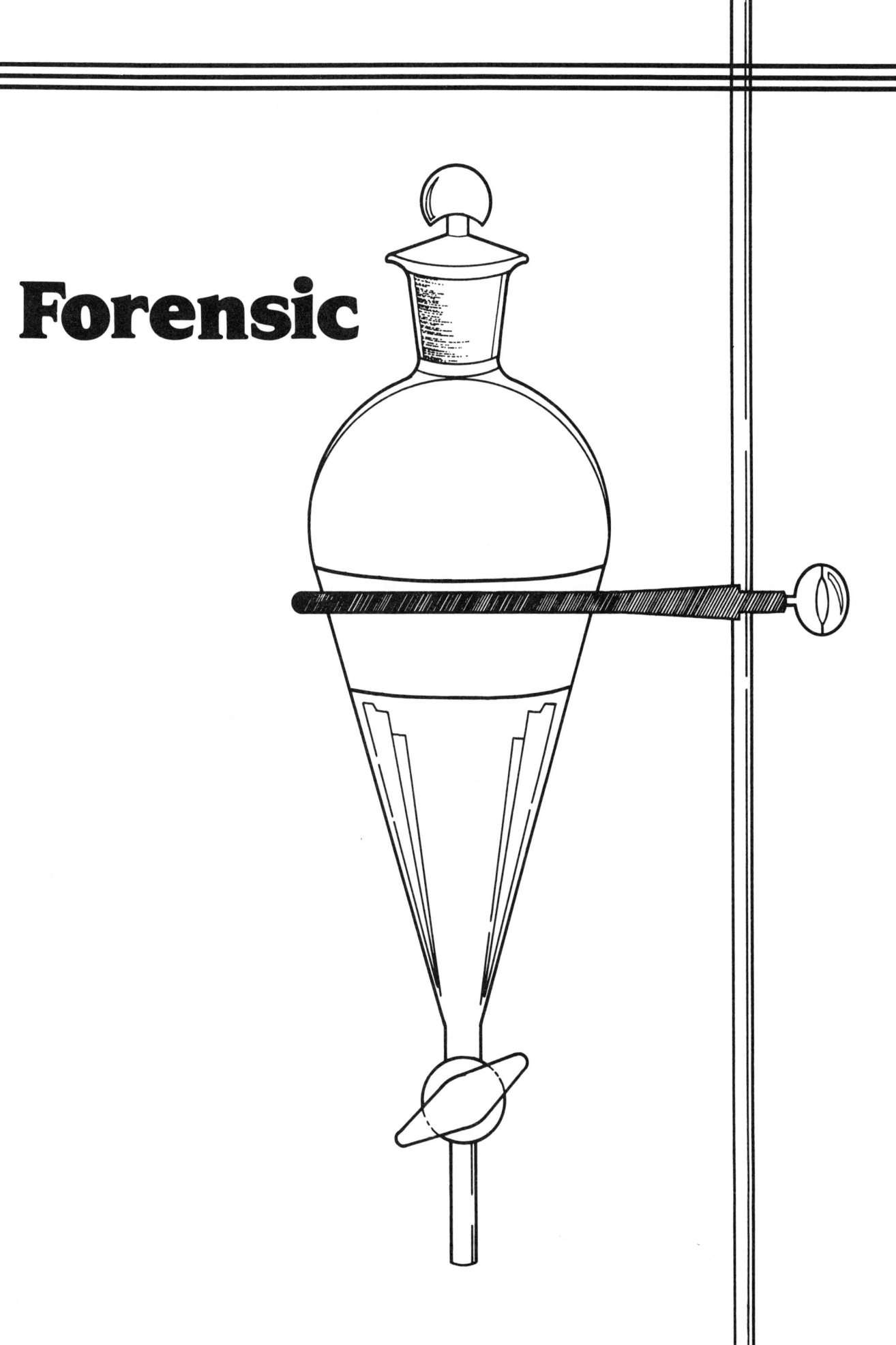

The forensic chemist is the ultimate analytical detective. In most instances he is involved from the beginning to the end of a crime. Evidence gathering, analysis of the clues, and interpretation of the data in a court of law are all parts of the job. The evidence presented to the jury as a result of the work of the forensic scientist must be convincing "without a reasonable doubt." This requires care in sampling, technical expertise, and a well-designed analytical plan.

"**Drugs of Abuse Detected in Hair by Radioimmunoassay**" discusses the detection of heroin, morphine metabolites, and PCP or "angel dust" in hair. It has been known for some time that hair may serve as an indicator of drug use if more specific and sensitive analytical procedures could be developed. Radioimmunoassay offers this possibility. Unlike the situation in body fluids, the drug levels in hair remain stable over long time periods, and it is possible to ascertain the approximate time of drug use from an analysis of the hair growth. This analytical method offers the possibility of being extended further to medicinal drugs.

"**FBI investigates Analytical Chemistry**" spotlights the scientific analysis section of the Agency in a series of complex case studies. The type of bullet and approximate distance from the firer to the target can be determined from analysis of a bullet hole in clothing. It is also possible to determine the sex of an individual from blood stains or from a single human hair. The sophisticated work of the Analysis Section will continue to improve the FBI's efficiency and range of investigative scope.

In "**The Case of the Great Yellow Cake Caper**," analytical chemists were able to develop sufficient evidence on the source of stolen uranium yellow cake to force a confession from the perpetrators. Detailed analyses of trace metals and processing chemicals from samples taken from six different production sources were required to stop this theft.

Arson is becoming more prominent as a crime. "**Analytical Techniques in Arson Investigation**" describes the forensic tools used by investigators to locate the source of a fire and the igniting material from evidence obtained at the site. Headspace chromatography of vapors from materials collected at the suspected fire source has been proved to be extremely powerful. Combined with GC/MS analysis, headspace chromatography should cause concern for any arsonist.

Legal manufacture of a controlled drug requires registry with the Drug Enforcement Administration and adherence to strict guidelines of the Controlled Substances Act. Illegal laboratories are the subject of "**Investigations of Clandestine Drug Manufacturing Laboratories.**" In one recent year, 221 clandestine laboratories were seized. Forensic chemists play a major role in all phases of these actions, including investigation, analysis of evidence, and prosecution. They work mainly at laboratory analytical work, with intermittent appearances as expert witnesses. They are present from seizure—to obtain samples of raw materials and products—to trial. A case history is discussed to offer background on the complexities of a case.

The forensic chemist, "**Analytical Detective**," has a difficult task. Society demands that samples be analyzed prior to accusation and prosecution of any violation of law. The forensic chemist must be both swift and completely reliable in his analyses. Two drug cases are cited, one in which a drug was shown to be harmless and one in which what appeared to be a drug overdose became a homicide investigation.

Conspiracy theories are highly important in American history. In "**JFK Assassination: Bullet Analyses**," the author applies a modern analytical technique (instrumental neutron activation analysis) to the fragments of bullets recovered in the JFK assassination case. In this fascinating study, he concluded to the Senate Select Committee reviewing the case in 1978 that all bullet fragments were from a specific 6.5 mm brand said to be owned by Oswald, that fragments from only two bullets were found, and that the bullet found on Governor Connally's stretcher was indeed the one that fractured his wrist. These results are all consistent with the original Warren Commission findings.

When a surgeon was accused by other doctors of murder by curare injection, a chain of events was set off ending in an acquittal after the second longest criminal trial in U.S. history. "**Detection of Curare in the Jascalevich Murder Trial**" outlines both the technical difficulties in analyzing for trace quantities of material in a body enbalmed and interred for a decade and the complexities of presenting highly sophisticated analytical data to a lay jury. In particular, the admissibility of data obtained from novel methods was an issue. Even after a decision allows admission of data, how does a jury evaluate conflicting evidence from experts? The jury acquitted the doctor in two hours, following a seven-and-a-half month trial.

"**Behind the Identification of China White**" is the story of a month-long search for the identity of a substitute heroin that had been implicated in several deaths. An analytical team conducted a painstaking analysis and verification of this very toxic material using a combination of spectroscopic techniques for identification. The irony is that after a month of concentrated effort the press wondered "why did it take so long to come up with an answer?" (á la Quincy).

Nonprofit research institutes often act as referee or corroborator witnesses in law suits because they have no vested interest in any private industry. In "**The Analytical Chemist as a Third Party Fact Finder**," the author presents three examples in which his institute effectively developed and presented data important to resolving significant problems through the use of analytical chemistry. A general analytical approach to acting as a referee or corroborator is presented.

Drugs of Abuse Detected in Hair by Radioimmunoassay

Stuart A. Borman

American Chemical Society, Washington, DC 20036

Originally published in ANALYTICAL CHEMISTRY, 1980, Vol. 52, No. 1.

A group of California researchers reported on detection of drugs in hair samples at the 1979 Pacific Conference on Chemistry and Spectroscopy held in Pasadena, Calif., Oct. 10–12. The group, consisting of Annette M. Baumgartner and Peter F. Jones of The Aerospace Corp., El Segundo, Calif., and Werner Baumgartner and Charles T. Black of the Veterans Administration Wadsworth Medical Center in West Los Angeles, had already been successful at detecting heroin and morphine metabolites in hair. Their Pasadena presentation reported the application of their methodology to the determination of phencyclidine, better known as PCP or "angel dust," in hair.

The idea is not a new one. As early as 1954 it was reported that barbiturate residues had been found in the hair of guinea pigs one month after administration of the drug. However, the methods of extraction and analysis and the large amount of sample required were unsuitable for routine analysis. It was evident that hair might serve as an indicator of drug use if more sensitive and specific analytical procedures could be developed.

Such a procedure was the radioimmunoassay (RIA), developed by Rosalyn S. Yalow and Solomon A. Berson in the 1950's. The Baumgartners, Jones, and Black carried out their analyses on methanol extracts of dried, crushed human hair by adapting commercial RIA kits designed for the determination of PCP and heroin and morphine metabolites.

To analyze PCP, for example, the researchers add 0.1 mL of the methanol extract (equivalent to approximately one human hair) to a test tube containing ^{125}I-labeled PCP. The labeled drug and the unlabeled drug are then allowed to compete for binding sites on PCP antibodies in the commercial RIA reagent. After an incubation period of 20 minutes, the bound antigen is precipitated and removed from solution. Free antigen in the supernatant is then counted with a gamma counter.

Drug abuse is presently monitored by analysis of body fluids such as serum and urine. But although drugs are rapidly cleared from body fluids, levels in hair are stable over long time periods. In addition, the researchers have had excellent results in determining heroin and morphine metabolites in sectioned hair, enabling them to ascertain the approximate time of drug use since the drugs are apparently detectable only in hair grown during or immediately after drug use. The sensitivity of the RIA method is indicated by the detection of a signal in the hair of an individual who reportedly had smoked only five PCP "joints" over a six-month period.

One of the technique's other applications may involve the assigning of hair to a given individual by drug profiling. Dr. Jones explains that if a murder has been committed and a hair found on the suspect resembles that of the victim, analysis of the hair could be performed for a more positive identification.

The Aerospace-VA group reports greatest interest in the new technique among probation officers, who anticipate determining not only drug use, but history of drug use in their parolees. But the Baumgartners, Jones, and Black are now working to extend the method to medicinal drugs. In addition, they are investigating the effects of hair treatments of various types on drug stability, and the variations from one part of the body to another and from one individual to another.

FBI Investigates Analytical Chemistry

Stuart A. Borman

American Chemical Society, Washington, DC 20036

Originally published in ANALYTICAL CHEMISTRY, 1981, Vol. 53, No. 3.

At the Federal Bureau of Investigation (FBI), analytical chemistry is being enlisted in the fight against crime. When most people think of the FBI Laboratory, they think of handwriting analysis, comparisons of typewriter lettering, and ultraviolet scans of bad checks. But the FBI also has a scientific analysis section, with responsibility in such areas as toxicology, explosives analysis, instrumental analysis, and research. In the Bureau's research unit, FBI scientists are involved in projects such as gunpowder analysis by gas chromatography/mass spectrometry (GC/MS), determinations of sexual identity from blood stains, differentiation between persons on the basis of variations in characteristic enzymes and antigens, and the determination of blood types from human hair samples, among others.

FBI research facilities are presently located in the Hoover Building in Washington, D.C. (above), but will soon be relocated to a new facility in Quantico, Va.

Gunpowder

Some months ago FBI research chemist Dennis Hardy was asked if he could develop a method to determine trace organic gunpowder components on garments involved in shooting incidents. When gunpowder is manufactured, three or four organic compounds are usually mixed into each manufacturer's formulation. These three or four compounds, added to the gunpowder, for example, to change the speed of burning, are selected from among perhaps 30–40 possible additives. Hardy's task was to find out if these organic compounds were deposited on target garments. If so, could they be determined by analytical methods, and could the analytical results then be used to associate a particular bullet hole with a particular weapon?

Hardy found that he was indeed able to extract the organic additives from around a bullet hole into acetone and could determine the components by GC/MS. "As it turned out," explains Hardy, "not only can most of these organic compounds be extracted from around the bullet hole, but the ratios of the various trace components appear to be preserved from the ratios of unburned trace organics in unfired bullets."

Taking the investigation one step further, Hardy also tried extracting spent bullet cartridges for the organic additives. Again, he detected the trace components, and again the ratios held. "We could relate a bullet hole with a spent cartridge and with powder from a bullet that was not fired," Hardy says.

Another facet of his research on gunpowder involved determinations of firing distance. "You would think if I fired a gun a foot away from the garment, more of these organic compounds would be deposited on the garment than if I fired from three feet away," Hardy explains. Indeed, that also turned out to be the case, under laboratory conditions. Hardy found that there were statistical limits within which he could draw conclusions about absolute firing distance, and he plans to publish his findings soon. Fortunately, there is no detectable time-dependence to the effect. The organic compounds apparently adhere strongly enough that Hardy obtains the same analytical results, whether the area around the bullet hole is extracted immediately after firing or after a delay of several weeks. "If I extract a 10-cm diameter cloth sample centered around the bullet hole," says Hardy, "I can tell you how far away that gun was fired, up to six feet away to within a six-inch tolerance, based on the mass spectral ion abundance ratios of these organic components."

Sex from Bloodstains

FBI research chemist Barry Brown has been busy with a different problem—determining sexual identity from bloodstains. The sex of an individual can be determined by examining the stained chromosomes of nucleated cells. The nuclei of the white blood cells, for example, contain either two X chromosomes (female) or one X and one Y chromosome (male). This technique, however, has limited application to the forensic sciences, since the white blood cells lose their integrity as blood dries out. The results of chromosome staining are thus quite time-dependent and often inconclusive. Red blood cells cannot be used in this test, since they do not have nuclei.

Brown devised a combination of column chromatography and radioimmunoassay (RIA) to measure steroid levels in the bloodstain, as an alternative to chromosome staining. Brown tests for three steroids in particular: testosterone, considered a male hormone, but also produced in the female in smaller quantities; and progesterone and estradiol, considered female hormones, but present in smaller amounts in males. Ratios of one steroid to another are utilized for the sex determinations, since the volume of blood that went into a bloodstain is almost impossible to determine due to differences in thicknesses and absorbances of various materials. Thus, it is not enough to determine absolute steroid levels in the stains, since the volume figures that could turn such data into concentration levels are not known.

Because the bloodstain volumes are unknown, Brown cuts a stain out, extracts it with an organic solvent, and uses column chromatography to separate the three steroids of interest from one another, and also from contaminants. He then uses RIA to quantitate the collected steroids, and divides one value by another to get ratios.

In one case Brown looked at testosterone/progesterone ratios in bloodstains from 112 females and 34 males. There was quite a difference between the male and female ratios, the average values being 3.5 (male) and 0.37 (female). Similar results were obtained by ratioing testosterone to estradiol. Brown even detected male/female differences in the progesterone/estradiol ratio, but here the variations were too great to reliably predict sexual identity in individual cases. This was partly due to the high variability of progesterone levels during the menstrual cycle and during pregnancy.

Brown's most conservative estimate is that he can predict sex from a bloodstain in probably 70% of the cases with the new method he developed. Less conservatively, using a range of one standard deviation from average values, Brown was able to predict 83% of the females and 91% of the males in one real sample. And there were no errors—the others were simply listed as "too close to call."

Now that Brown has completed the methods development and testing, the next step is to apply the method to a real case and testify on it in court. It can take years from the point a new method is first conceived until it is established firmly enough to be used in court testimony.

But Brown is already working on his next idea—developing an antibody to the Y chromosome. "If we had that," he says, "we could run an antibody–antigen reaction to test for the presence of the Y chromosome." As hybridoma technology for the production of monoclonal antibodies comes of age in the next few years, such procedures may become much more popular.

Polymorphic Enzymes

FBI research chemist Paul Mied is interested in other information that can be obtained from the analysis of bloodstains. A number of electrophoretic techniques have been developed for the separation of polymorphic enzymes in blood. Red blood cells contain many different enzymes, and some of these enzymes are polymorphic—that is, they appear in different forms in different individuals.

To get evidentiary information from a bloodstain, the sample is electrophoresed to separate the component enzyme systems. The pattern formed by a particular enzyme is characteristic of the polymorphic form of that enzyme found in that bloodstain, and, indeed, in that individual.

For instance, one enzyme might exist in three forms (phenotypes) and another might take five forms, in different individuals. If a bloodstain were analyzed for, say, 10 polymorphic enzymes, the product of the probabilities of occurrence of each of the enzyme phenotypes would type the individual involved very specifically. He or she might be one in 10 000, or even one in 500 000, based on his or her enzyme phenotypes.

Such a specific characterization can obviously be very useful in a criminal case. A particular suspect, for instance, could be associated with a bloodstain at the scene of a crime with an extremely high probability. In another case, it might be determined that a suspect's blood was definitely *not* present at the scene of the crime.

Mied has been studying the application of higher resolution electrophoretic techniques to the determination of these blood enzyme phenotypes. Phosphoglucomutase (PGM) is one such enzyme found in red blood cells. With conventional starch gel electrophoresis, the procedure commonly used, three phenotypes of PGM can be distinguished. But with the more sensitive isoelectric focusing, 10 PGM phenotypes can be resolved. In starch gel electrophoresis, the different enzymes migrate on the basis of their net charge in a buffer. This procedure tends to produce broad diffuse bands wherein overlapping of phenotypes may occur. Isoelectric focusing, on the other hand, separates the molecules into very narrow bands on the basis of their isoelectric points, which are characteristic of the specific amino acid composition of each enzyme molecule. So isoelectric focusing provides higher resolution than starch gel electrophoresis.

The problem is, isoelectric focusing is more expensive and more difficult

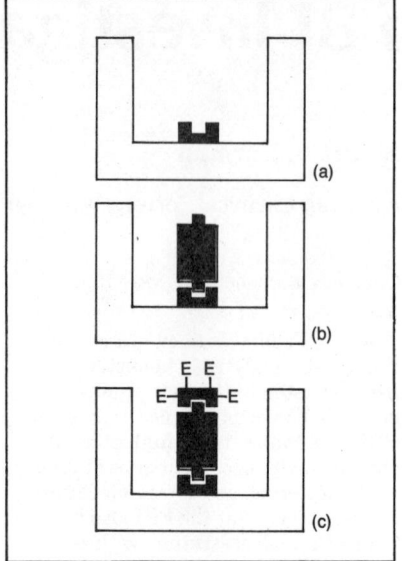

This ELISA method for detection and measurement of antigen is called the double antibody sandwich method. Antibody is immobilized on polystyrene (a). After wash, test solution containing antigen is added (b). After second wash, enzyme-labeled specific antibody is added (c). After final wash, enzyme substrate is added. Amount of hydrolysis catalyzed by enzyme indicates amount of antigen in sample

to perform than starch gel electrophoresis. So the disadvantages of isoelectric focusing still outweigh its advantages for most polymorphic enzyme determinations. But Mied is investigating situations in which the additional expense and difficulty of isoelectric focusing may be worth it. For example, isoelectric focusing improves the limit of detection for various enzymes. This may be an important advantage with aged stains, since enzymes can seriously degrade with time and with poor storage conditions.

ELISA for Antigens

Another project Mied is involved in is the development of new assays for human leukocyte antigens, or HLAs. HLAs are found on the surfaces of all nucleated cells in the body, and are often referred to as histocompatibility antigens. These are the antigens that cause rejection of foreign tissue in operations such as kidney transplants. As in the case of the polymorphic enzymes, the HLA phenotype is highly characteristic of a particular individual, and thus of potential value as forensic evidence. Unfortunately, the immunological tests commonly used to determine an individual's HLA phenotype are time-consuming and laborious. So Mied is in the process of adapting the enzyme-linked immunosorbent assay (ELISA) to HLA determinations.

In one form of ELISA Mied has devised (figure), antibody to a particular antigen being sought is immobilized on a polystyrene tube. If HLA-A_1 is being sought, for instance, anti-HLA-A_1 is first immobilized. The bloodstain is extracted, and the extract is added to the polystyrene tube. If A_1 antigen is present in the bloodstain, it will bind to the immobilized anti-HLA-A_1. In the next step, a different antibody labeled with an enzyme, is added to the tube. When that enzyme's substrate is added in the final step, the extent of enzyme-catalyzed reaction (such as hydrolysis) that occurs indicates the amount of antigen originally present in the bloodstain.

A number of other projects are under way at the FBI laboratories, including determination of blood type from single human hairs and ion chromatography for the identification of explosives. Research chemist James Mudd is busy adapting various immunological techniques to the determination of blood group antigens in hair. And researcher Dennis Reutter has been working on the use of ion chromatography to identify dynamites and explosive residues. "Dynamite residues have been done here by boiling down an aqueous solution of the residue to dryness and doing X-ray diffraction," Reutter explains. "But there are problems with this. The samples are often hygroscopic, and they contain predominantly low atomic number elements like hydrogen, carbon, and nitrogen, which don't show up well in X-ray diffractometry. Ion chromatography is much faster, more sensitive, and devoid of artifact problems associated with drying the sample."

Besides innovative research for the development of new methods, the FBI can also handle some of the most sophisticated instrumental work. Whether it be ion microprobe analysis, scanning electron microscopy, or neutron activation analysis, the FBI can handle the job.

The FBI laboratories are currently located in Washington, D.C., but Bureau researchers are looking forward to relocation to a new facility, the $8 million Forensic Science Research and Training Center in Quantico, Va., which is scheduled to open in April or May of this year. At Quantico, opportunities will be available for scientists from academia, industry, and from other government agencies to work with FBI researchers on projects of mutual interest. With its expanded research center at Quantico, the FBI will no doubt remain preeminent in the effort to successfully adapt analytical chemistry to the fight against crime.

Stuart A. Borman

The Case of the Great Yellow Cake Caper

P. A. Budinger, T. L. Drenski, A. W. Varnes, and J. R. Mooney
Standard Oil Co., Cleveland, OH 44128

Originally published in ANALYTICAL CHEMISTRY, 1980, Vol. 52, No. 8.

In the fall of 1978, five barrels of yellow cake, a complex salt of probable composition $(NH_4)_2[(UO_2)_2SO_4(OH)_4(H_2O)_n]$ (1), valued at $200 000, were stolen from the Standard Oil Co. of Ohio's (Sohio) uranium mill facility in New Mexico. The Federal Bureau of Investigation (FBI) tracked down and arrested two men, both Sohio employees, and another person, commonly known as the "fence," in El Paso, Tex., and the stolen yellow cake was recovered. The research department of Sohio was requested to determine scientifically to whom the yellow cake belonged. Samples of the stolen yellow cake were acquired from the FBI. Other samples were collected from Sohio's mill and neighboring mills.

Trace Metals Analysis

The integrated analytical approach obviously promised to provide the most support for proof of ownership. A trace metal analysis, which might act as a "fingerprint" for the stolen yellow cake, was expected to furnish the strongest evidence.

There were two possible sources of unique trace metal distribution in samples from various mills.
- The ores could be mined from different formations with different trace metal distributions.
- Processing could affect the amount of residual metals.

Fortunately, the Sohio operation uses ammonium hydroxide to precipitate uranium oxide from an organophosphorus extract of the ore acid leach. Most other mills use sodium hydroxide. Ammonia may form soluble complexes with several metal ions, thereby depleting them from the solid uranium oxide. Furthermore, sodium may be coprecipitated when sodium hydroxide is used to neutralize the acid leachate.

Because speed was essential, inductively coupled plasma (ICP) emission spectroscopy seemed ideally suited for the task. Our instrument, a Jarrell-Ash 975 AtomComp, was one of the first ones manufactured, and was acquired three-and-a-half years ago. ICPs are capable of simultaneously analyzing up to 40 elements, with an analysis time of 1 min per sample. At present, our instrument is equipped to handle 28 elements. It has detection limits of sub-ppm for most elements, and usually only a small amount of sample is required.

The principle behind ICP is as follows. A radio frequency generator provides energy to a plasma torch and creates an rf field. A stream of argon gas passes through the field and is ionized to form the plasma (about 10 000 °K). A liquid sample directed into the plasma emits radiation which is focused onto a grating. The grating diffracts the emitted light through precisely aligned exit slits to photomultiplier tubes, which convert the light to electrical energy proportional to the intensity of the spectral lines. These electrical pulses are fed into a computer, which processes the data and prints out the concentrations of the elements.

The yellow cake sample preparation involved an extraction procedure to remove the uranium from the samples in order to eliminate spectral interferences. The method used was essentially the one described by Ward and Marciello in the Jarrell Ash Plasma Newsletter (2). The sample (0.5 g) was digested on a hot plate with acid, and the acidified sample then extracted with tri-n-butyl phosphate to remove the uranium. The aqueous phase was combined with acid washings of the TBP phase. It was then evaporated to near dryness and diluted with acid and water.

The samples were examined for 22 elements. Data from the various samples are recorded in Table I. The ranges reported are the spread of results from two to five samples of the given material. Rigorous null-hypothesis testing was not performed, but a qualitative approach did reveal significant differences.

The various columns were compared and the results are shown in Table II. A "yes" means that the extremes of the spread for the suspect sample either included or were included by the extremes of the subject material. "Maybe" means that one extreme of the suspect range fell in the range of the known sample. "No" means that the ranges did not overlap.

As can be seen from Table II, all metals except magnesium were found in the "yes" or "maybe" category for the Sohio yellow cake. This result was a strong indication that the two materials had a common origin, but it was not proof, because it could be claimed that any yellow cake would yield good agreement, especially because the ranges exhibited for the five barrels of purloined yellow cake were quite broad for most elements.

Several samples of yellow cake produced from competitors' ore by Sohio's process were also analyzed. Much to our consternation, we found almost the same level of agreement for the competitors' material as for our own! The United Nuclear material agreed even better than our own. Thus, our fervent prayer was that the processing variables would be more significant than the origin of the yellow cake.

The rest of the data in Table II show that, indeed, there were significant differences when the material was subjected to competitor processes. All four samples gave negative results for one-half to two-thirds of the elements we determined. Boron, iron, and nickel tended to be lower than the suspect, while calcium, magnesium, and manganese were higher than the stolen material. These results provided more convincing evidence that there were meaningful differences between Sohio's yellow cake and that from other mills.

It should be noted that the thieves' modus operandi contributed to the spread in analytical results. The miscreants took advantage of an oversight in Sohio's accounting practices. The mill runs a batch operation, and at the end of the day, only completely filled barrels were accounted for. The residue should have been combined with the next day's run, but instead was diverted by the suspects. Partial drums were combined, so the resultant mixture might be expected to show a somewhat broader range of trace elements than would material from a single batch.

The key element for establishing a clear difference between the materials was sodium. As noted earlier, competitors were alleged to use a sodium hydroxide neutralization, rather than ammonium hydroxide. Thus, coprecipitation might result in higher sodium in competitors' products. Sure enough, Kerr-McGee, Anaconda, and United Nuclear processed samples were two to six times higher in sodium than the suspect. Unfortunately, United Nuclear Church Rock material showed the same sodium level as material from the Sohio process and the suspect material. It was with much relief that we were informed that Church Rock utilizes ammonium hydroxide neutralization.

Organics Analysis

Concurrently, as the trace metals examination was proceeding, another analysis was in progress to detect and identify any unique organics that the stolen and the Sohio yellow cake might contain. It happened that at the time of the theft Sohio used tridecanol during the processing of the ore. Tridecanol is used as an additive to increase miscibility of an organo-phosphorus solvent that is used to extract uranium from the acid leach of the ore. Other mining concerns use isodecanol. It was decided an attempt should be made to detect any traces of the tridecanol, even though a 1600 °F calcining step for water removal was involved after its use, as well as any other unique organics that might have been introduced in some way after the calcining step. The analysis was not done on yellow cake from the neighboring mills because of lack of time and not enough available sample.

Table I. Trace Elemental Analysis of Yellow Cake (ppm)

	Sohio	Stolen	UN[1] Sohio process	ANA[1] Sohio process	KM[2]	UN[2]	ANA[2]	UNCR[2] Ammonia process	Detection limit
Aluminum	1400–1600	800–1400	400–2200	400–1000	1000–1300	750–1100	450–600	500–600	2
Arsenic	80–200	60–100	30–70	50–200	40–60	70–80	45–50	30–35	10
Boron	30–80	20–27	20–40	30–80	5–15	9–13	10–15	<1	1
Barium	4–11	3–16	2–4	3–6	15–25	100–130	3–4	60–65	0.4
Beryllium	2–9	1.6–2.6	1–4	2–8	0.3–1	0.8–1.0	0.1–0.3	0.4–0.5	0.4
Calcium	130–210	100–250	80–130	170–240	480–600	1300–1600	2000–2500	300–450	0.1
Cadmium	5–20	3–5	4–8	4–15	1–3	4–5	1–2	1.5–2.5	0.8
Cobalt	<15–30	<15	<15	0–20	<15	<15	<15	<15	15
Chromium	30–200	40–70	30–80	50–200	10–40	25–30	10–15	15	0.5
Copper	20–70	13–20	10–30	17–65	6–12	200–500	5–10	5–10	0.2
Iron	1500–2600	2100–2800	200–2000	500–700	1350–1500	2200–4500	1400–1700	1500–2000	3
Magnesium	0–5	10–60	<5–40	10–40	70–80	180–200	1600–1800	30–35	.06
Manganese	13–21	16–20	7–40	9–18	20–30	140–260	200–250	14–17	0.1
Molybdenum	60–180	100–350	50–120	80–110	450–500	20–30	25–35	150–550	5
Sodium	400–700	500–1300	100–500	1300–1700	8000–8200	3000–4000	4000–7000	600–650	2
Nickel	90–300	60–100	50–130	80–340	20–40	45–55	5–10	10–20	2
Lead	60–170	30–50	40–80	50–160	13–30	50–60	15–20	10–15	7
Selenium	30–100	30–50	10–40	20–100	30–60	15–25	20–30	14–17	5
Silicon	700–900	450–750	260–1000	500–900	110–140	500–1200	250–350	50–120	2
Titanium	8–30	6–12	5–14	8–30	2–4	80–100	5–10	1.2	0.2
Vanadium	70–90	40–80	20–100	30–80	70–90	1000–1300	15–30	20–25	0.3
Zinc	10–25	10–40	3–16	10–40	10–15	20–30	15–25	25–60	0.2

Abbreviations used are: UN, United Nuclear; ANA, Anaconda; KM, Kerr-McGee; UNCR, United Nuclear Church Rock.
(1) Yellow cake from ore of neighboring mills processed by Sohio
(2) Yellow cake from ore of neighboring mills processed by those mills

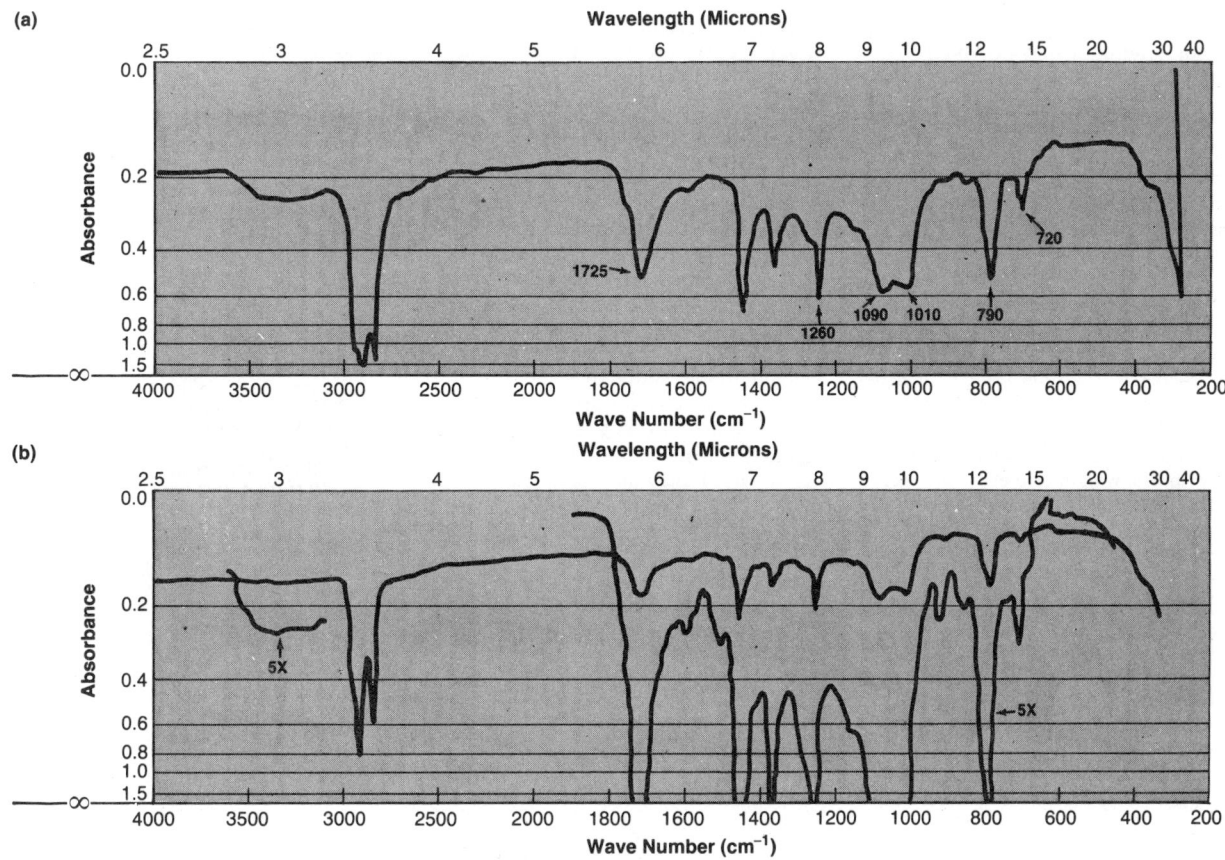

Figure 1. Infrared spectra of the pentane solubles from the Soxhlet isopropanol extract of the stolen sample (a) and the Sohio sample (b)

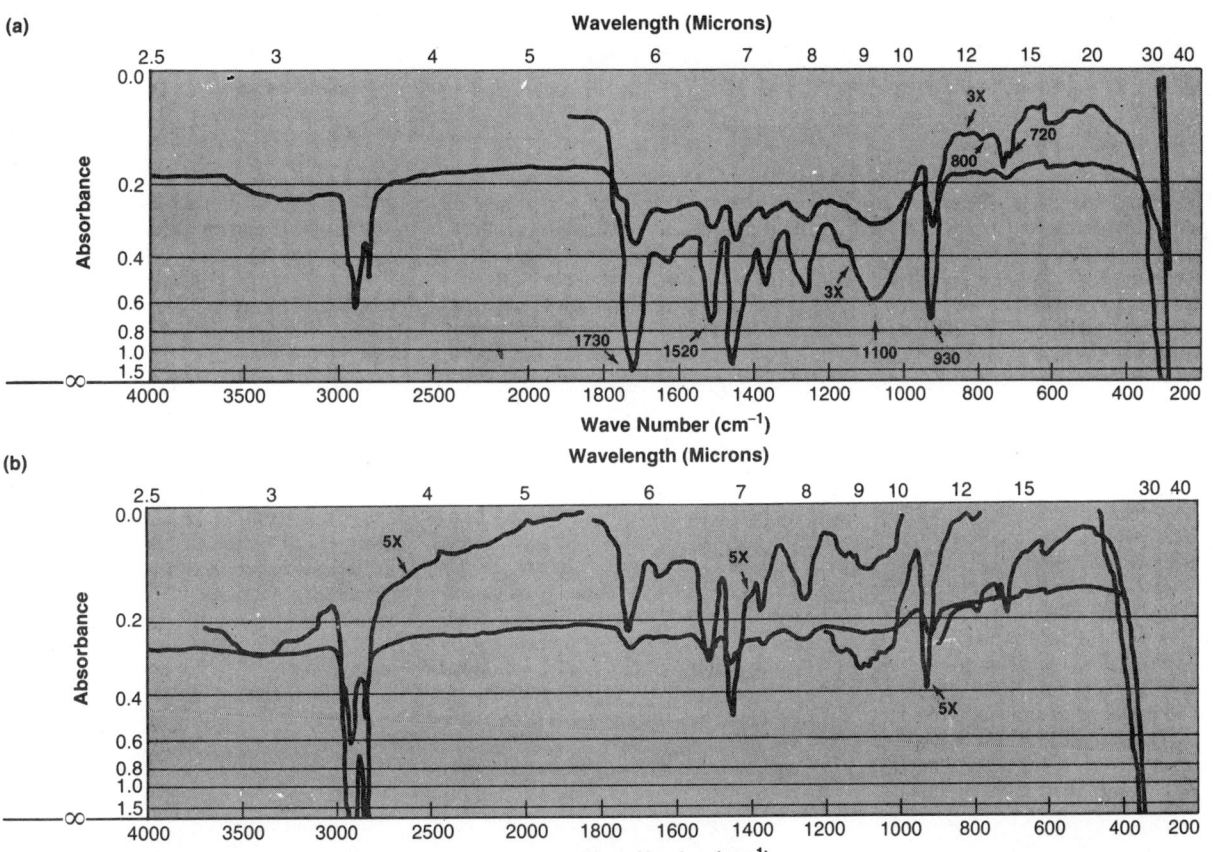

Figure 2. Infrared spectra of the carbon tetrachloride solubles from the Soxhlet isopropanol extract of the stolen sample (a) and the Sohio sample (b)

sist of the following. There was a major amount of silicone which was probably copolymerized with ethylene glycol. Also present was a large amount of an assorted mixture of carbonyl-containing materials; a small amount possibly belonged to the silicone/ethylene glycol component. The rest appeared to consist of major amounts of carboxylic acids, carboxylic acid salts, and smaller amounts of other carbonyl-containing materials. Their complexity of mixture and the fact that they were observed in all the extracts suggests that they were either humic acids or oxidation products created from processing alcohols during the calcination. A small amount of crystalline wax or polyethylene was indicated. A very minor quantity of an amide was detected. Very small amounts of inorganics were detected in the extract. They were found to be composed largely of silicate and a uranate, some carbonate, and a very small amount of sulfate. (The latter was detected in the FBI sample only.) Tridecanol, an alcohol unique to Sohio's process, was not detected in either sample, because it was removed by a calcining step.

Had time permitted, the silicone/ethylene glycol agent would have been examined more closely. ^{1}H NMR references of these materials show the glycol methylene to silicone methyl ($Si(CH_3)_2$) ratios vary for the assorted agents (4). It is possible that the silicone found is unique to Sohio. Investigation of these ratios for the samples on hand, in addition to more background information on the silicone agents used, may have provided us with further indicative evidence.

The evidence presented in this paper was never taken to court. It was presented to the suspects and induced them to plead guilty and admit that the yellow cake was indeed from the Sohio mill.

Acknowledgment

The authors would like to thank P. S. Fay for his input and coordination in this effort, which permitted the analysis to be done in a timely fashion.

References

(1) H. F. Mark, et al., "Kirk-Othmer Encyclopedia of Chemical Technology," 2nd ed., Vol. 21, p 1, Interscience Publishers, a division of John Wiley & Sons, Inc., New York, N.Y., 1970.
(2) A. F. Ward, and L. Marciello, Jarrell-Ash Plasma Newsletter, 1, (4), 10 (1978).
(3) "Sadtler Commercial Infrared Spectra—Surface Active Agents," Sadtler Research Laboratories, Inc., Philadelphia, Pa., 1977.
(4) "Sadtler Commercial NMR Spectra—Surface Active Agents," Sadtler Research Laboratories, Inc., Philadelphia, Pa., 1977.

Analytical Techniques in Arson Investigation

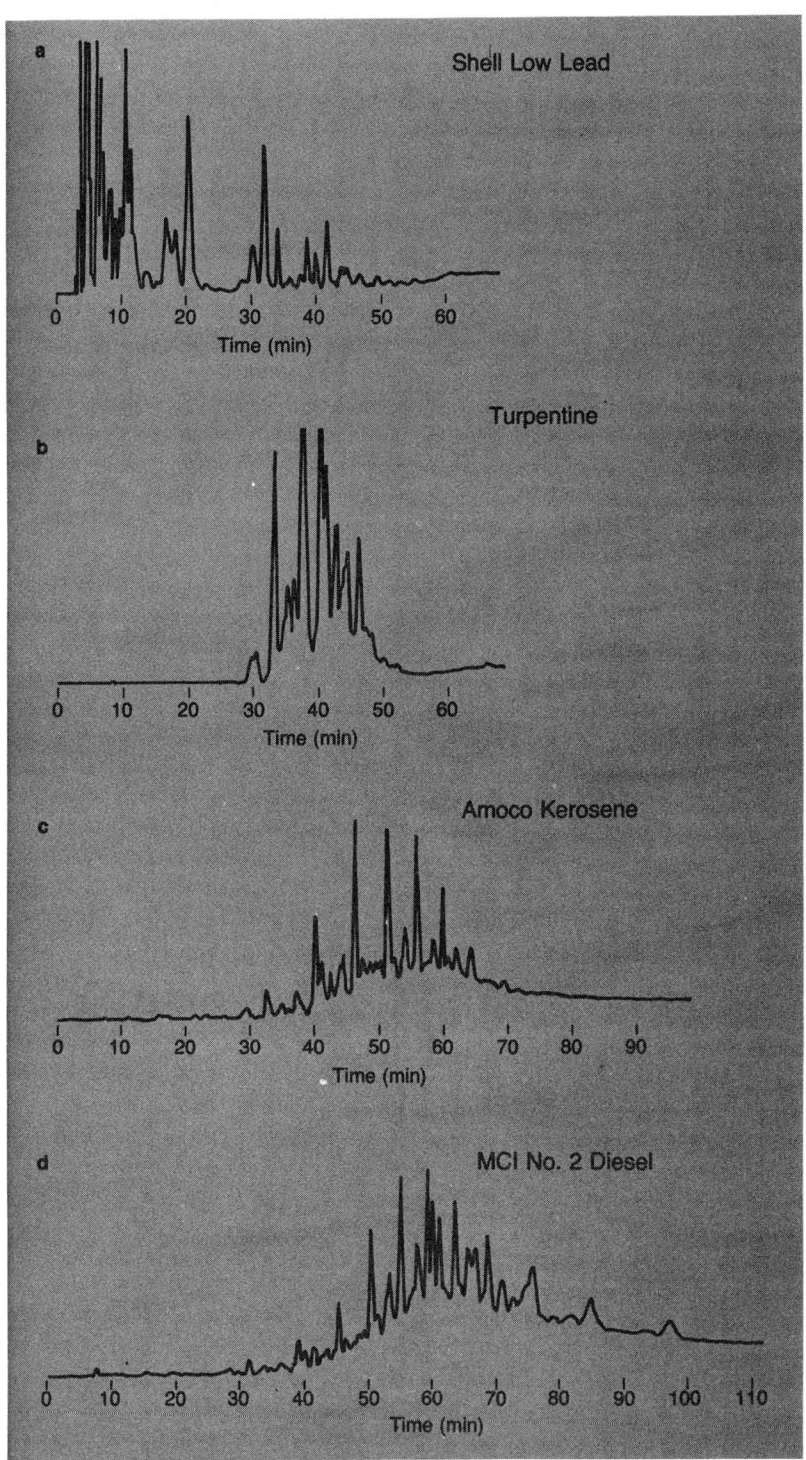

Figure 1. Typical chromatograms of different accelerant classes
(a) Gasoline (b) Turpentine (c) Kerosene (d) Higher Fuel Oil

Michael J. Camp

Northeastern University
Institute of Chemical Analysis
Applications, and Forensic Science
Boston, MA

Cambridge Analytical Associates, Inc.
Watertown, MA

Originally published in
ANALYTICAL CHEMISTRY, 1980
Vol. 52, No. 3.

Arson is the willful destruction of property by fire. In recent years this crime has received increasing public awareness. After any major fire has occurred, it is common to hear or to read the phrase "the arson squad is investigating." This article will discuss the role of the forensic laboratory in determining if arson has indeed occurred and how it was accomplished.

The analysis of an allegedly arsonous fire begins at the scene. Any chemicals used to initiate or to spread the fire, and any mechanical or electrical device used to contain or to control these chemicals will leave some trace at the location where they were used. Even if the fire was accidental, the cause will still be located at the fire's origin. Thus, the exact location where the fire started must be found.

In most jurisdictions a qualified investigator bears this responsibility. The details of his method are not within the scope of this article. However, some basics will be given as they will affect the laboratory's results. The investigator will use eyewitness reports of what part of the building was first on fire, descriptions of the burning characteristics from the firefighters, knowledge of the combustibility of construction materials, and educational and personal experience to locate the region of first involvement. He may also use a hydrocarbon sniffer to aid in his investigation. After finding the place of origin and reconstructing the burning sequence, the investigator has to determine if it was an accidental or an unnatural fire. Laboratory tests may be used in either case to substantiate or to disprove his theory.

83/4463-0179 $07.00 © 1983 American Chemical Society

The laboratory is entirely dependent upon the investigator to provide it with proper samples—proper in both the chemical sense, i.e., that they are from the fire's origin, and in the legal sense, i.e., that the legal requirements have been satisfied. The most commonly encountered accelerants are gasoline, kerosene and home heating fuel oil. For this reason the collection, preservation and analysis of arson samples are geared toward detection of volatile materials as the first step.

The samples of debris should be sealed at the scene in airtight containers. New, unlined metal paint cans are preferred since there is no organic matter to cause a background. Rubber-ring-sealed glass jars are also acceptable. Large items can be sealed in plastic bags at the scene and then cut up and resealed in cans as soon as possible. Paper bags and also plastic ones will slowly breathe, causing a loss of volatile vapors. Even worse, they will allow vapors to enter the sample. This is of concern since the plastic bags may be transported with or stored in areas with liquids. Cross-contamination would negate any later result.

The main technique used for the detection of volatile materials is headspace vapor analysis. The goal is not to determine the brand and grade of gasoline, but to determine if any is present in the debris. Petroleum products are complex mixtures of several hundred aromatic and aliphatic compounds. They are sold for specific purposes and their composition and properties vary with the end-use purpose. Thus, they can be classed into groups by any convenient measurable property such as flash point, boiling point range or distillate fraction.

In chromatographic analysis each class has a similar distribution of peaks and peak heights. Typical chromatograms are shown in Figure 1. These are fresh samples injected as liquids. Chromatograms of headspace vapor residues will be slightly different. The more volatile components may be completely absent and the next peaks reduced in intensity when compared to the later, less volatile materials. The heat of the fire will cause a differential volatility of the residue which is absorbed onto some debris material. It is, however, still possible in the majority of cases to discern the characteristic pattern.

Before discussing the chromatographic techniques, the concept of class identity should be examined. Modification to the basic methods will depend upon the class of vapors present. The definition of a class of petroleum hydrocarbons may vary from one lab to another. It is usually based upon the patterns seen on the chromatogram (Figure 1). The author uses the following guidelines. All brands and grades of automotive gasolines belong to one class. All brands of kerosene and fuel oil number one belong to a second class. The higher fuel oils are a third class, even though it may be possible to estimate the grade. The centroid of the peaks shifts to longer times as the grade number increases. Distillates with a narrow boiling range are distinguished from the fuels by the compactness of the chromatogram. The peaks are seen over a narrower time range. This class is called a distillate material and contains thinners, charcoal lighter, patio torch fuel, turpentine (even though it is a natural product) and the various products with distillate solvents available over the counter. Very light fractions such as lighter fluids and white gas do not have the heavier, higher boiling components of the automotive gasolines. Nonpetroleum products present a problem since they contain very few components. These usually require an alternative method since the pattern is not complex enough to classify it directly.

A standard reference file run under your conditions is needed for aiding in the analysis of the chromatograms. In addition to fresh samples, partially evaporated and trial burn samples with and without an accelerant should be run. Blank or control samples are a must. In some actual cases a unique class may not be determinable. In this event it is only possible to report the detection of a volatile hydrocarbon residue.

Typical chromatographic conditions for a basic headspace vapor analysis of arson debris are shown in Table I. Variations are common and are usually tailored to a specific instrument and column. To obtain a sample, the sealed container is heated in an oven or on a water bath to a temperature which will desorb the volatile materials. Depending upon the amount of water present, temperatures of 60–110 °C for 5–20 min will be sufficient. The more water there is, the longer it will take to reach the final temperature and the greater the internal pressure will be. Experience is the best teacher. A chromatogram with weak peaks should be rerun after further heating before choosing a second method.

The sample (1–5 cc of headspace vapor) is removed through a septum. Two devices are shown in Figure 2. Both are attached before heating. Either a gas-tight glass syringe or a disposable plastic syringe can be used. The former must be cleaned thoroughly between samples; the latter introduces a medium intensity peak in the first 2 min. No interference with the analysis has been noted when the plastic syringes were used. On some gas chromatographs the large volume of air and moisture may extinguish the flame. The resultant chromatogram is then compared to the standard reference file, to other items in the same case, or to liquid samples submitted for reference.

Since analysis time for fuel oils would be over an hour, the investigator may attempt to characterize these materials by relying on the use of different columns (1). An Apiezon L column would be used first. This column would detect and classify the lighter hydrocarbons up to kerosene. If a higher material was suspected, a second run would be made using a Dexsil-300 column at a higher temperature. On this column the gasoline components are not separated; however, the higher boiling materials can be classified in a short time. Results are satisfactory if sufficient sample amounts are available.

In most cases the headspace technique leads to a positive or negative conclusion, especially if there is sufficient sample and the background does not cause problems. Occasionally a chromatogram is obtained which does not yield a definite conclusion. The peaks may be just larger than the base line and not well resolved. More sample is needed before a distinction can be made.

The peaks may be large but too few to conform to a typical class, or they may have very short retention times. Could it be background material or a special class of accelerant? Liquid samples may be submitted when the purpose is to determine if sample A was stolen from gas station B or gas station C to the exclusion of all other sources. And there is the instance when the suspect accelerant is a higher fuel oil whose vapor pressure may be insufficient to give an appreciable headspace concentration. It is to these questions that current work is directed.

Much effort has gone into preconcentration of the headspace vapors prior to injection. Historically, steam distillation and solvent extraction methods were the first to be used. Distillation using water, ethylene glycol or other high boiling materials produces a visible layer floating on top of the distillate. This material can be injected directly into the gas chromatograph (2). In addition, instrumental methods such as infrared spectroscopy or mass spectrometry can be utilized for a more complete analysis (3). Results are again satisfactory. However, distillation times may exceed 48 h (4), and the rate of distillation may be important. When the debris contains wood, especially pine, the chance of distilling a large amount of background material is high.

Solvent extraction techniques include a soaking and rinsing or longer refluxing with a Soxhlet system. In the soaking method the volume of solvent (hexane, heptane, or petroleum ether) must be concentrated before injection. The process is time-consuming, and the concentration steps may cause loss of part of the sample. Soxhlet extraction also takes time and leads to possible background peaks from wood or plastics.

Another classical separation method which leads to a concentration of sample is vacuum distillation (5). One report describes a simple approach using a sealed can, a chunk of dry ice, and a heat lamp (6). Reported results are satisfactory for gasoline and kerosene, but long times are required for fuel oils.

All of these classical methods will work in some cases. Most are labor- or time-intensive. In addition, the distillation and extraction methods may generate a background level of materials which could be taken as a false positive or which could interfere with the basic pattern classification procedures.

The search for better methods has led to the adoption of concentration procedures similar to those developed for environmental air sampling and bomb detection. The vapors are concentrated by absorption onto a substrate and later desorbed. The absorbant material, process of absorption, and the process of desorption are currently being studied.

Charcoal is known to be an excellent absorbant. It has a high capacity but bonds materials strongly (7, 8). Other materials such as Poropak Q, Tenax-GC, and the Amberlite XAD resins are also suitable (9). Alumina and silica do not absorb appreciable amounts of hydrocarbons.

Absorption can be done at the scene by drawing air through a packed tube. This allows an investigator to collect air samples at specific sites. Volatile materials from a larger area will also be collected. In the laboratory, absorption is done by sweeping dry nitrogen through the sealed can and onto a packed tube. Background is limited to the actual sample material. The can may be placed in a water bath to increase the vapor pressure of the accelerant.

Desorption is achieved by heating or by solvent stripping. For thermal desorption from charcoal, a high temperature is required. Lower temperatures can be used for the other absorbants (7). Commercial units are available which are attached directly to the injection port of the gas chromatograph. On instruments which have a moderately fast injection port heating rate, the absorbant can be packed into a spare injection port liner. Desorp-

Table I. General Chromatographic Conditions for Headspace Vapor Analysis of Arson Residues

Carrier gas: Nitrogen or helium at 30–40 mL/min

Injection port: 150–200 °C

Column: 1/8 or 1/4 in, glass or metal, 6 to 12 ft long, packed with Apiezon L, OV-1, OV-101, SE-30, or SP1200/Bentone 34

Oven: Programmable, 50–70 °C initially for 0–5 min, heat to just below the column's maximum temperature at 5–20°/min, and hold there until no further peaks are seen.

Detector: Flame ionization

Sample: 1 µL of liquid, 1–5 cc of headspace vapor

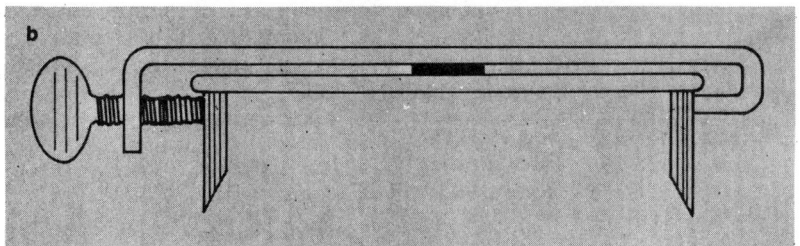

Figure 2. Two headspace sampling devices
(a) An apparatus which pierces a paint can lid and screws tight via an o-ring. The top has a septum under a Swagelok nut. (courtesy of the Wisconsin State Crime Lab)
(b) Simple device to hold a septum in place with a tension bar. Bar has a hole in it which must mate with hole in can lid. (courtesy of Demers Laboratory, Springvale, Maine)

tion can then be accomplished inside the injection port by rapidly heating from room temperature to 250 °C (10).

If charcoal is used as an absorber, elution with a suitable solvent will strip off all the volatiles. The solvent must be pure and have a very short retention time or be nondetectable by the flame ionization detector. A detectable solvent which tails considerably may mask the initial peaks of the accelerant.

Carbon disulfide has been found to fulfill these requirements. The flame response is very low and elution is quantitative. (8, 9) It can not be used with Tenax-GC as it dissolves the polymer.

More work is being conducted to optimize absorption/elution techniques. One major concern is the presence of background vapors. They will be concentrated at the same time as the accelerant vapors. A suitable blank sample is desirable, although it may be hard to determine what is suitable at the scene. Test burns of various materials without the use of an accelerant can be concentrated and used to form a reference library. If any doubt arises in a specific case, the use of mass spectrometry is recommended.

A GC/MS combination has several useful applications in arson analysis (11, 12). By monitoring selective ions, one can determine the presence and distribution of materials, such as the polycyclic aromatics, which are common to petroleum products. The mass spectrometer's sensitivity is greater than the flame ionization detector, enabling very weak samples to be analyzed. Bones from bodies which may have been burned have been successfully studied (11). In this instance it was necessary to add some water to the bone before appreciable vapor was detected.

The mass spectrometer is currently the best method for determining if observed peaks may be a commercial solvent or may have originated from the background. In those cases where a nonpetroleum accelerant is used there are too few peaks to yield any classification. Mass spectral identification is required. Similarly the decomposition of wood and plastic will yield a number of peaks but their pattern is not consistent with known accelerants except for turpentine. Identification of these materials will assist the analyst in deciding that no accelerant was present. It is not known whether GC/IR will be of assistance in this area.

A preliminary study reports the use of the ratios of the lead alkyls ($Me_xEt_{4-x}Pb$) as determined by GC/MS (13). Another area of possible GC/MS involvement would be to determine the proprietary additives used by each manufacturer. Detection of these additives may lead to the long-sought goal of brand and/or grade identification.

When liquid samples are submitted, the class is usually known and more specific identification on direct comparison of samples is required. There is more information in a liquid than in the vapor and not being sample limited is a definite advantage. Traditional classification by flash point or the percent of the sample which distills in a given temperature range can be determined. The dyes can be examined and compared by thin layer chromatography, and the lead and bromine content can be quantitated.

Beyond these basic steps one wants to determine the relative amounts of each component. This requires much greater resolution than the packed column can provide. Capillary columns will resolve gasoline into 200+ peaks (14). When coupled with steam distillation, capillary columns have been used on test burns. Identification is based on the use of Kovats indices as well as visual comparison (4).

This brings up the last topic in arson analysis. The final decision is currently made by visual examination of a complex chromatogram. How reliable is this step? For most instances it is quite good. Could it be better? The use of computer-based pattern recognition has been applied to many forensic topics. Paper, glass and whiskey adulteration are just a few of these areas. A preliminary report suggests that these methods may well be applicable to arson analysis (15).

The application of new methods to arson analysis is a rapidly growing field. It will become easier for the laboratory to detect trace levels of accelerant with more confidence. The goal of identification of brand or grade of gasoline may be attained in the near future. This may have an impact in a related criminal area if the theft of fuels becomes a major problem.

References

(1) H. Evans, Unpublished Master's Paper, Northeastern University, 1979.
(2) R. W. Clodfelter and E. E. Hueske, *J. Forensic Sci.*, **22**(1), 116 (1977).
(3) I. C. Stone, J. N. Lomote, L. A. Fletcher, and W. T. Lowry, *J. Forensic Sci.*, **23**(1), 78 (1978).
(4) A. T. Armstrong and R. S. Wittkower, *J. Forensic Sci.*, **23**(4), 662 (1978).
(5) R. Hrynchuk, R. Cameron, and P. G. Rodgers, *Can. Soc. Forens. Sci. J.*, **10**(2), 41 (1977).
(6) H. G. Linde, *Crime Laboratory Digest*, **79-6**, 3 (1979).
(7) F. J. Conrad, T. A. Burrows and W. D. Williams, Sandia Laboratories Report SAND78-1246, August, 1979.
(8) D. V. Canfield and S. J. Irwin, private communication.
(9) W. Graziano, Unpublished Master's Paper, Northeastern University, 1980.
(10) H. Evans and P. Demers, private communication.
(11) H. Harris, "Identification of Accelerant Peaks by Gas Chromatography—Mass Spectroscopy," presented at the 5th Annual Meeting, Northeastern Association of Forensic Scientists, Albany, N.Y., October, 1979.
(12). M. H. Mach, *J. Forensic Sci.*, **22**(2), 348 (1977).
(13) L. T. Lytle, L. C. Ford and P. T. Williamson, "GC-Mass Spectroscopy: Its Use in Determining Gasoline Lead Alkyl Ratios and Their Use in Forensic Examinations," presented at the 31st Annual Meeting of the American Academy of Forensic Sciences, Atlanta, Ga., February, 1979.
(14) K. A. Oakes, "The Use of Glass Capillary Columns in Arson Analysis," presented at the 31st Annual Meeting of the American Academy of Forensic Sciences, Atalnta, Ga., February, 1979.
(15) B. R. Kowalski and W. R. Gresham, "Use of 'Pattern Recognition' in Arson Cases: a Progress Report," presented at the 31st Annual Meeting of the American Academy of Forensic Sciences, Atlanta, Ga., February, 1979.

Investigations of Clandestine Drug Manufacturing Laboratories

Terry A. Dal Cason

Drug Enforcement Administration, North Central Regional Laboratory, Chicago, IL 60607

Richard Fox and Richard S. Frank

Drug Enforcement Administration, Forensic Sciences Division, Washington, DC 20537

Originally published in ANALYTICAL CHEMISTRY, 1980, Vol. 52, No. 7.

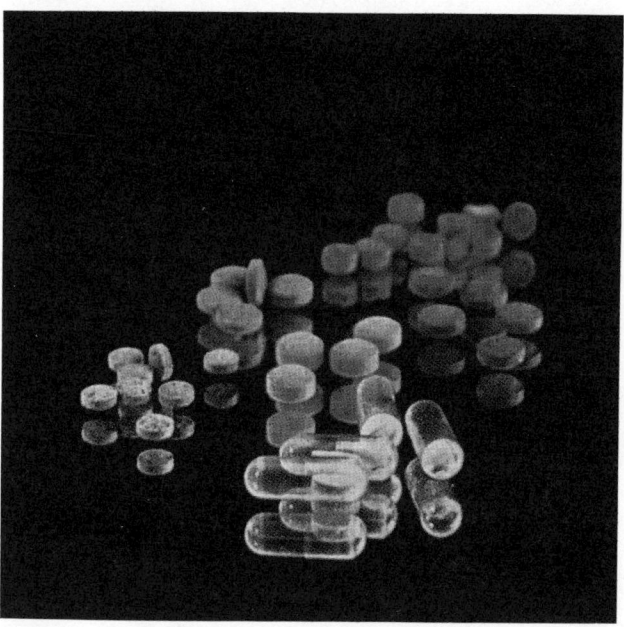

In order to legally manufacture or synthesize a controlled drug substance, a firm or individual must become registered as a manufacturer with the Drug Enforcement Administration (DEA) and follow a strict set of guidelines specified in the Controlled Substances Act (Public Law 91-513). Clandestine, or illegal, laboratories have been established by the criminal element to circumvent these requirements with the goal of supplying drugs of abuse to the illicit market. Investigation of clandestine drug manufacturing laboratories is a high priority activity of DEA because elimination of these laboratories will thereby eliminate the drugs of abuse at their source.

During 1979, a total of 221 clandestine laboratories were seized by DEA, with the following breakdown by drug:

Methamphetamine	121
Phencyclidine (PCP)	47
Amphetamine	20
Methaqualone	7
All Others	26

Forensic chemists from DEA's eight field laboratories play a very important role in all phases of DEA's clandestine laboratory activities, such as investigation, analysis of evidence, and criminal prosecution. A previous article in THE ANALYTICAL APPROACH (1) described the duties of a forensic chemist, likening the duties to that of an "analytical detective." A forensic chemist's duties normally entail analytical work in the laboratory, punctuated by appearances in court as an expert witness. The duties associated with clandestine drug manufacturing laboratories, however, directly involve DEA chemists in the active criminal investigation, in addition to analytical laboratory and court duties.

Participation by DEA chemists frequently begins when an investigator identifies some chemical compounds and wishes to determine what drugs are most likely being produced, the quantities involved, an appropriate time for scheduling seizure (from surveillance of activities at the site), and technical information necessary for preparation of the search warrant.

At the time of the seizure, the forensic chemist accompanies the raiding party. The chemist's role is to ensure safety of people and property, as well as to provide scientific advice. Operating illegal laboratories must be carefully shut down in order to avoid fire or explosion, and hazardous chemicals must be identified and disposed of. Chemists then assist in identifying what should be seized as evidence (essential precursors, finished and in-process materials) and identify other useful evidence that will help to prove that a controlled drug was being synthesized in violation of the Controlled Substances Act.

Following the analysis of the evidence, the chemist must develop a concise report, detailing what was found, what was being synthesized or could have been synthesized at the site. Important points are the chemical reactions or synthesis routes used, chemicals, equipment, intermediates, and final products.

During a recent investigation, a DEA laboratory was notified that a large purchase (approximately 2000 lb) of anthranilic acid was delivered to a company allegedly involved in the textile and dye industry. Additional information was available that this company had also purchased acetic anhydride. The chemist assigned at the initial stage of this investigation informed DEA special agents that through the use of acetic anhydride, it is an easy process to convert anthranilic acid to N-acetyl anthranilic acid or N-acetyl-anthranil (2). Either of these two substances can be used as a precursor for the synthesis of methaqualone. He also suggested that the agents provide information as to any subsequent purchases by the suspect company of o-toluidine, toluene, or phosphorous trichloride, all of which may be used in the synthesis of methaqualone.

Although no o-toluidine, toluene, nor condensing agent was purchased, nine 55-gal drums of o-chloroanaline were acquired by the company. The following equipment was also procured:

- three tabletting machines,
- two 200-gal glass-lined vessels,
- three 40-L flasks with accompanying heating mantles,
- a 6 ft × 5 ft × 4 ft electric drying oven,
- and scales with 100-lb, 310-lb, and 140-kg maximum capacities.

The size of the operation was apparent from the large capacity of the equipment.

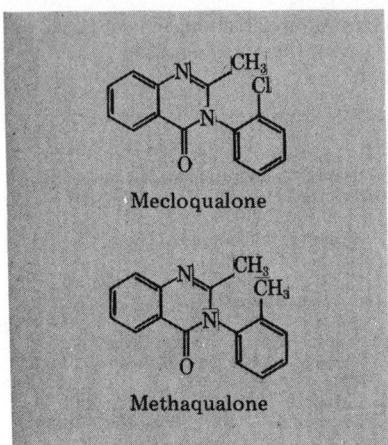

Figure 1

The chemist was able to ascertain, after a review of the scientific literature, that by substitution of o-chloroanaline for o-toluidine, it was possible to produce mecloqualone, rather than methaqualone (3, 4, 5) (Figure 1). Mecloqualone, placed in Schedule I of the Controlled Substances Act on July 10, 1975, is a nonbarbiturate sedative and hypnotic agent of the quinazolinone family. It is not marketed for legitimate medical purposes in the U.S.

A search warrant was executed for seizure of the suspected clandestine laboratory. Approximately 30 exhibits of seized chemicals and equipment were inventoried as evidence. Only one exhibit contained mecloqualone, which was confirmed by both ultraviolet and infrared spectroscopy.

Notes found at the clandestine laboratory site gave no indication as to the synthetic methods utilized. Most of the previously known syntheses required the use of a condensing agent, such as phosphorous trichloride, phosphorous oxychloride, or ferric chloride, or required special conditions or equipment (6). No condensing agent or special equipment was found at this laboratory. An extensive literature search was made to find a procedure for the synthesis of mecloqualone that would require only those chemicals and equipment found in the laboratory at the time of seizure. In a patent issued to a firm in Paris, France, methaqualone is prepared by adding excess acetic anhydride to anthranilic acid, heating and maintaining this for a period of time. The product is recrystallized from alcohol and gives a 70% yield. The DEA chemist performed this procedure by substituting o-chloroanaline in place of o-toluidine, and confirmed the product as mecloqualone (Figure 2).

At this stage, all elements were present from the scientific viewpoint for successful prosecution:

- All reagents and equipment necessary for production of a controlled substance were present on the premises at the time of seizure.
- A valid synthetic method for producing the drug is recognized in the literature, requiring only the items seized in the laboratory.
- Finished product was found on the premises.
- The projected synthesis used in the clandestine laboratory was successfully duplicated by the DEA.

The teamwork of the criminal investigator and forensic chemist led to the seizure of this clandestine laboratory and prevented a potential 2 700 000 mecloqualone tablets from entering the illicit drug market.

As discussed earlier, the most technically demanding aspect of the entire chain of events involved in the forensic chemist's role in clandestine laboratory activities is preparation for, and participation as an expert witness in, the trial of the alleged violator(s). The chemist must have a full understanding of the reaction pathways, side reactions, reaction byproducts, and intermediates. Alternate synthesis routes must be known. The reagents and products that are important to describe the amount of drug being synthesized should be known, as well as other purposes for the chemicals found at the site. All this, in addition to the intimate knowledge of the analytical techniques used to examine the evidence, is necessary.

In one case, an alleged clandestine methamphetamine laboratory was seized and 50 exhibits of evidence were collected at the site. Ten of the exhibits were subsequently identified as methamphetamine hydrochloride, another 11 as being either chemicals involved in the process of synthesizing methamphetamine or its immediate precursor—phenyl-2-propanone.

During jury trial, the defense introduced the argument that the DEA, having used a GC/MS for identification of the methamphetamine, failed to prove beyond reasonable doubt that the substance was methamphetamine. Defense introduced literature stating that quadropole mass spectrometry cannot resolve the difference between methamphetamine and phentermine.

The forensic chemist involved in the case responded that by comparing ratios of m/e 65, 119 and 134 peaks, differentiation was possible. Graphic explanation of peak ratio calculations was presented (Table I). The ratios

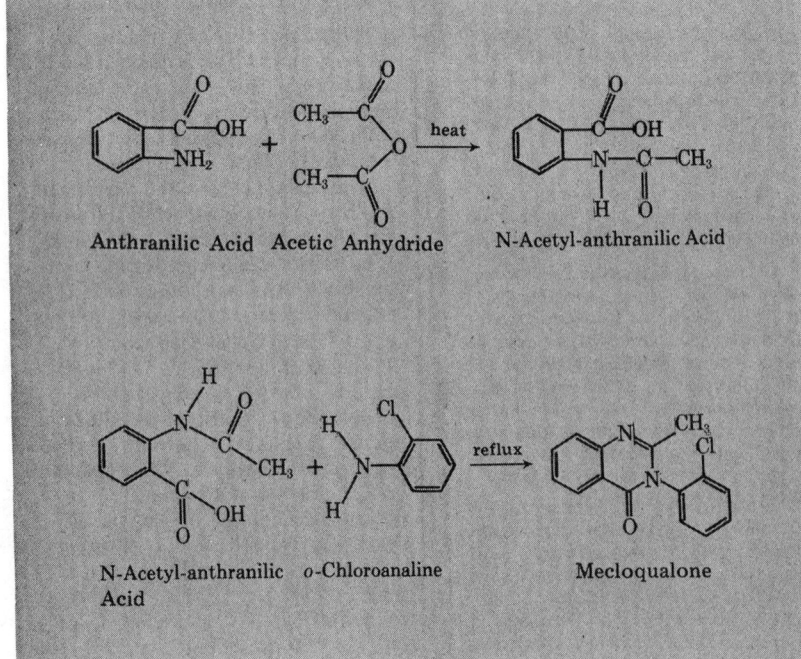

Figure 2

Table I. MS Peak Ratios

Spectra	(Relative to m/e 65 = 100%)	
	m/e 119	m/e 134
Methamphetamine standard #1	28%	42%
Methamphetamine standard #2	26%	39%
Phentermine standard	6.5%	81%
Evidence	27%	40%

clearly demonstrated that the sample was methamphetamine.

The defense then cited a reference from the scientific literature in which the two compounds were differentiated using m/e 56, 58, 91, 119, and 134, ratioing to the m/e 58 peak. They argued that this published procedure had not been used in the DEA investigation. It was then up to the DEA chemist to convince the court that both the published procedure and the DEA procedure were adequate to accomplish the analysis. The DEA forensic chemist must thus have confidence in his instrumental conditions, analytical technique, and spectral interpretations, and must have the ability to explain the role of various chemicals in the synthetic process.

Thus, the DEA forensic chemist is confronted with various problems and situations which require expertise in many areas of chemistry, a knowledge of law, the ability to present findings in court, and the ability to use the correct analytical approach.

Acknowledgment

The authors would like to thank Arthur W. Davidson, Drug Enforcement Administration, San Francisco, Calif., and John W. Gunn, Jr., Drug Enforcement Administration, Washington, D.C., for their contributions to this article.

References

(1) Eichmeier, Larry S., and Caplis, Michael E., "The Forensic Chemist—An Analytical Detective." *Anal. Chem.*, **47**, 841 A–844 A (1975).
(2) Bogert and Chambers, *J. Am. Chem. Soc.*, **27**, 649 (1905).
(3) "*Cuttings Handbook of Pharmacology,*" Windsor Cutting, 5th ed., p 561, Meredith Corp., 1972.
(4) "Merck Index," 8th ed., pp 647 and 672, Merck & Co., Inc., Rahway, N.J., 1968.
(5) Grimmel, H.W., Guenther, A., and Morgan, J.F., *J. Am. Chem. Soc.*, **68**, 542–3 (1946).
(6) Bogert and Chambers, *J. Am. Chem. Soc.*, **27**, 649 (1905).

The Forensic Chemist An "Analytical Detective"

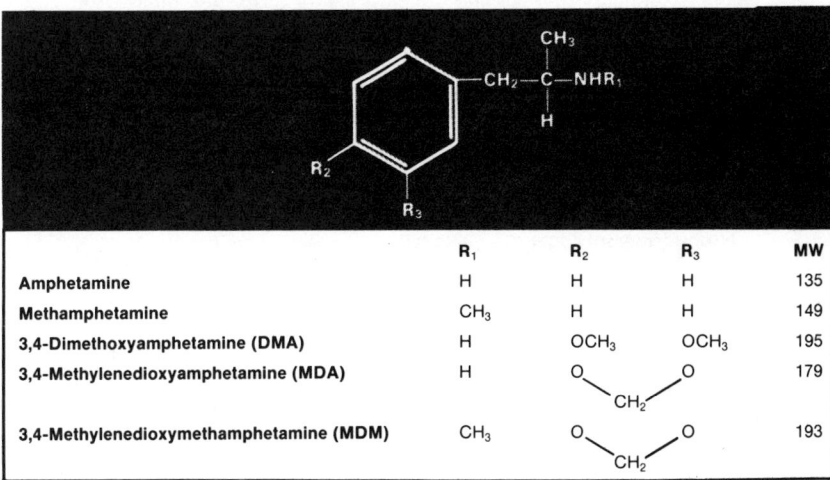

Figure 1. Structurally related amphetamines

Larry S. Eichmeier and Michael E. Caplis

Northwest Indiana Criminal Toxicology Laboratory, Gary, IN 46402

Originally published in ANALYTICAL CHEMISTRY, 1975, Vol. 47, No. 9.

The forensic chemist has the acute problem of constantly being required to make a rapid assessment of the nature of suspected illicit materials or to demonstrate the presence of poisons, drugs, and toxic chemicals in biological samples. Society demands that samples be completely analyzed prior to accusation and prosecution of any violation of its laws. Thus, the forensic chemist must be both swift and absolutely reliable in his assessment. To this end, it is his responsibility to identify and determine, without a reasonable doubt, illicit or toxic substances within a relevant time.

By virtue of this responsibility, the forensic chemist becomes an "analytical detective" who must deal with complex mediums from pharmaceutical and illicit preparations to samples of body fluids and tissues. Determination of toxic compounds in a biological matrix is complicated by the low concentration levels of the compounds and by their possible conversion to metabolites. Complete analysis of such materials requires a multitechnique approach. Techniques utilized include thin-layer chromatography, gas chromatography, high-pressure liquid chromatography, fluorospectrophotometry, gas chromatography–mass spectrometry, UV–VIS spectrophotometry, and infrared spectrophotometry, as well as chemical methods. Equipped with this armamentarium, there can be little justification for an incorrect analysis from the forensic laboratory.

Two examples in which a judicial combination of techniques was utilized are cited. One involved the identification of a new street drug closely resembling a known illicit preparation, and the other involved a drug overdose which led to a homicide investigation.

Identification of Suspected Illicit Drug

A major abused class of drugs is amphetamine and its derivatives. In an attempt to combat this problem, undercover narcotic agents are constantly striving to determine the source and trafficking route of these drugs. In one particular instance, agents purchased capsules suspected to contain amphetamine or methamphetamine from a major distributor (Figure 1). These capsules containing a pink powder were submitted to the forensic laboratory for analysis. Preliminary screening of a $1M$ sulfuric acid solution of the powder by UV spectrophotometry indicated a dioxyamphetamine derivative, such as 3,4-dimethoxyamphetamine (DMA) or 3,4-methylenedioxyamphetamine (MDA). Amphetamine and methamphetamine were ruled out since the indicative pattern for monosubstituted benzenes (252, 257, 263 nm) was absent. Thin-layer and gas chromatographic screening after chloroform extraction from a basic solution showed chromatographic properties similar to 3,4-methylenedioxyamphetamine (MDA).

Differential chemical visualization following thin-layer chromatographic separation was used to determine if the substance was a primary, secondary, or tertiary amine (Table I). The material did not react with ninhydrin–acetone, but reacted with ninhydrin–isopropanol–acetic acid and iodoplatinate, indicative of a secondary or tertiary amine.

On-column gas chromatographic derivatization with acetic anhydride and benzaldehyde was employed to confirm the amine structure. The on-column derivatization of the amine with acetic anhydride gave a change in peak retention time indicative of the acid amide formation expected with a primary or secondary amine.

However, on-column derivatization with benzaldehyde resulted in loss of any gas chromatographic activity, indicating formation of the chromatographically inactive substituted secondary amine.

Table I. Summary of Tests and Conclusions for Suspected Illicit Drug

Examination	Results	Conclusion
UV (1M H_2SO_4)	252 nm, 257 nm, 263 nm — Absent	No monosubstituted benzene present
	287 nm, 233 (max) — Present	Dioxyamphetamine-like derivative
TLC		
Ninhydrin–acetone	No reaction	
Ninhydrin–isopropanol–acetic acid	Reaction	Secondary or tertiary amine
Iodoplatinate	Reaction	
GC		
Acetic anhydride	Peak shift	Secondary amine
Benzaldehyde	No peak	
Mass spectrometer		
193 m/e	Weak—recognizable (molecular ion)	Secondary amine
135 m/e	Strong	Aryl methylenedioxy substituent
58 m/e	Intense (base peak)	Alkyl amine
151 m/e	Absent	No aryl dimethoxy substituent present
91 m/e	Absent	No monosubstituted benzene present

$$R-CH_2\text{-}N(R^1)\text{-}H + \phi CHO \longrightarrow R-CH_2\text{-}N(R^1)\text{-}C(OH)(H)\phi$$

Mass spectral analysis showed fragments at 58, 135, and 193 m/e. Even though the ion 193 m/e was recognizably weak, it was determined to be the molecular ion. Characteristically, a weak parent ion is common to aryl-substituted secondary amphetamine-like derivatives. The base peak at 58 m/e can be attributed to either of two fragments:

(A) $(CH_3)_2 C=NH_2^+$

(B) $CH_3\text{-}CH=N^+(H)(CH_3)$

Greater significance can be assigned to fragment B because previous data indicated a secondary amine. Additionally, the absence of any ion at 151 m/e negated the possibility of a dimethoxy substituent, whereas the presence of the 135 m/e ion indicated a methylenedioxy substituent:

(151 m/e) — dimethoxybenzyl cation (CH_3O, OCH_3)

(135 m/e) — methylenedioxybenzyl cation

Inspection of the analytical data allowed identification of the material as 3,4-methylenedioxymethamphetamine (MDM). The data and conclusions are summarized in Table I. This substance, similar in structure and general analytical properties to MDA, is exempt from federal control, whereas MDA is controlled. Thus, the rapid utilization of general and specific techniques provided valuable information on a suspected illicit preparation and averted embarrassment and unwarranted prosecution.

Drug Homicide

Blood from a woman found by a relative was submitted to the forensic laboratory for toxicological analyses. There was no apparent cause for death. The woman and two of her male friends had been drinking heavily outside her apartment complex earlier that evening. Police were called to quell a disturbance, and the trio had retired to her apartment. Later a witness testified that cocktails had been forcibly administered to the woman in an attempt to subdue her. Medical reports showed the woman was under a doctor's care with a prescription for phenobarbital.

Examination for blood volatiles involved gas chromatography and UV spectrophotometry (Figure 2). Screening was accomplished by injecting 10 μl of whole blood into a gas chromatograph. Results indicated the presence of ethyl alcohol. The ethanol concentration was determined by an enzymatic procedure with alcohol dehydrogenase (ADH).

$$CH_3CH_2-OH \xrightarrow[\text{NAD} \rightarrow \text{NADH·H}]{\text{ADH}} CH_3-C(=O)H \xrightarrow{\text{Semicarbazide}} \text{Semicarbazone}$$

In this procedure a mole of ethanol is oxidized to a mole of acetaldehyde, which is removed as the semicarbazone, with the concurrent reduction of a mole of nicotine adenine dinucleotide (NAD). Monitoring of the reduced NAD at 340 nm with a UV spectrophotometer gave an ethanol concentration of 0.23% (0.1% = intoxication).

Preliminary screening with a chloroform extract of an acidified blood sample (Figure 2) gave conflicting results. Gas liquid chromatography suggested the presence of three barbiturates: amobarbital, secobarbital, and phenobarbital. However, thin-layer chromatography indicated only the presence of amobarbital and secobarbital. Gas chromatography–mass spectrometry resolved the inconsistency with identification of amobarbital, secobarbital, and n-dibutylphthalate. Dibutylphthalate has the identical gas chromatographic properties of pheno-

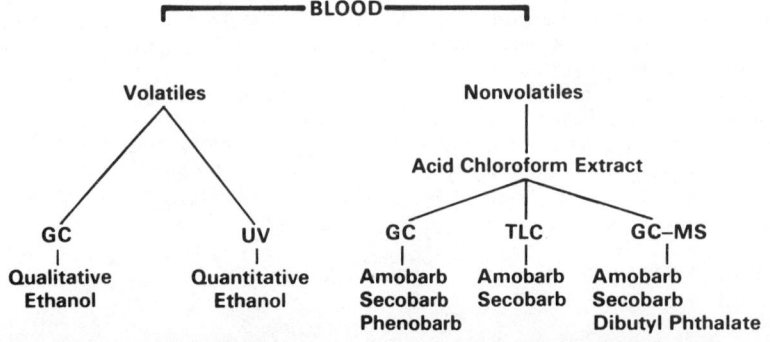

Figure 2. Isolation and identification scheme for volatiles and nonvolatiles in blood

barbital but without the chemical visualization utilized in thin-layer chromatography.

The secobarbital and amobarbital were quantitated by gas chromatography as their dimethyl derivatives via an on-column reaction with trimethylanilinium hydroxide (Figure 3). This provides sharper chromatographic peaks with a minimum amount of tailing. The secobarbital and amobarbital levels determined were greater than each drug's minimum toxic level, 1.0 and 1.5 mg/100 ml, respectively.

Each of the drugs [alcohol (0.23%), secobarbital (1.1 mg/100 ml), and amobarbital (1.6 mg/100 ml)] was present at toxic levels and individually could have caused intoxication. The synergistic effects of alcohol and barbiturates are well documented, and in this case, an acute intoxication and overdose death were quite probable. The forensic chemist's analysis indicating adverse blood barbiturate levels, to which the subject had no known access, prompted further investigation.

Apprehension of the male companions showed one to have traces of white powder in the pockets of the pants worn on the fatal evening. Preliminary screening of the trace powder with UV spectrophotometry indicated the presence of a barbiturate (Figure 4). Further examinations of the powder by gas chromatography, thin-layer chromatography, and gas chromatography–mass spectrometry showed the powder to contain both amobarbital and secobarbital.

Thus, a suspected overdose death was transformed into a drug homicide investigation and eventual criminal trial as a consequence of the forensic chemist's analyses.

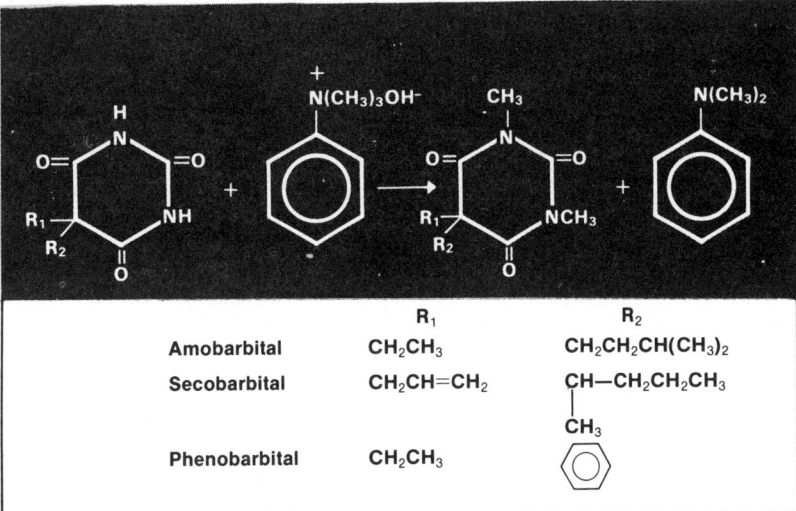

Figure 3. Methylation of barbiturates

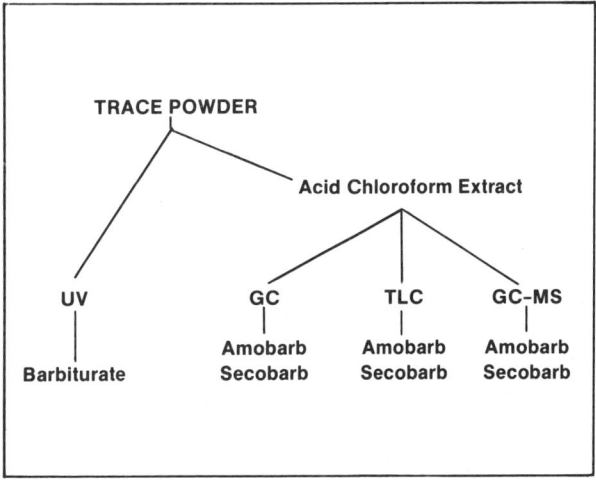

Figure 4. Identification scheme for unknown powder

JFK Assassination: Bullet Analyses

Vincent P. Guinn

University of California, Chemistry Department, Irvine, CA 92717

Originally published in ANALYTICAL CHEMISTRY, 1979, Vol. 51, No. 4.

The world was shocked on November 22, 1963, when President John F. Kennedy was assassinated by rifle fire in Dallas, Tex. It rapidly became clear that either some or all of the bullets were fired at the President's motorcade limousine from a corner room on the sixth floor of the Texas School Book Depository, just after the motorcade turned onto Elm Street in Dealey Plaza. Within about 2 h after the shooting, Lee Harvey Oswald was captured a few miles away, but only after he had fired four revolver shots into Officer J. D. Tippit, killing him instantly. Initially arrested for the killing of Officer Tippit, it soon became evident that Oswald was probably also the assassin of President Kennedy. A search of the Book Depository sixth floor room resulted in the finding of three spent Western Cartridge Co. (WCC) 6.5-mm Mannlicher-Carcano (MC) cartridge cases on the floor of the room, and an Italian-made Mannlicher-Carcano military rifle with one unfired cartridge of the same type still in the gun. Oswald, an employee of the Book Depository, had been seen there that morning and also a few minutes after the assassination—disappearing soon thereafter.

During the next 48 h, Oswald was interrogated repeatedly, but consistently denied having killed either the President or Officer Tippit. While the investigation was still in progress, Oswald was shot and killed by Jack Ruby while Oswald was being transferred to other quarters.

Warren Commission Investigation

Within days, numerous rumors were afoot that Oswald was part of a conspiracy to assassinate the President, and that he had one or more confederates, firing from the same location or from another location. The facts, soon unearthed, that Oswald was an avowed Marxist, had recently lived for about three years in the Soviet Union where he married a Russian (Marina), and was an outspoken supporter of Fidel Castro's Cuba, led many people to suspect an international conspiracy. To investigate all of these possibilities in depth, and to provide the American public with a factual analysis of the assassination, the new President, Lyndon B. Johnson, appointed a Presidential Commission to conduct a thorough investigation of all aspects of the assassination. Headed by then Chief Justice Earl Warren, the 7-member Commission became known as the Warren Commission. Constituted only one week after the assassination, it conducted a 10-month investigation assisted mainly by the FBI, and published its findings in September of 1964 in the famous Warren Commission Report.

In this investigation, the Mannlicher-Carcano rifle was definitely proved to belong to Oswald (his palmprint was found on it), the three recovered cartridge cases were proved to have been fired from it, and the almost undamaged copper-jacketed bullet found on Governor Connally's stretcher at the Parkland Memorial Hospital in Dallas and two large pieces (a nose portion and a base portion) of copper-jacketed bullet(s) found in the President's limousine were all shown to have been fired from that particular rifle. Clearly, Oswald appeared to be either the lone assassin or at least one of the assassins.

1963–1964 Analyses

All of the analytical measurements conducted by the Dallas police (inconclusive dermal nitrate tests run on paraffin casts taken of Oswald's hands and his right cheek soon after he was apprehended) and by the FBI Laboratory shed very little light on the subject. The FBI took the Oswald paraffin casts to the Oak Ridge National Laboratory and analyzed them by neutron activation analysis (NAA) for the possible presence of primer residue (barium and antimony) (1) still there even after the Dallas dermal nitrate tests. This effort was thwarted by the fact that the casts were badly contaminated, essentially as much Ba and Sb being found on the outside surfaces of the casts as on the inside surfaces—which had been in contact with Oswald's skin. The right cheek cast, if it had not been contaminated by improper previous handling, might have established that Oswald had very recently fired a rifle. The FBI Laboratory also analyzed the various bullet fragments recovered from the Dallas limousine, President Kennedy's brain, and Governor Connally's wrist plus the bullet recovered from his stretcher. These specimens were analyzed by emission spectrography. The results showed that all of the bullet-lead specimens were qualitatively generally "similar" in elemental composition,

192 The Analytical Approach

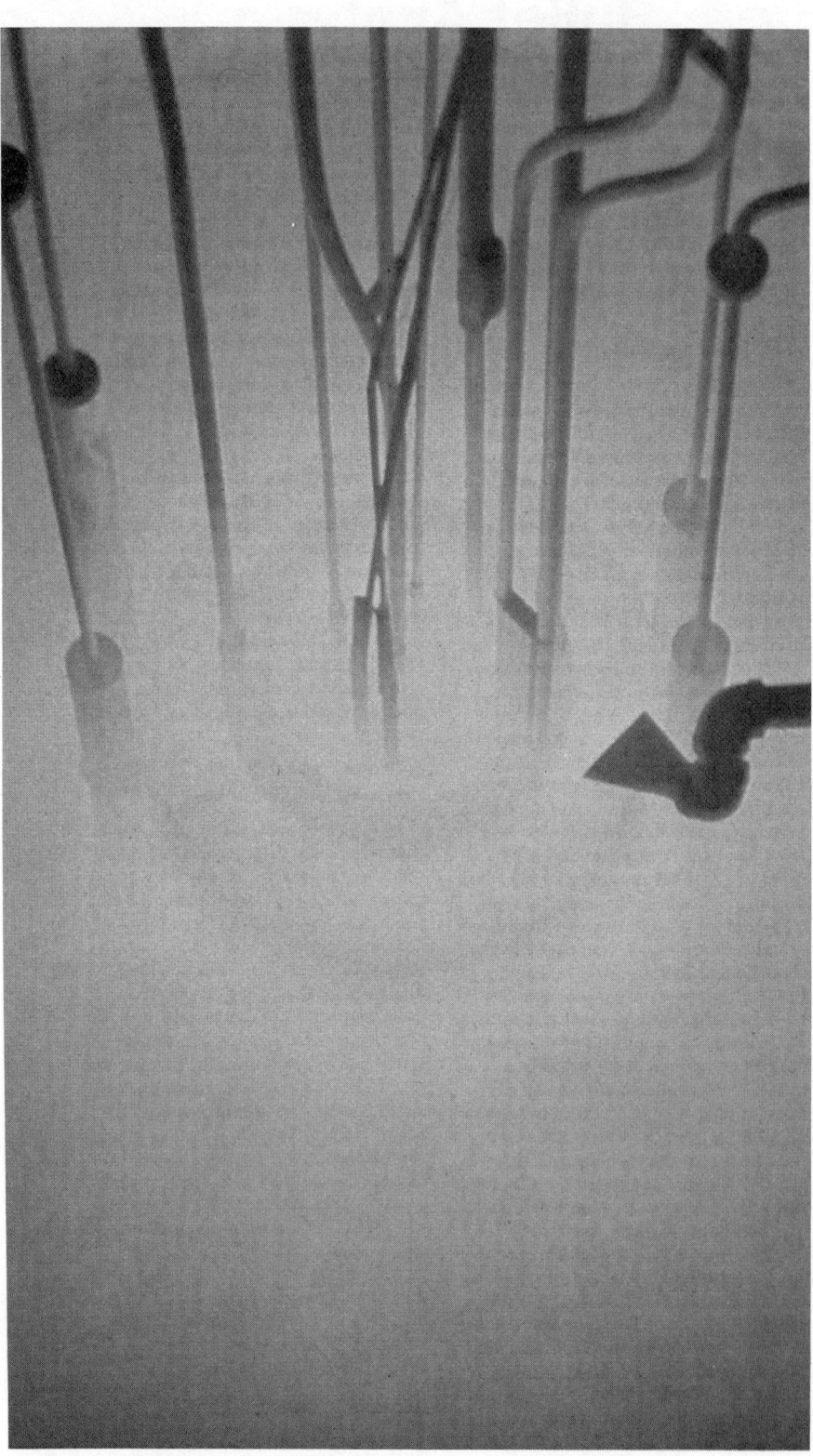
The core of the U.C. Irvine TRIGA Mark I nuclear reactor

and "could be" all of the same brand of ammunition. That was all of the chemical analysis work cited in the Warren Commission Report.

Surprise Letter Found in the National Archives

For a number of years after publication of the Warren Commission Report, the author, several other forensic scientists, and various critics of the Commission urged that the Dallas bullet-lead evidence specimens should be examined in more quantitative detail by some more powerful method, such as NAA, to test the Commission's conclusion that all of the specimens recovered were fired only by Oswald. The FBI, under J. Edgar Hoover, declined or ignored all such suggestions. And then a great surprise occurred late in 1973, almost 10 years after the assassination—a letter from J. Edgar Hoover to the Warren Commission, dated July 8, 1964, turned up in the National Archives. This hitherto unknown letter disclosed the fact that, after the generally not very informative emission spectrographic analyses, the bullet-lead specimens had been analyzed by the FBI, using the NAA method. The letter contained no numerical results at all, but merely stated that the NAA results were also inconclusive, and did not allow one to discern how many bullets were represented by the various recovered fragments, although it did state that some compositional differences were found. Why Mr. Hoover chose not to reveal that these NAA measurements had been made, and why there is no mention of them in the Warren Commission Report, is still a mystery.

1964 FBI NAA Data

Working with John Nichols, a forensic pathologist at the University of Kansas Medical School, I joined in efforts to obtain a copy of the 1964 NAA data obtained by the FBI on the bullet-lead specimens. This also proved to be a slow uphill battle. Finally, and only by taking legal action under the amended Freedom of Information Act, Dr. Nichols succeeded in obtaining a copy of the FBI data—in April of 1975. Dr. Nichols immediately flew out to California with the data, and I began a detailed examination of the 70 pages of the raw NAA data and calculated results that had been obtained at the Oak Ridge National Laboratory in May 1964 by John F. Gallagher of the FBI Laboratory (now retired). The author's initial examination of these data tended to agree with Mr. Hoover's statement that the results were inconclusive. But I will discuss more about these data later.

WCC Mannlicher-Carcano Bullet Lead

At this point, Dr. Nichols and I urged a reexamination of the bullet-lead evidence specimens—this time using nondestructive instrumental neutron activation analysis (INNA) with a high-resolution Ge(Li) semiconductor gamma-ray detector instead of the low-resolution thallium-activated sodium iodide [NaI(Tl)] scintillation detector that had been used in 1964. During the period 1972–1976, I had analyzed a number of samples of WCC/MC 6.5-mm bullet lead, from all four of the production lots made by WCC, using instrumental NAA with Ge(Li) gamma-ray spectrometry. These known samples were supplied by Dr. Nichols and gave surprising results. The results showed that this type of ammunition was quite different from virtually all other brands of bullet lead I had ever analyzed before (2, 3). Although individual bullets were fairly homogeneous in their antimony and silver contents, they exhibited a great heterogeneity from bullet to bullet—even within the same production lot and even within an individual box of 20 cartridges. The range of Sb values was especially large, all the way from around 20 ppm up to 1200 ppm Sb. Although still in the range of unhardened lead, they clearly were not made from virgin lead but instead obviously contained appreciable and variable amounts of recycled lead—some of which was antimony-hardened lead. The silver levels ranged from about 5 ppm to around 15 ppm Ag. These background results showed that a more detailed analysis of the Dallas bullet-lead specimens, by INAA with Ge(Li) gamma-ray spectrometry, might be able to establish whether all of the specimens were or were not in the range of WCC/MC bullet lead, and whether they corresponded to two, or more than two, bullets.

The Warren Commission, it should be noted, had concluded that Oswald fired three WCC/MC bullets; that one of them had missed completely; that one of them had struck the President in the back, exited from his throat, gone on to strike Governor Connally in the back (he was sitting in a jump seat right in front of the President), exited from the Governor's chest after fracturing one of his ribs, struck the Governor's right wrist—shattering it, exited from the under side of his wrist, and then came to rest in his left thigh after penetrating the flesh only slightly—finally to fall out on the Governor's stretcher at the hospital; and that a later (or last) one had struck the back of the President's head, exiting near the right front of his head, causing a massive and fatal wound. After issuance of the Commission's Report in 1964, numerous critics scoffed at their conclusion that a single bullet could cause the President's back wound and the Governor's back, wrist, and leg wounds—and still end up with only a slight dent in it and with only about a 1% weight loss. They dubbed it the "Magic Bullet." In addition, partly based on the Zapruder film and on eyewitness accounts of gunfire from a region in front of the limousine (the so-called "grassy knoll"), rather than from the rear, some critics claimed that the fatal head shot did not come from Oswald's location.

A Phone Call from the House Select Committee

In the early summer of 1977, the author received a phone call from a staff member of the U.S. House of Representatives Select Committee on Assassinations—inquiring (1) whether the author thought it might yield new information if the Dallas bullet-lead evidence specimens were reanalyzed, using improved INAA techniques; (2) what kind of information might be generated by such measurements; and (3) whether the author would be willing to conduct such analyses for the Select Committee. To (1) and (3) the answer was "yes," and the answer to (2) was a summary of the information possibly obtainable from such measurements, based upon the author's earlier studies of WCC/MC bullet lead. It was then agreed that the author would reanalyze the Dallas specimens, using the sensitive and nondestructive INAA method, with Ge(Li) gamma-ray spectrometry.

New INAA Measurements

In mid-September of 1977, James L. Gear of the National Archives flew out to California with the Dallas bullet-lead specimens. During a three-day period, the author examined the specimens, prepared them for analysis, analyzed the samples in two reactor runs under different conditions, and returned the samples to him. An amusing aspect was that the measurements had to be conducted under tight security and secrecy—during every working hour of those three days Mr. Gear and the author were accompanied by two armed, uniformed federal guards. Needless to say, speculation amongst the U.C. Irvine students ran high! Later, reading the Ge(Li) pulse-height data from the samples and the Sb, Ag, and Cu standards back from the magnetic tape, I proceeded to calculate the ppm levels of these three elements in each sample.

Preparation of the samples for analysis was itself somewhat of a problem. The various samples ranged in size from about 1 mg on up to one large portion of a jacketed bullet (Q2) on up to an almost whole bullet (the Connally stretcher bullet, Q1). The smaller pieces were taken each in its entirety. From the large Q2 specimen, a piece of the bullet lead free of copper jacket was cut off with a stainless steel scalpel, for analysis. A sample of bullet lead was drilled out from the base of the stretcher bullet (Q1), using a 0.5-mm diameter carbon-steel drill. Each sample was examined under magnification to be sure no specks of imbedded jacket material could be detected. To remove as much as possible of any external contamination from the samples, each was washed three times, alternately, with distilled/deionized water and reagent grade acetone. Each sample was weighed into a specially cleaned small polyethylene vial on an analytical balance. Several small standard samples each of Ag, Sb, and Cu were prepared in the same size vials, the solutions evaporated to dryness, and the materials cemented to the inside of the bottom of the vial by paraffin (using a few drops of 10% paraffin/CS_2 solution, followed by air drying). The conduct of the analyses was somewhat restricted by the security/secrecy requirements, and by the short time available to carry them out (precluding the possibility of doing replicate determinations of each sample, except for the silver determinations, which could be repeated, due to the short half life of ^{110}Ag). Each night, all of the evidence specimens and the analytical samples had to be taken away for overnight security storage at the Laguna Niguel branch of the National Archives.

In the first run, the 1–50-mg samples and the standards were activated and counted one at a time, using the pneumatic-tube facility of the U.C. Irvine TRIGA Mark I nuclear reactor. The conditions used were those of our rapidscreening method (4), giving precision measurements for silver and less precise values for antimony and copper: irradiation, decay, and counting times of 40 s each. The samples were activated in a thermal-neutron flux of 2.5×10^{12} n cm^{-2} s^{-1}, and each was transferred to a fresh polyvial during the 40-s decay period. Each activated sample was counted with a 38 cm^3 Ge(Li) detector/4096-channel gamma-ray spectrometer, and the pulse-height data were promptly stored on magnetic tape. The most prominent induced activities detected were 24.4-s ^{110}Ag, 93-s $^{124m_1}Sb$, and 5.10-min ^{66}Cu. Quantitative results were based upon the largest photopeak of each (658, 498, and 1039 keV,

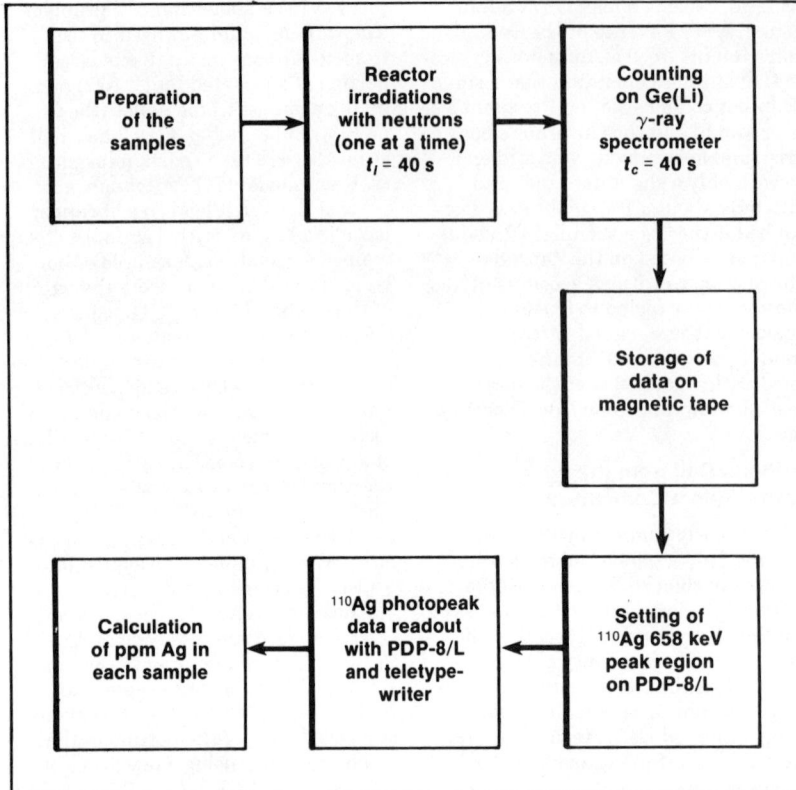

Figure 1. Schematic of INAA determination of silver in bullet-lead specimens via 24.4-s ^{110}Ag

respectively). The analytical sequence is shown schematically in Figure 1.

In the second run, the same samples (and standards) were activated and counted again, but this time all together in the 40-tube rotary specimen rack of the reactor—for 1 h at a thermal-neutron flux of 1.0×10^{12} n cm^{-2} s^{-1}. After about a 1-h decay, the samples and standards were counted as before, but this time for 5 min each. The most prominent induced activities were 2.80-day ^{122}Sb and 12.8-h ^{64}Cu, providing more precise values of the antimony and copper levels. Quantitative results were based upon the largest photopeak of each (564 and 511 keV, respectively).

Results of New INAA Measurements

As can be seen in Table I, samples Q1 and Q9 (the Connally stretcher bullet and fragments from Connally's wrist, respectively) are indistinguishable from one another in their Sb and Ag concentrations, but are clearly distinguishable from the Q2, Q4, 5, and Q14 samples (Q4, 5 being fragments recovered from two different areas in the Dallas limousine)—these latter three samples, in turn, being indistinguishable from one another. The results for copper follow the same pattern except that no reliable value for Cu could be obtained for sample Q9, because it was greatly contaminated with Cu from imbedded jacket material. From the induced 66.9-min ^{204m}Pb activities [an (n,n') fast-neutron product of lead] (5), it was also established that all of the samples were at least 90% lead. The sample Q designations, by the way, are those originally assigned to the evidence specimens by the FBI. The CE designations, which are also shown in Table I, are the Warren Commission Exhibit numbers assigned to them.

The conclusions derived from these results—interpreted in the context of my earlier measurements on background WCC/MC bullet-lead samples—are definite and straightforward: all of the Dallas samples are in the unusual (though not necessarily unique) concentration ranges of WCC/MC bullet lead; and the specimens show clearcut evidence for the presence of two, and *only* two, WCC/MC bullets—one of a composition of 815 ppm Sb and 9.3 ppm Ag, the other of a composition of 622 ppm Sb and 8.1 ppm Ag.

A Second Look at Earlier FBI Data

After I had obtained these new results, it seemed to me that the presence of two different compositions should have been discernible from the FBI's 1964 NAA data, in spite of the complication of the 20-fold poorer energy resolution of the NaI(Tl) scintillation detector that Mr. Gallagher used [the high-resolution Ge(Li) detector was not generally available in 1964]. My previous examination of the FBI data had revealed that the results had been obtained for silver (via the 24.4-s ^{110}Ag induced activity, as in the newer measurements) under one set of irradiation/decay/counting time conditions. His values for silver agreed closely with the new values. The complication, however, was that Mr. Gallagher measured antimony (via the 2.80-day ^{122}Sb and 60.4-day ^{124}Sb induced activities) under four different sets of irradiation/decay/counting time conditions—unfortunately obtaining four rather widely different Sb values for each sample. The wide spread of Sb values for each sample obscured any distinction between the Q1 and Q9 samples, on the one hand, and the Q2, Q4, 5, and Q14 samples, on the other—if all the results were viewed simultaneously. This confusion no doubt led Mr. Hoover to state that the results were inconclusive.

However, my second review of the FBI data (6) [benefiting, of course, from the hindsight gained from the newer Ge(Li) results] resolved this anomaly. Although not heretofore realized, the old FBI data also showed that samples Q1 and Q9 were similar to one another in their Sb and Ag contents (Cu was detected, but not measured), and distinct from samples Q2, Q4, 5, and Q14—which, in turn, were similar to one another. This conclusion was reached by examining the Sb results obtained by the FBI for all the samples under all four conditions—but with the data for each condition

Table I. Author's 1977 INAA Silver and Antimony Results[a]

Sample no.	ppm Ag	ppm Sb	Sample description
Q1 CE-399	8.8 ± 0.5	833 ± 9	Connally stretcher bullet
Q9 CE-842	9.8 ± 0.5	797 ± 7	Fragments from Connally's wrist
Q2 CE-567	8.1 ± 0.6	602 ± 4	Large fragment found in car
Q4, 5 CE-843	7.9 ± 0.3	621 ± 4	Fragments from President Kennedy's brain
Q14 CE-840	8.2 ± 0.4	642 ± 6	Small fragments found in car

[a] The measurement precisions shown in Tables I and II as ± values represent 1 standard deviation. Studies by the author on background samples of WCC/MC bullet lead show that the Sb variability within an individual bullet, particularly, is usually several times larger than the measurement precision on an individual sample.

Table II. 1964 FBI Antimony NAA Results from Four Sets of Measurements (and the One Set of Silver Results)

Sample no.	ppm Ag	ppm Sb			
		Set 1	Set 2	Set 3	Set 4
Q1	9.4 ± 0.3	945 ± 16	1002 ± 13	813 ± 43	705 ± 54
Q9	9.2 ± 0.1	977 ± 24	1090 ± 37	773 ± 22	676 ± 14
Q2	7.9 ± 0.9	745 ± 16	747 ± 20	626 ± 57	534 ± 30
Q4, 5	8.5 ± 0.4	783 ± 5	858 ± 46	614 ± 37	561 ± 32
Q14	8.5 ± 0.2	793 ± 10	879 ± 33	629 ± 18	562 ± 21

compared only with one another, rather than intercomparing the results obtained for each sample under all four conditions. Examined in this fashion, it was revealed that, for each of the four FBI measurement conditions, samples Q1 and Q9 matched closely, and were quite different from samples Q2, Q4, 5, and Q14, which in turn matched one another closely. The FBI results are displayed in this fashion in Table II. Apparently, some errors in some of the standards used, and/or in the counting conditions used in the four different measurements led to some consistent (determinate) errors that resulted in four different Sb values being obtained on the same sample. The FBI values for one of their conditions (set 3) agree closely with the newer Ge(Li) values, but their results for the other three conditions are numerically considerably different from the Ge(Li) results.

Presentation of Results Before the Select Committee

After I had submitted a detailed report to the Select Committee on my INAA results and conclusions concerning the Dallas bullet-lead evidence specimens, the Committee requested that I present these at their public hearings in Washington, D.C. On September 8, 1978, I presented a 90-min summary of my findings and conclusions (nationally televised on public service TV), the 90 min including questioning by various of the 12 Congressmen who constitute the Select Committee.

Thus, analytical chemistry—which in 1963–64 had not shed much light on the assassination—finally succeeded in producing significant useful information. The nondestructive instrumental neutron activation analysis results have demonstrated that, to a high degree of probability, all of the bullet-lead evidence specimens are of WCC/MC 6.5-mm brand, that there is evidence for the presence of portions of two—and only two—such bullets, and that the Connally stretcher virtually intact bullet indeed caused the fracture wound of Governor Connally's wrist—a previously hotly disputed part of the Warren Commission's theory. The back wounds of President Kennedy and Governor Connally involved essentially no damage to the bullet (or bullets) causing them, and thus produced no fragments for possible analysis. The new results cannot prove the Warren Commission's theory that the stretcher bullet is the one that caused the President's back wound and all of the Governor's wounds, but the results are indeed consistent with this theory.

And What Now?

In due time, my report to the Select Committee will be made public and available, and I will be submitting a series of several papers to the *Journal of Forensic Sciences* (of the American Academy of Forensic Sciences) that will cover this investigation in greater detail. These papers will also include additional studies in my laboratory, some of them still in progress at this writing, on background WCC/MC bullet-lead samples (further homogeneity studies) that will enable one to calculate actual numerical probabilities.

References

(1) R. R. Ruch, V. P. Guinn, and R. H. Pinker, *Trans. Am. Nucl. Soc.*, **5**, 282, subsequent papers, AEC report (1962).
(2) H. R. Lukens and V. P. Guinn, *J. Forensic Sci.*, **16**, 301 (1971).
(3) H. R. Lukens, H. L. Schlesinger, V. P. Guinn, and R. P. Hackleman, "Forensic Neutron Activation Analysis of Bullet-Lead Specimens," U.S. AEC Report GA-10141, 48 pages, 1970.
(4) V. P. Guinn and M. A. Purcell, *J. Radioanal. Chem.*, **39**, 85 (1977).
(5) T. Izak-Biran and V. P. Guinn, *Trans. Am. Nucl. Soc.*, **28**, 94 (1978).
(6) V. P. Guinn and J. Nichols, *ibid.*, p 92.

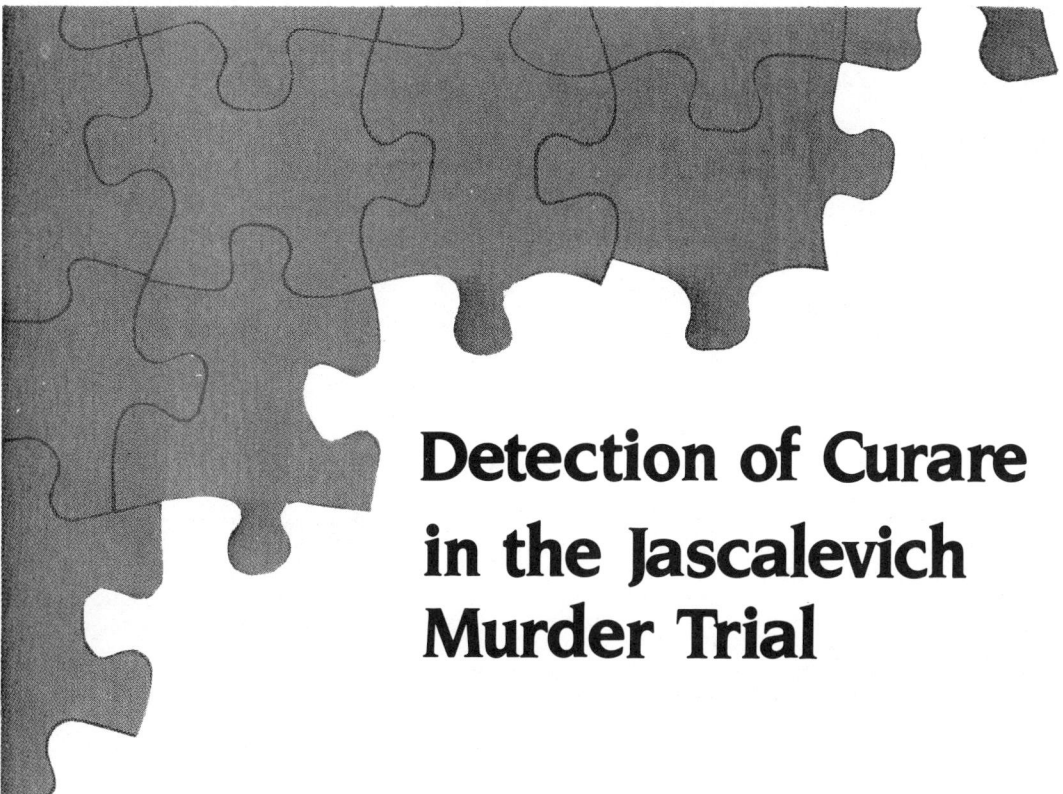

Detection of Curare in the Jascalevich Murder Trial

Lawrence H. Hall

THE STAR-LEDGER, Newark, NJ 07101

Roland F. Hirsch

Seton Hall University
Chemistry Department
South Orange, NJ 07079

Originally published in ANALYTICAL CHEMISTRY, 1979, Vol. 51, No. 8.

The murder trial of Dr. Mario E. Jascalevich was one of the most complicated criminal proceedings ever tried in an American courtroom. The 34-week trial before a Superior Court judge in New Jersey resulted in a not-guilty verdict for the Englewood Cliffs, N.J., surgeon. The questions concerning analytical chemistry raised in the trial will continue to be discussed in years to come.

Not since the controversial trial of Dr. Carl Coppolino—convicted in a Florida courtroom in 1967 of murdering his wife with succinylcholine chloride—have so many forensic experts of national and international stature labored so long over the scientific questions at issue in the case:

• What happens to human tissue, embalmed and interred for a decade?

• Assuming lethal doses of a drug such as curare were given to hospital patients, would the drug have changed chemically or have been destroyed entirely over a 10-year period?

• Assuming again that the drug had been injected, what analytical techniques could be employed to trace submicrogram amounts of it?

• Could components of embalming fluids or bacteria in the earth react chemically, forming substances giving a false positive reading in the analytical procedures used?

Forensic scientists first grappled with these questions during the latter part of 1966. Two of Jascalevich's colleagues at Riverdell Hospital in Oradell, N.J.—Dr. Stanley Harris, a surgeon, and Dr. Allan Lans, an osteopathic physician—suspected him of murdering their patients with curare. There were no eyewitnesses to the alleged murders, but Drs. Harris and Lans discovered 18 vials of curare in Jascalevich's surgical locker after breaking into it.

They took their suspicions to the Bergen County Prosecutor's office in November 1966, and a brief but unpublicized investigation was launched. Items taken from the surgeon's locker, including the vials of curare and syringes, were sent for analysis at the New York City Medical Examiner's office.

In the interim, Jascalevich told authorities he used the muscle-relaxant drug in animal experiments at the Seton Hall Medical College. The surgeon presented the prosecutor his medical research papers and other documentation to support his contention. In addition, he reviewed the medical charts of the alleged murder victims and told the prosecutor there was no need for the operations the patients received. Malpractice and misdiagnosis were the causes of the deaths, Jascalevich stated at that time.

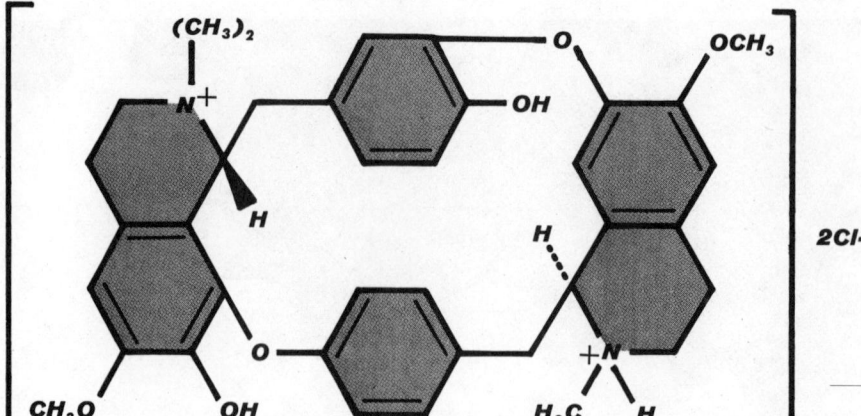

(+) tubocurarine chloride

Dr. Milton Helpern, chief of the New York City Medical Examiner's office, and his staff in early 1967 concluded their testing on the items taken from Jascalevich's locker. Dog hair and animal blood were detected on the vials of curare and syringes.

The prosecutor's office terminated its investigation and stated there were more reasons to look into allegations of malpractice than murder at the small osteopathic hospital.

In January 1976 a series of articles about a "Doctor X" suspected of murdering patients at Riverdell Hospital appeared in the *New York Times,* and the Bergen County Prosecutor's office reopened its case.

A month prior to the case being officially reopened, however, New York Deputy Medical Examiner Dr. Michael Baden supplied an affidavit with the Superior Court in Bergen County, stating that at least a score of patients who died at Riverdell in 1966 succumbed from other reasons than those stated on death certificates.

In his affidavit in support of exhumation of the patients' remains, Dr. Baden stated, "It is my professional opinion that the majority of these cases reviewed are not explainable on the basis of natural causes and are consistent with having been caused by a respiratory depressant."

The deputy medical examiner continued, "I am aware that because unexplainable respiratory arrests have been involved in many of these deaths, the possibility of poisoning by a curare-like substance (specifically d-tubocurarine) was considered and investigated at the time of the initial inquiry in 1966.

"The ability to identify d-tubocurarine, often referred to as curare, in human tissue was limited at the time of the initial investigation.

"It is my professional opinion that recent technological advances now permit the detection of very minute amounts of d-tubocurarine in tissues removed from dead bodies. This is because d-tubocurarine is a chemically stable compound that can exist unaltered for many years.

"Therefore, the aforementioned new techniques to detect curare-like compounds can be applied to tissues removed from bodies that have been interred for long periods of time."

A Superior Court judge signed the order in January 1976, granting the prosecutor's office the right to exhume the bodies of Nancy Savino, 4; Emma Arzt, 70; Frank Biggs, 59; Margaret Henderson, 27; and Carl Rohrbeck, 73.

All these patients entered Riverdell Hospital between December 1965 and September 1966 for routine surgical procedures and succumbed days afterward.

In mid-January 1976 the body of the Savino child was exhumed from a gravesite in Bergen County and taken to the medical examiner's office in New York City.

There, Dr. Baden, in the presence of New Jersey State Medical Examiner Dr. Edwin Albano and others, began performing the almost 4-h examination of the child's body, which was said to be well preserved.

Assisting Dr. Baden in the analytical studies carried out on the tissues were Dr. Leo Dal Cortivo, chief toxicologist for Suffolk County, N.Y., and Dr. Richard J. Coumbis, chief toxicologist for the New Jersey Medical Examiner's office. The defense experts, headed by former Westchester County (N.Y.) Medical Examiner Dr. Henry Siegel, were not permitted to be present at the reautopsies.

The state began its work. In March, a week before the grand jury met, a newspaper article declared that curare had been detected in the Savino child. However, in his grand jury testimony weeks later, Dr. Baden stated his experts could not be certain if curare could be detected: ". . . we have to look and see whether or not we can develop adequate procedures."

On May 18, 1976, Dr. Jascalevich was indicted for five murders.

A little more than a year later, the state's forensic experts began using radioimmunoassay (RIA) and high-performance liquid chromatography (HPLC) on the tissue specimens. In the fall of 1977, the defense received from Drs. Baden and Dal Cortivo samples of tissues and embalming fluids of the alleged murder victims.

For the remainder of the year, both the defense and the state experts worked to develop analytical procedures to settle the question of detection of curare in human tissue.

In addition, there were numerous pretrial hearings at which time the defense, headed by Jersey City attorney Raymond Brown, requested medical slides, reports, and patient charts relating to the alleged murder victims, as well as the methodologies used in treating the specimens.

On February 28, 1978, a panel of 18 jurors was chosen for what was to become the second longest criminal trial in the nation's history. At the outset, the defense wanted a hearing to ascertain the validity of the scientific procedures employed by the state to reportedly detect curare.

The defense contended that RIA and HPLC were relatively new procedures and could not be used to detect curare in human tissue. RIA, for example, could only be used to detect drugs in blood and body fluids, according to defense experts.

The defense motion for a hearing outside of the presence of the jury was denied by Superior Court Judge William J. Arnold, who maintained the motion could be made later in the trial when the evidence obtained by the analytical techniques would actually be scheduled for presentation to the jury.

The trial got underway with testimony by osteopathic physicians, nurses, and other hospital personnel employed by Riverdell during the time the alleged murders were committed. The physicians told Assistant Prosecutor Sybil Moses that in each in-

stance the patient had been recovering from surgery when he succumbed.

However, on cross-examination, the physicians admitted they had misdiagnosed their patients' conditions and that there was inferior postoperative care. For example, in the case of the Savino child, the defense experts held that the little girl died of acute diffuse peritonitis—the source of her abdominal pain when she was brought into Riverdell after having been diagnosed as having acute appendicitis.

After the prosecution completed presentation of the medical aspects of its case, the defense renewed its request for a special hearing on the admissibility of the evidence obtained by radioimmunoassay, liquid chromatography, and other analytical techniques. This request came as Dr. Baden took the witness stand to explain why he had recommended reautopsy of the bodies. The prosecution was opposed to a hearing:

"The techniques used by the State are not new toxicological methodologies, but are standard methods, used widely throughout the field. These methodologies include radioimmunoassay and high-pressure liquid chromatography. . . .

"Since the methodologies used to detect the curare are widely accepted in the scientific community, there is no necessity for the Court to conduct a hearing as to their reliability."

Nevertheless, Judge Arnold ruled that a hearing should be held. Arguments began, in the absence of the jury, on June 10. Both sides presented statements by their technical experts and affidavits from other scientists regarding the validity of the analytical methods. The prosecution cited various cases in support of its position:

"Practically every new scientific discovery has its detractors and unbelievers, but neither unanimity of opinion nor universal infallibility is required for judicial acceptance of generally recognized matters [*State v. Johnson,* 42 N.J. 146, 171 (1964)].

"The law in its efforts to enforce justice by demonstrating a fact in issue, will allow evidence of those scientific processes, which are the work of educated and skillful men in their various departments and apply them to the demonstration of a fact, leaving the weight and effect to be given to the effort and its results entirely to the consideration of the jury [*State v. Cerciello,* 86 N.J.L. 309, 314 (E&A 1914)]."

The prosecution stated, "Federal courts have held that newness or lack of absolute certainty in a test does not require its inadmissibility." In one case involving neutron-activation analysis, a federal appellate court held in part:

"Every useful new development must have its first day in court. And court records are full of the conflicting opinions of doctors, engineers, and accountants, to name just a few of the legions of expert witnesses" [*United States v. Stifel,* 433 F. 2d. 431, 437, 438 (6th Cir. 1970)].

The prosecution noted, "The Florida Appellate Court in *Coppolino v. State* . . . held that not only established techniques but methods developed specifically for that case could be used to detect a previously undetectable drug in the body of the decedent . . . .

"The tests by which the medical examiner sought to determine whether death was caused by succinylcholine chloride were novel and devised specifically for this case. This does not render the evidence inadmissible. Society need not to tolerate homicide until there develops a body of medical knowledge about some particular lethal agent. The expert witnesses were examined and cross-examined at great length and the jury could either believe or doubt the prosecution's testimony as it chose" [*Coppolino v. State,* 223 So. 2d. 75 (Fla. App. 1968)].

Finally, the prosecution noted the following holding of the New Jersey Superior Court Appellate Division:

"The general rule in New Jersey regarding the admissibility of scientific test results is that, if the equipment or the methodology used is proven to have a high degree of scientific reliability, and if the test is performed or administered by qualified persons, the results will be admissible at trial" [*State v. Chatman,* 101 N.J.L.S. index 307, 308 (App. Div. 1973)].

The defense contended, "that the methodologies of thin layer chromatography (TLC), high pressure liquid chromatography, ultraviolet spectrophotometry, and radioimmunoassay which have been utilized by the State do not meet the required level of acceptance under the circumstances of the tissues in this case . . . . Since there have never been any attempts to demonstrate the presence of d-tubocurarine in embalmed, buried tissue . . . the State cannot even assert that the techniques it wishes to utilize to demonstrate this have been generally accepted."

The defense presented affidavits from a variety of forensic scientists, from which we present one example:

"It should be noted that even though the newer analytical methods and some of the sophisticated equipment are extremely sensitive for drug detection, the sensitivity of some method is not a criterion of its specificity. Sensitivity is the minimum amount of an unknown substance below which a test gives a negative result. Specificity is the ability of a test to establish the individual characteristics and/or configuration of a particular substance by differentiating it from all other substances, especially in a biologic mixture.

"Currently, the reported analytical methods, which include ultra-violet absorption spectrophotometry, thin layer chromatography, high pressure liquid chromatography and radioimmunoassay, alone or in conjunction lack such a degree of specificity with any degree of scientific certainty required to support the opinion that they identified the isolated material as d-tubocurarine in embalmed, decomposed and skeletonizing tissues that have been in the ground for ten years under varying climatic conditions" [Abraham Stolman, Chief Toxicologist, State of Connecticut Department of Health].

On June 20 the judge ruled that the analytical evidence was admissible. He stated; "All I'm saying is under the law the evidence is admissible. I'm not going to comment on the value or trustworthiness of the witnesses [who testified]. The ultimate decision must be made by the jury."

Following this decision, the jury began listening to the scientific evidence, with the State's and the defense's witnesses in the process explaining such points as: What is curare, and specifically d-tubocurarine? What is radioimmunoassay? What is an antibody, and how is the antibody for d-tubocurarine created? What is high-pressure liquid chromatography?

Dr. Richard Coumbis testified about his finding tubocurarine in tissues from four of the five patients: "... can only state there is presumptive evidence" that curare was discovered in the fifth patient. Under cross-examination by defense attorney Raymond Brown, Coumbis maintained that the RIA and HPLC procedures were valid methods of detecting curare because "on the basis of my personal experience, I did not find any other substance interfering with curare."

The toxicologist admitted that the counting efficiencies of the instruments he used to get the RIA displacement values varied from day to day and were subject to error. Brown disagreed with the displacement figures Coumbis arrived at, and wanted to know whether there was a "cut-off point" whereby he arrived at the conclusion that curare was or was not present in tissues. The RIA results ranged from as low as 77 counts all the way up to 700. Somewhere within that range, Brown argued, was a point at which Coumbis arrived at the decision that the drug was detected or not. Where, he asked, was that point? The toxicologist responded by saying that the higher the figure, the more likely curare was present. He said in many instances, however, he had to use his discretion to determine the cut-off point.

Dr. David Beggs of Hewlett-Packard then testified that he found curare in the Savino lung and liver samples using mass spectrometry. He said the Biggs and Arzt samples contained possible traces of curare; however he could not be scientifically certain of this. He stated that mass spectrometry "is not an absolute test" for curare, but "just indicated that it is probably there." He did carry out a solvent blank as a means of eliminating false positives. He held under cross-examination that the electron impact technique used by him resulted in a spectrum with 12 major peaks and that 10 were sufficient for "fingerprint" identification of curare.

Dr. Sidney Spector of the Roche Institute testified about how he had developed the antibody for d-tubocurarine and applied it in RIA analysis of body fluids such as urine and blood. He had not himself run any tests for curare in human tissue samples and stated, "If there were curare in tissues, there is the possibility it could be detected." He said that the State's RIA experiments were "inadequate" in relying on aqueous solutions of curare to develop a standard curve. He held that the RIA procedure could give an indication that curare was present, but that the finding would only be presumptive evidence and not sufficient to say that the muscle-relaxant drug was positively present. He made the same point about HPLC and said that even if the two techniques were used together, there still would only be presumptive proof that the drug was present.

Dr. Leo Dal Cortivo then took the witness stand and testified that he had found curare in tissue remains of three of the patients using HPLC. He also had measured curare in vials found in the defendant's locker at Riverdell Hospital in 1966, which the defense contended had been used in animal experiments conducted by Jascalevich at the College of Medicine in Jersey City. It was necessary to use RIA for the detection of curare in the HPLC eluates. The samples were prepared for LC analysis by an extraction procedure which Dal Cortivo stated gave a 75% recovery. He rejected the contention that the extraction and LC method might have allowed positive results because of an interfering substance.

The prosecution then completed its case. At this point Judge Arnold dismissed two counts of murder and stated that the prosecution had not presented scientific evidence for the presence of curare in the bodies of Emma Arzt and Margaret Henderson. The defense then began presentation of its case with testimony about the medical aspects.

In September, attention returned to the analytical data. Drs. Frederick Rieders and Bo Holmstedt testified about the experiments they carried out on the samples provided by the prosecution. The major question they addressed was that of the long-term stability of curare under the conditions to which the bodies were subjected between 1966 and 1976.

Dr. Rieders maintained that, in his opinion, the RIA was not specific enough and "could only raise suspicions that something is there but it might not be there." The only procedure he found specific enough to be confident of identification of curare is mass spectrometry, using the entire spectrum, not just selected ion monitoring. In critical analyses, a four-step extraction procedure was used to isolate d-tubocurarine from the samples: The sample was crushed while frozen, then—in the cold—1. The sample was homogenized with acid buffer and dichloroethane, discarding the organic layer containing acidic and neutral substances; 2. The buffer was made alkaline and extracted to remove organic bases; 3. KI solution was added, and the curare extracted as an ion pair with iodide; and 4. The organic layer was back extracted with a small amount of HCl solution. The latter solution was evaporated down in the mass spectrometer probe, after which the spectrum was run.

Rieders tested for the stability of curare and found that both embalming fluids and tissue juices (from the patients) had destructive effects on this compound. He added curare to these liquids and could detect it by TLC initially, but after a few days could find no trace of it or other nitrogenous bases. These liquids altered curare chemically to the point where it was no longer recognizable as such. He concluded that the rapid rate of decomposition meant that to detect curare in the specimens in 1976 would have required huge, medically impossible amounts to have been present in 1966.

Rieders tested the samples for curare and found it only in the liver specimen of Nancy Savino. He stated that mass spectrometry indicated that the curare in this sample was highly pure and could not have been present in the ground for 10 years. Furthermore, if curare was present in the liver, it should also have been found in the child's muscle tissue. That it was not detected in the latter specimen was a "tremendous inconsistency."

Dr. Bo Holmstedt then stated that curare could not survive in embalmed bodies for 10 years, especially because of the effects of bacteria and repeated fluctuations in temperature of the bodies. He reviewed experiments which showed that curare, upon injection, shows levels of the same order of magnitude in liver and muscle tissues. After 10 min, "40% of the drug is to be found in the muscle and 3% in the liver."

On October 14 the defense rested its case. On October 23, after both sides had presented summations of their cases, Judge Arnold gave his charge to the jury. The next day, October 24, 1978—seven and a half months after the trial had begun—the jury received the case. After just over 2 h of deliberations, the jury returned a unanimous verdict of not guilty on all three remaining counts of murder. Two years and five months after the indictments against him had been returned, Dr. Mario Jascalevich was free.

Acknowledgment

Comments of several of the participants in the trial are appreciated, as well as the assistance of the *Star-Ledger* in providing photographs.

Bibliographic Note

This article is not a technical paper; hence, there are no references cited. As articles relating to the technical evidence in the trial are published, we will prepare a bibliography that will be furnished to interested readers.

Behind the Identification of China White

Theodore C. Kram, Donald A. Cooper, and Andrew C. Allen

U.S. Department of Justice, Drug Enforcement Administration, Special Testing and Research Laboratory, McLean, VA 22101

Originally published in ANALYTICAL CHEMISTRY, 1981, Vol. 53, No. 12.

Many forensic scientists who are avid viewers of the TV series "Quincy" cannot help cringing each time the protagonist peeks through a microscope and, without a moment's hesitation: "Eureka!" We sink a little lower in our seats as he rips a gas chromatogram from a recorder, studies it intensely for several nanoseconds, then slaps his right temple with the palm of his hand: "Of course! I should have known it from those yellow spots on his big toe!"

The speed and bull's-eye accuracy displayed by Dr. Quincy each week, although intended to keep the show moving at a pace suited to the average TV audience, unfortunately reflects—and may even reinforce—the image that much of the community has of forensic scientists: that with little more than a microscope and a reagent that produces specific colors with every compound imaginable, the forensic scientist can produce a positive identification in just a few minutes.

Although we in the real world are frequently exposed to pressures not unlike those depicted on "Quincy," we have learned only too well that painstaking and time-consuming effort is often required to solve a tough analytical problem. Even after we piece together a molecular structure that appears to have analytical properties consistent with our data, we must subject it to what is oftentimes a laborious process of verification.

We at this laboratory are often asked how we go about identifying an unknown. In our college days, unknowns generally consisted of several grams of a single substance, identifiable by use of a variety of classical procedures, and for which an abundance of reference data was readily accessible. Forensic drug exhibits, however, usually consist of powders, liquids, and amorphous masses that are seldom obtained from reagent bottles, but, more likely, from scrapings of glassware in illicit laboratories or from matchbooks that have changed hands under cloak-and-dagger circumstances.

We consider ourselves fortunate if we have gram quantities to work with, and extremely fortunate if we discover that we have but a single ingredient to consider.

Since the majority of unknown exhibits that we examine consist of illicitly manufactured drugs or reaction intermediates, we can be assured of much work in determining which ingredients of forensic significance may be present, either as finished products or as substances that may be reacted to form them.

The initial approach to the identification of the sample components depends primarily upon the following factors: (1) the quantity of material submitted; (2) its known history; and (3) the preferred analytical procedures, instrumental and otherwise, of those assigned to the case.

We have to maintain complete flexibility in our approach as we gather information on the nature of the exhibit. Where a sizable amount of sample is available, for example, attempts are generally made to analyze separate portions by microscopy; thin layer chromatography; and mass, magnetic resonance, and infrared spectrometries. In many instances, the compounds of forensic interest are identified within a matter of hours, sometimes minutes, by a combination of interpretation and comparison of data with those obtained from reference substances.

With no suitable reference data, however, identification becomes increasingly difficult with increased molecular complexity. It is at this point that the data already obtained must be scrutinized in greater depth. What more can be gleaned from the mass spectrum? Is it possible that the ion of highest mass that we observe is not the molecular ion? What about the simple ingredients that might be present? Could they provide some clues as to what was being synthesized? Might they constitute building blocks to be found in the larger molecules? If the sample was obtained in an illicit laboratory, do we have information on any of the other exhibits found at the laboratory site? Perhaps the identity of commercial packages of chemicals might provide some useful clues.

The "China White" Case

In the matter of "China White" we were confronted with an impure, complex material with which the forensic community had no prior experience. It was available, initially, in only minute concentrations, and its identification was of the greatest urgency. Among drug users word had begun to spread of a synthetic substitute for heroin. Among law enforcement people a macabre aspect became prominent; people were dying and no one knew why.

Our first encounter with China White occurred last fall with a sample arriving from California. Although it had been implicated in a death where symptoms of analgesic overdose were exhibited, no drug could be detected either in it or in the victim's body fluids. Our initial examination of the powder revealed lactose and nothing more. Subsequent analysis by GC/MS of a highly concentrated extract produced a weak spectrum that was total-

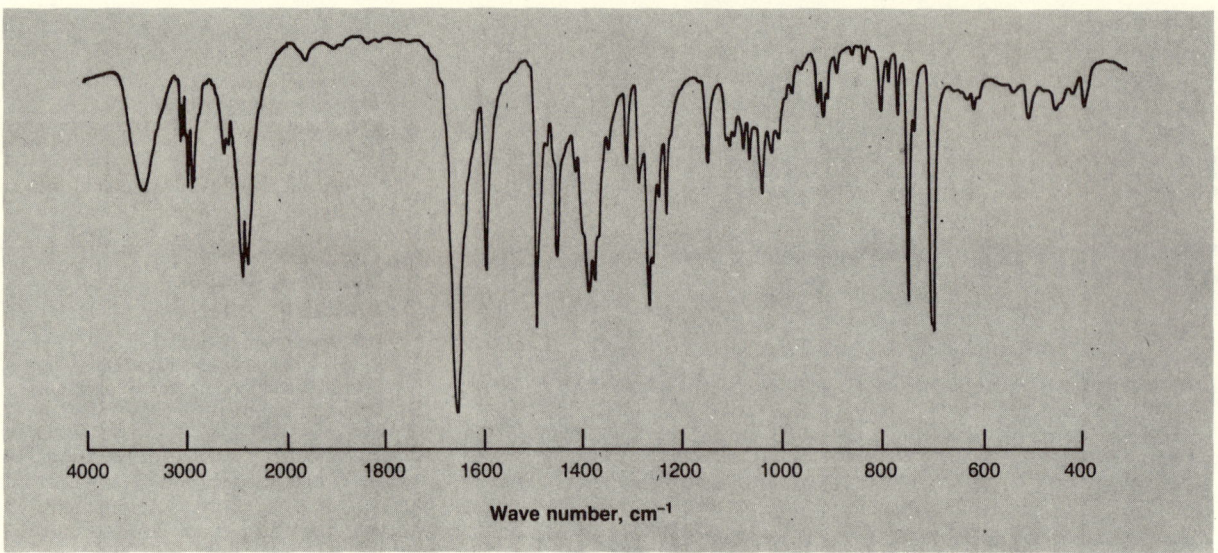

Figure 1. IR spectrum of hydrochloride salt of compound isolated from "China White" and prepared in a KBr pellet

ly unfamiliar and, unfortunately, whatever produced this spectrum was present only at very low concentration levels.

A short time later, however, an exhibit was referred to us that consisted of about 200 mg of powder having a greater concentration of this same substance. It was accompanied by a mass spectrum from GC/MS analysis of a methanolic extract of the exhibit and by an infrared spectrum of a hydrochloride salt prepared from the suspected active principal.

The mass spectrum resembled that obtained from the earlier exhibit. The infrared spectrum bore similarities to that of amphetamine; however, a band attributed to carbonyl was present, its location suggesting a tertiary amide (Figure 1).

The illicit laboratory from which the powder was thought to originate was believed to be engaged in the manufacture of methamphetamine. Indeed, a toxicological study conducted on an apparent drug overdose victim associated with that laboratory failed to reveal anything but amphetamine or methamphetamine. However, the quantities of these drugs were well below overdose levels.

Upon receiving the powder, we dry-extracted a substantial portion with deuterochloroform, then filtered it to remove the lactose. Examination by ^{1}HNMR revealed the presence of at least three ethyl groups, a possible cyclic group, and aromatic protons nearly equal in number to all the others (Figure 2). Following extractions from the chloroform with deuterated water and backwashes with additional chloroform, it was clear that we were dealing with three major components, one of which was a small compound with a partitioning preference for water. It contained an ethyl group, but gave no indication of aromaticity. The other compounds, however, exhibited a distinct preference for chloroform, although not in the same degree, and each contained both ethyl and aromatic groups. In addition, citrate was observed, present either as a mixture of citric acid and an inorganic salt, or in loose association with one of the major components.

It was evident from the weakness of

Figure 2. 200-MHz ^{1}HNMR spectra obtained from exhibit of "China White": (a) CDCl$_3$ extract of the powder; (b) D$_2$O extract from (a); (c) CDCl$_3$ extract of (b); (d) D$_2$O layer after CDCl$_3$ extraction; (e) CDCl$_3$ extract of (d)

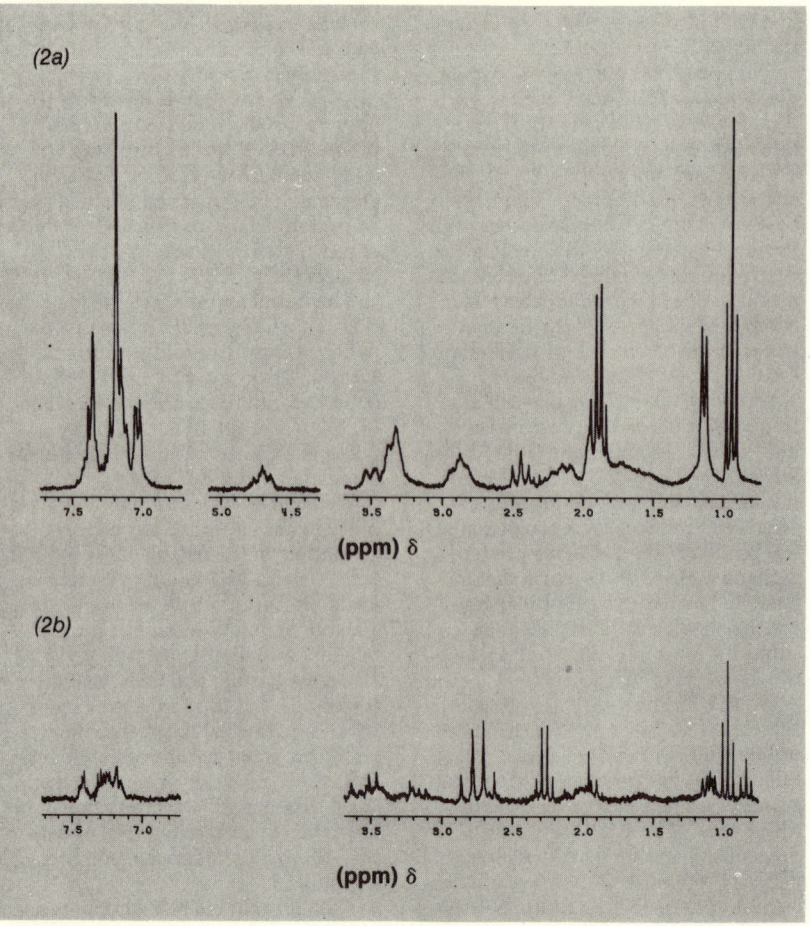

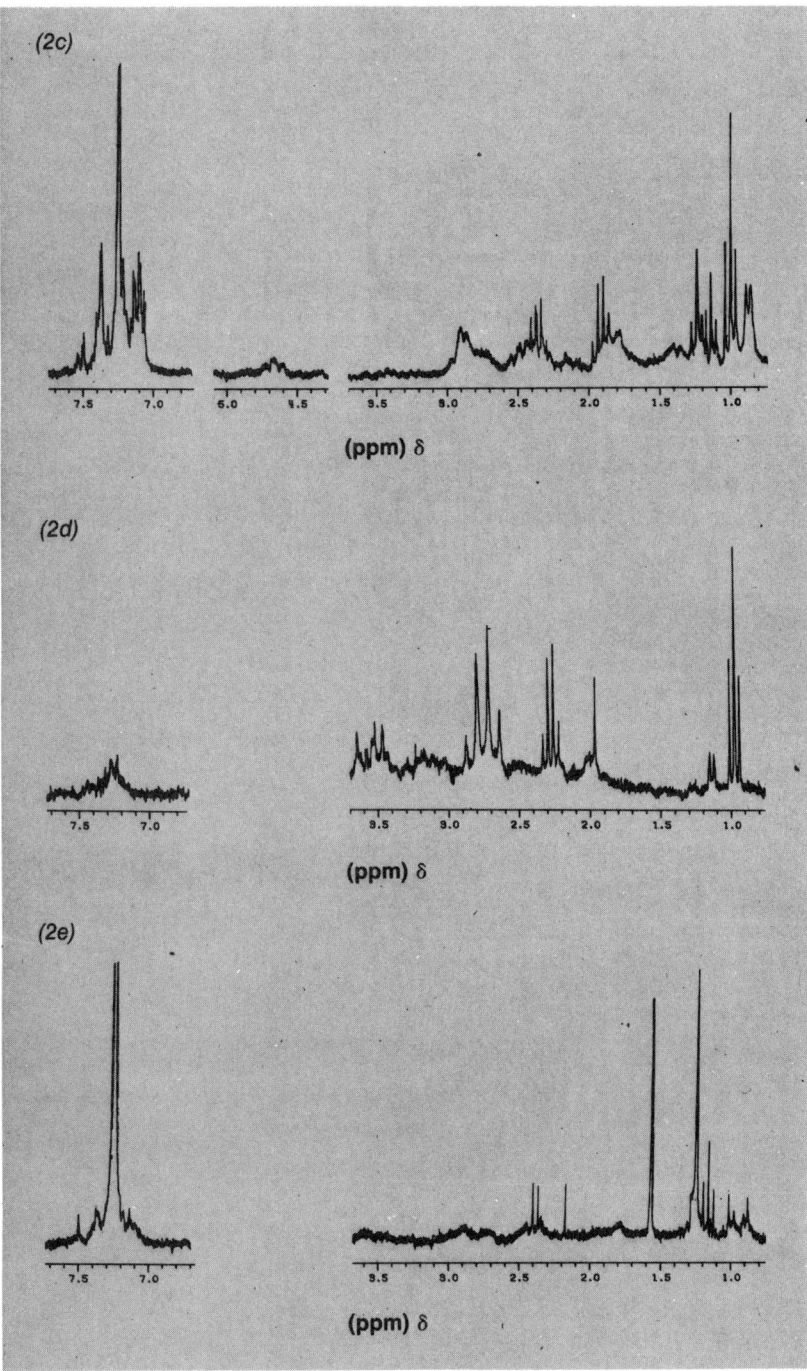

the ¹HNMR spectrum that, although our exhibit had a greater concentration of suspected active principal than the one received originally, we had a total of no more than about 1 mg of this compound to work with. The limitations this would place on our analytical flexibility would undoubtedly delay the desired outcome.

Another portion of sample was subjected to GC/MS. Two major chromatographic peaks were observed, with several minor ones appearing between them. The early major eluant produced a fairly uncomplicated spectrum (Figure 3). With a base peak at m/z 93, an apparent molecular weight of 149, a small but significant fragment at m/z 120, and fragmentation characteristic of a phenyl group (m/z 77, 65, 51, 39), very little was required to complete the picture. The other major component, however, produced the mass spectrum that had stymied us previously (Figure 4). We felt certain that this was the one to tackle. Assuming a molecular weight of 259 from the fragment of highest observable m/z, we spent some time in trying to find clues in the fragmentation that would allow us to assemble building blocks adding up to 259. At this point it was clear that we had to make a more meaningful determination of molecular weight. The technique most readily available to us for this purpose was that of chemical ionization mass spectrometry (CIMS). Attempts would also be made to see if the major components could be separated by simple extraction procedures so that we might obtain more meaningful ¹HNMR spectra and also to see if we could learn more about the structural functionalities of the major components from partitioning characteristics.

GC/CIMS provided a major breakthrough for us: Although the early eluter was indicated to have a molecular weight of 149, the late eluter appeared to have one of 350. The loss of 91 mass units to produce the 259 fragment in the E.I. mass spectrum might tentatively be ascribed to loss of a benzyl group. The presence of an intense m/z 91 ion in the E.I. spectrum further supported this hypothesis. A proposal that the m/z 110 fragment resulted from a loss of 149 from the m/z 259 ion and an indicated relationship to the early eluting compound was especially attractive despite the fact that charged and neutral 149 fragments of great structural variety are not uncommon in the mass spectrometry of organic molecules.

The mass 149 compound was found to extract as a neutral. The mass 350 compound, already known to have an amine function from its observed tendency to form a hydrochloride salt, was, predictably, found to be extractable from alkaline solution and also as a hydrochloride ion pair from acid solution.

The remaining pieces of the mass 149 compound could now be determined with greater assurance. With an odd molecular weight of 149 there had to be at least one nitrogen atom. We knew we had an ethyl group; a nitrogen had to be attached to the phenyl ring in order to produce an m/z 93 ion (assuming hydrogen transfer). Since the original ¹HNMR spectrum favored attachment of an aprotic carbon atom to the ethyl group and because of the presence of an m/z 57 ion, the existence of a carbonyl was all but assured. The pieces fit together nicely to form propionanilide.

CH₃CH₂CONH—⟨phenyl⟩

Further support for the structure was obtained from the ¹HNMR spectrum of the isolated compound (1). With a reported LD_{50} of 1100 mg/kg (2), however, this compound could hardly qualify as a toxic substance, much less be the potent drug we were looking for.

In the ¹HNMR spectrum of the hydrochloride salt prepared from the mass 350 compound (Figure 5), the presence of an anilido group was further supported. With most of the molecule tentatively accounted for, our attention was now focused on the m/z 110 fragment. We knew from the even molecular weight of the compound that, with only one nitrogen accounted for, there had to be another one here, or, with decreasing likelihood, three, five, or seven. Integration of the ¹HNMR spectrum told us that we probably had approximately 30 hydrogen atoms. Further consideration of the mass spectral and NMR data strongly suggested the presence of a piperidine ring, and a methyl group attached to CH, possibly at the 3-position of the piperidine ring.

The pieces were now beginning to come together. The next step was to search *Chemical Abstracts* for information on highly toxic compounds that could conceivably fit the proposed structural data. Armed with empirical formulas derived from the tables of Beynon and Williams (3) that were consistent with presumed molecular weight and proton count, the search commenced. After much diligence, a likely candidate emerged: 3-methylfentanyl.

CH₃CH₂CON(⟨phenyl⟩)—⟨piperidine with 3-CH₃⟩—N—CH₂CH₂—⟨phenyl⟩

This compound seemingly fit the bill from the standpoint of structural as well as toxicological characteristics. The piperidine structural moiety was sufficiently consistent with mass and ¹HNMR spectral data that it was felt that a tentative identification had been made. It appeared that our search had just about come to an end.

However, the following remained for us to do:
• Confirm the proposed structure with the aid of spectral comparison with a reference compound;
• develop methodology that could be employed by other forensic laboratories for rapid screening, isolation, and identification of the active component; and
• obtain all available information from the literature on the proposed compound.

Upon further study of the data, however, an inconsistency with the proposed structure became apparent. A weak ¹HNMR signal at about 4.7 ppm, representing a single proton believed to be in the 4-position of the piperidine nucleus, produced a pattern that was seemingly inconsistent with the assigned structure. In both salt and basic forms of the compound, it appeared as a "triplet of triplets," a result of two large coupling constants and two small ones. Since this could arise only from a pair of axial–axial (and/or geminal) proton interactions, and a pair of axial–equatorial proton interactions, this meant that substituents could not be present in both the 3- and 4-positions. The possibility of overlapping "doublets of triplets," so spaced to produce an apparent "triplet of triplets" from a mixture of cis and trans isomers (axial and equatorial 3-methyl, respectively), had to be discounted after serious consideration because of the theoretically vast differences in coupling.

By now we had some standards available for comparison and some informative reprints (4–6). The 3-methylfentanyl standard, following purifi-

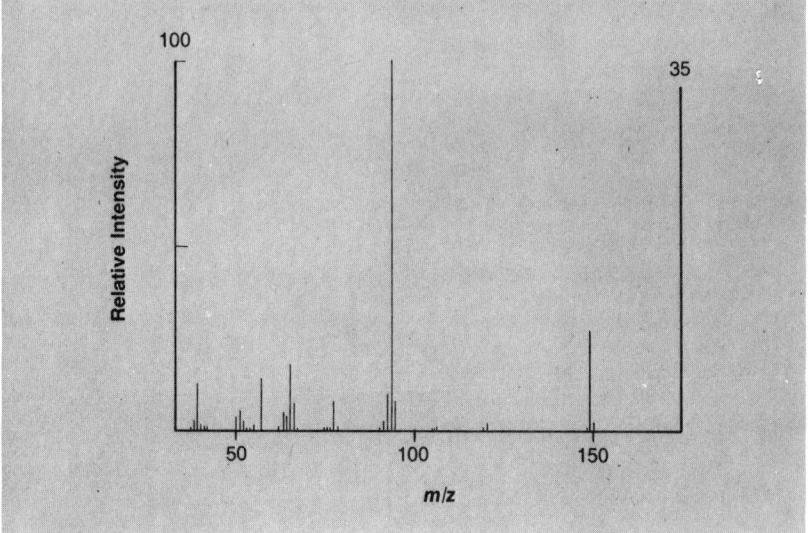

Figure 3. Electron impact mass spectrum of first major GLC eluant

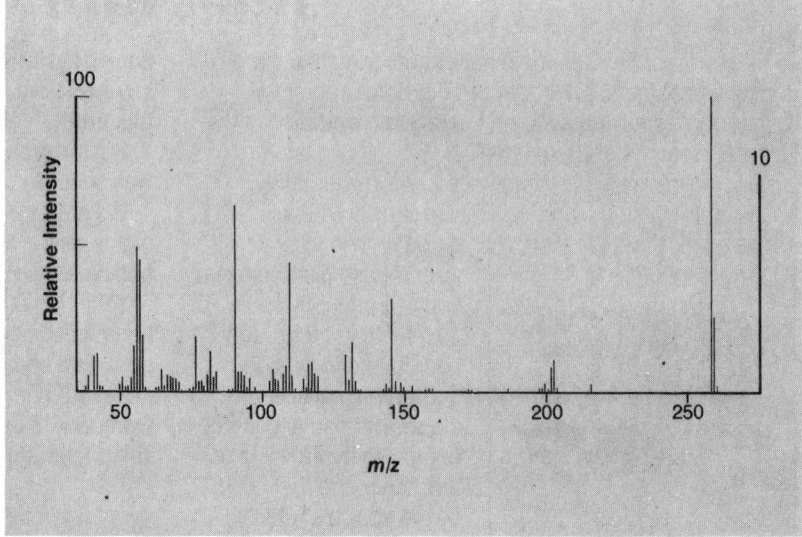

Figure 4. Electron impact mass spectrum of second major GLC eluant

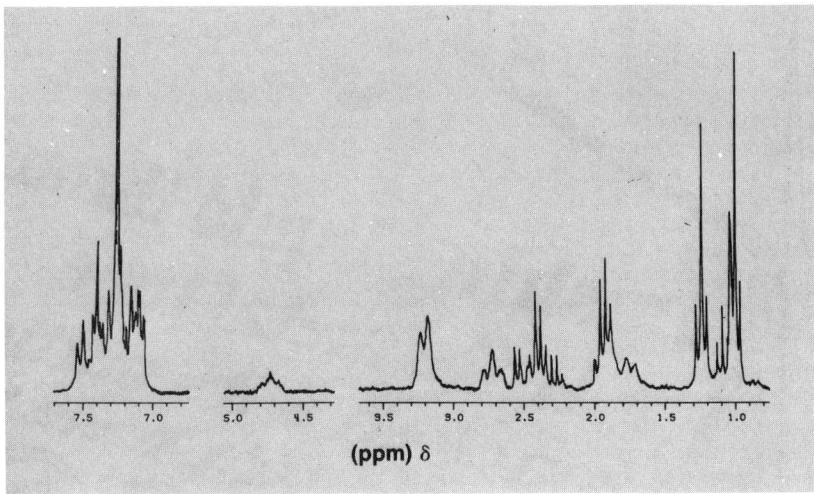

Figure 5. 200-MHz ^{1}NMR spectrum of hydrochloride salt prepared from alkaline extract of the exhibit (in $CDCl_3$)

cation by HPLC, did not compare with the unknown, although some structural similarities were evident. The 2-methylfentanyl, although it produced a ^{1}HNMR pattern for the 4-position proton similar to that of the unknown, was not the compound of interest. In view of the structural assignments for which we had solid support, where could we reassign this methyl group? Only two places remained: alpha and beta to the piperidine nitrogen. The beta position seemed highly unlikely because of the predicted effects on the mass spectral fragmentation: intense ions at m/z 105 (in addition to the m/z 91 ion) and at m/z 96 (at the expense of the m/z 110 ion), neither of which was evident.

This left us with the alpha isomer, which we could find no reason to discount.

All efforts now were concentrated on obtaining a reference standard of this material as quickly as possible. Within days it was synthesized, purified, and subjected to spectrometric examination. It was examined as a base and as a hydrochloride salt. The outcome? A match with the sample in all respects!

What of the water-soluble component that contained an ethyl group and was not observed by GC/MS? It was identified as propionic acid. It was believed to have arisen either as a by-product of the synthesis or as a decomposition product.

As the case drew to a close, we became subjected to a steady stream of reporters and photographers from the various media. The last of them, an established science reporter from a radio network, discussed the case with us at great length. Toward the end of the interview, however, I (T.C.K.) was abruptly disturbed by a reminder of the legacy of Dr. Quincy and his fictional predecessors. "Why," asked this otherwise seasoned reporter, "did it take nearly a month to come up with an answer?"

References

(1) "Sadtler Nuclear Magnetic Resonance Spectra," No. 14246, Sadtler Research Laboratories, Philadelphia, Pa. 1966–1980.
(2) "Registry of Toxic Effects of Chemical Substances," Christensen, H. E.; Luginbyhl, T. T., Eds., National Institute for Occupational Safety and Health, Rockville, Md., 1975, p 974.
(3) Beynon, J. H.; Williams, A. E. "Mass and Abundance Tables for Use in Mass Spectrometry"; Elsevier Publishing Company: New York, 1963.
(4) Janssen, P. A. J. U.S. Patent 3 164 000, 1965.
(5) Riley, T. N.; Hale, D. B.; Wilson, M. C. *J. Pharm. Sci.* **1973,** *62* (6), 983–6.
(6) Van Bever, W. F. M.; Niemegeers, C. J. E.; Janssen, P. A. J. *J. Med. Chem.* **1974,** *17* (10), 1047–51.

The Analytical Chemist as a Third Party Fact Finder

Joseph G. Montalvo, Jr.

Gulf South Research Institute, New Orleans, LA 70186

Originally published in ANALYTICAL CHEMISTRY, 1978, Vol. 50, No. 14.

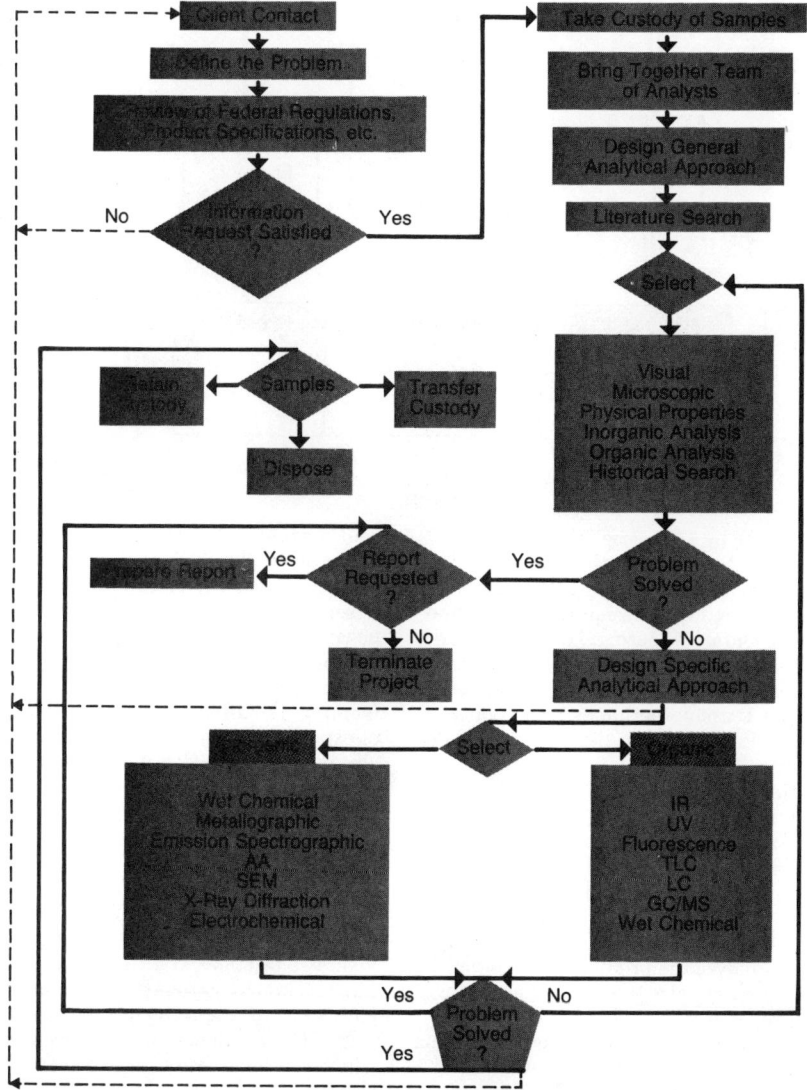

Figure 1. Analytical approach to referee and corroborative analyses

Gulf South Research Institute (GSRI) is a nonprofit research institute chartered and domiciled in the state of Louisiana and is engaged in contract research in the life, physical, and engineering sciences. This diversified base requires that analytical chemists work on a wide variety of problems. Situations encountered range from providing analytical support in process engineering and polymer membrane studies to developing new instrumentation.

As an independent laboratory, GSRI has no vested interest in private industry. Consequently, the Institute frequently engages in referee (third party) and corroborative (for first or second party) problem solving, which are the principal topics of this paper. The general scheme employed in referee and corroborative problem solving is shown in Figure 1. The three examples that follow serve to illustrate the general approach. The first involves a study of personnel hazards associated with a supposedly harmless commercial corrosion preventative coating. The second example shows that the solution of the corroborative problem associated with a ruptured pipe was beneficial to the sponsor in on-going litigation. The third problem cited is a classic—a shipload of molasses was suspected of being contaminated during transport aboard ship. In this case, the analytical results are important to the vendor, shipper, and customer. The examples cited are intended to illustrate that the "best" methods are not necessarily the most accurate or precise ones but the ones which will most quickly and easily provide answers to the real questions at hand.

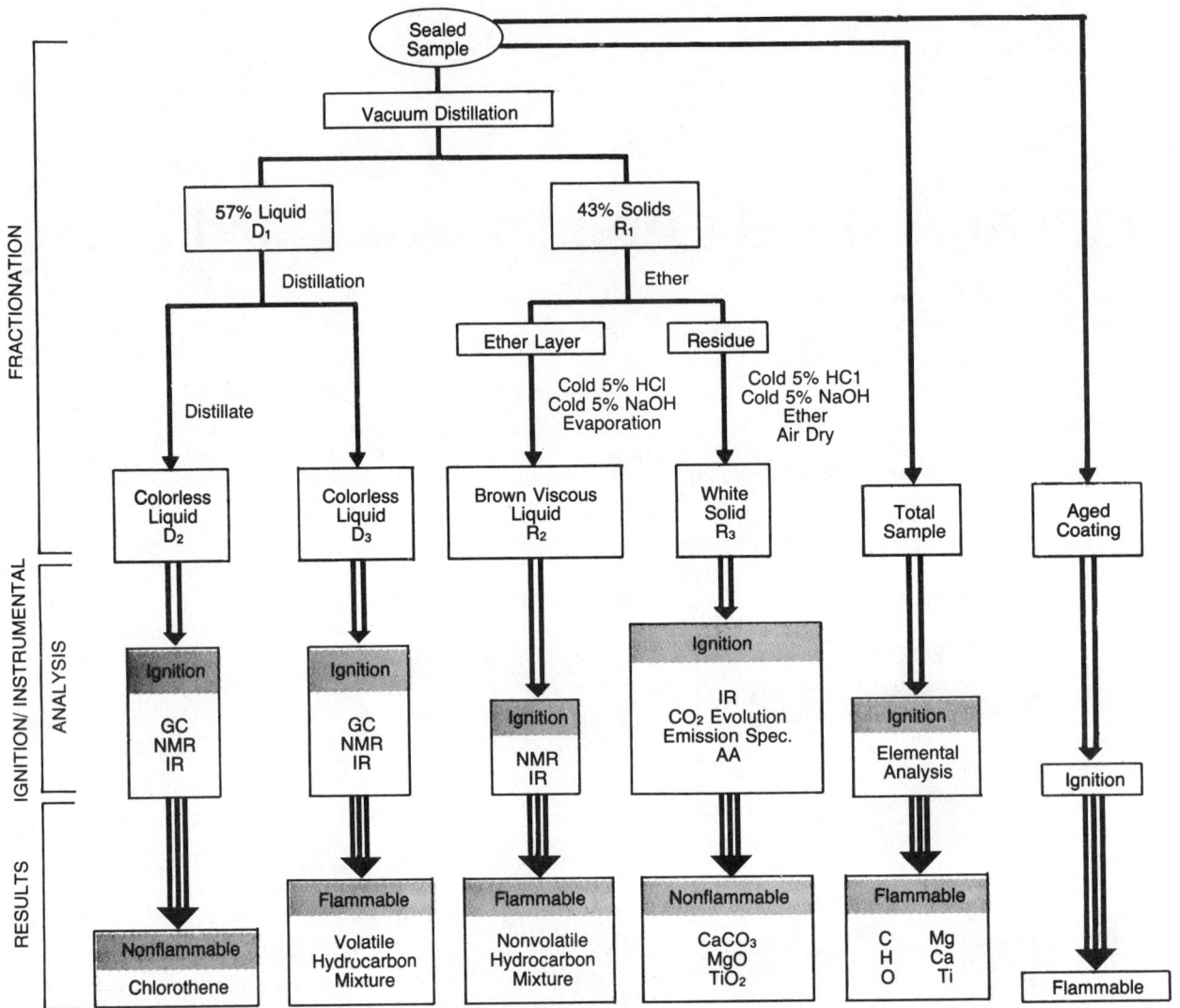

Figure 2. Analysis scheme for All-Safe

Corrosion Preventative Coating—Harmless Product?

Offshore drilling rig fabrication is a large business in the Gulf Coast area and has worldwide implications. It is not uncommon for rigs built in Louisiana to be towed hundreds and even thousands of miles in the search for oil. These metal structures are coated with unique formulations to minimize corrosion. Rigid specifications are required for such coatings that are to be applied to interior walls, particularly in areas of limited accessibility. Interior coatings must be durable, repel moisture, and not present hazardous conditions. Two types of time-dependent hazards are of importance: those associated with application and those related to on-going maintenance, welding, etc.

Not surprisingly, rig fabricating companies are always in the market for improved internal coatings. One such coating, identified fictitiously as All-Safe for this paper, was considered for use by a large rig fabrication company. According to the supplier's literature, All-Safe was a white, organic based, ready-to-use internal corrosion preventative coating. However, information concerning the safety of the formulation was sketchy. Chlorothene (1,1,1-trichloroethane), a component of the coating, was listed as a nonflammable hazardous ingredient, with a threshold limit value (TLV) of 350 ppm; the mixture was reported not to flash at the boiling point. An organic based formulation might be expected to flash and burn within the temperature limits encountered in welding operations. Therefore, GSRI was asked to analyze the product, with emphasis placed on investigation of hazardous properties.

Preliminary examination of All-Safe revealed that the product was a solid suspended in a liquid. The formulation was insoluble in water and contained a readily volatile solvent. The aroma was characteristic of petroleum. Upon addition of dilute hydrochloric acid, a gas evolved. An ignition test was positive. These data guided the analytical team in mapping a strategy to separate, identify, and quantify the hazardous components of the mixture. The analysis scheme is shown in Figure 2. Isolated fractions are coded to facilitate the discussion. Comparison of experimental infrared (IR) and gas chromatographic (GC) data with reference materials was qualitative, since the primary objective was to discern hazardous properties rather than to define composition.

The solid phase was isolated from the liquid by vacuum distillation. Distillate D_1 was redistilled at 1 atm to yield two fractions of different volatility. The more volatile fraction, D_2, gave a negative ignition test. GC analysis of D_2 indicated one major peak which was tentatively identified as chlorothene via nuclear magnetic resonance (NMR) and IR spectroscopy. The NMR assignment was confirmed by analysis of an authentic sample of chlorothene. Unexpectedly, the less

Table I. Characterization of All-Safe

Fraction	Chemical composition	
	Ingredient	%
Volatile	Chlorothene[a]	36
	Hydrocarbons	21
Nonvolatile	Hydrocarbons	17
	$CaCO_3$	21
	MgO	1
	TiO_2	4

	Fire and explosion hazard data		
	Flammable	Flash point (°F)	Fire point (°F)
Volatile hydrocarbons	Yes	186	200
Nonvolatile hydrocarbons	Yes	480	480
All-Safe		158–160	161–165

Will also flash and burn above boiling point after chlorothene evaporation. Is nonflammable at boiling point (167–176 °F)

[a] Hazardous compound (TLV = 350 ppm).

---

volatile fraction, D_3, gave a positive ignition test and had an aroma characteristic of petroleum. GC analysis of D_3 revealed unresolved peaks. NMR and IR data suggested a petroleum hydrocarbon oil. GC, NMR, and IR analysis of a reference mixture, Tampelene, a commercial petroleum-based paint thinning agent, gave similar instrumental responses.

The nonvolatile portion of the mixture (R_1) was shaken with ether. The ether (R_2) and residual (R_3) phases were washed with cold aqueous acid and base; R_3 was subsequently washed with ether, and the ether evaporated from both fractions. R_2 had an aroma characteristic of petroleum and gave a positive ignition test. Residue R_3 evolved CO_2 when treated with concentrated mineral acid. IR, emission spectroscopy, and atomic absorption spectroscopy studies, together with the results of a literature search, suggested the white pigments in the mixture were $CaCO_3$, MgO, and TiO_2. Elemental analysis of the total sample gave a positive test for C, H, O, Mg, Ca, and Ti.

Table I summarizes the chemical composition, hazardous ingredients, and fire and explosion hazards of All-Safe. Both volatile and nonvolatile fractions contain petroleum hydrocarbons which ignite and burn at temperatures considerably below the maximum encountered in welding. Flammability of the total sample is intriguing—flammability is observed above and below but not at the boiling point. This anomaly is apparently due to dispersion of the hydrocarbon vapors by the boiling chlorothene.

Demonstration of the fire hazard associated with an aged coating of All-Safe revealed a spectacular chain of events. One side of a 3-ft² section of sheet metal was coated with a 33-mil thick coating and allowed to age at atmospheric conditions for one week. The coated test plate was mounted outdoors in a vertical position and heated on the lower noncoated side with a welding torch. The coating blistered, melted, flowed down the metal surface, and rapidly burst into a flame that consumed the entire test strip.

Needless to mention, the rig fabrication company discontinued evaluation of the oil-based coating and initiated a paint quality control department.

Water in Urethane Foam Misleading

As events were described to GSRI, an underwater coaxial pipe carrying a molten product in the central line had ruptured. The outer pipe contained polyurethane foam insulation; it was suspected that traces of water in the foam had triggered a corrosion reaction which led to the rupture. The insulating material was applied in situ, and reactants were pumped in place and allowed to polymerize. The sponsor requested that the Karl Fischer (FK) technique be used to screen for water content, followed by a gravimetric confirmation. The case was in litigation; within 48 h after GSRI accepted custody of unofficial samples, representatives of the private company would visit the laboratory to observe tests on official samples submitted at that time. Unofficial samples were subsamples of official specimens.

A team of five investigators was assembled to develop and validate both procedures in the allotted time. Each member was asked to devise a general approach to the problem, and a consensus was obtained. Initial tasks were to conduct a literature search, purchase supplies, prepare laboratory space and equipment for the project, and develop theoretical approaches to the problem, if necessary. Published procedures for the specified assay methods were not found; therefore, the team was reassembled to develop specific analytical approaches.

Urethane foams are cellular products that may vary from softly resil-

---

Table II. Methyl Ester Composition of Fats and Oils

	Methyl ester composition (weight %)								
Fat and oil	Myristic 14:0	Palmitic 16:0	Palmitoleic 16:1	Stearic 18:0	Oleic 18:1	Linoleic 18:2	Linolenic 18:3	ΣD (%)	DI
Unknown tallow	3.5	33.0	1.6	14.6	42.9	3.6	0.8	...	...
Fat and oil reference mixture (lard, beef, tallow, and palm oil)	2.4	31.3	2.6	13.1	41.1	6.5	3.0	12.2	0.06
North American beef	6.3	27.4	0	14.1	49.6	2.5	0	19.1	0.10
Malaya palm oil	1.4	40.1	0	5.5	42.7	10.3	0	27.6	0.14
Congo palm oil	2.4	41.6	1.8	6.3	38.0	9.5	0.4	29.4	0.15
Belgian Congo palm oil	1.3	41.4	0	4.7	42.9	9.7	0	29.0	0.15
Liberian palm oil	0.6	37.6	1.4	3.7	50.3	6.4	0	29.6	0.15
English mutton	4.6	24.6	0	30.5	36.0	4.3	0	35.4	0.18
Cameroons palm oil	1.1	45.1	0.8	4.1	38.6	10.3	0	37.6	0.19
Fat and oil reference mixture (linseed, perilla, hempseed, and rubber seed oils)	0	7	0	5	18	36	34	131.2	0.66

ient to hard, rigid structures. Water might be held loosely or bound in urethane foams. To expedite the problem solving, it was assumed that traces of water in the samples under consideration existed primarily as loosely held water. The analytical approach examined the validity of this hypothesis. The sample handling procedure in the open air laboratory environment also had to be validated.

Two approaches to sample preparation for the KF titration were evaluated: solvent extraction and direct introduction of urethane foam into the titrating vessel. Direct introduction of a freshly cored sample (1 × 3 cm) was the method of choice, although about 45 min was required to reach the end point. This time lag is attributable to a diffusion limited process. A coulometric plot provided a practical means for obtaining a true end point. In this technique, facilitated by a digital readout, milligrams of water titrated were plotted vs. time. Readings were taken until the linear regression slope on the plateau of the titration curve was independent of time. This residual slope represented the leakage rate of water into the KF apparatus. The intercept of the regression line was a measure of water content in the sample. Splitting the core sample lengthwise immediately prior to KF assay provided a simple means for determining if a portion of the water in the intact core was unavailable for titration. The confirmatory gravimetric technique was designed to elucidate moisture equilibrium rates in air. Loss in sample weight upon heating or drying over P_2O_5 was equated to moisture content; regain in weight of the dried samples in the laboratory environment shed light on moisture equilibrium rates. Results indicated that water content of the unofficial foam samples was equivalent to that obtained by equilibrating samples in air.

The percent water in the unofficial samples measured by the KF technique was 1.8% (coefficient of variation = 6.4% for three replicate determinations). Splitting the core sample lengthwise immediately prior to assaying gave results which were not significantly different at the 95% confidence level when compared with the whole core values. Freshly cut foam cores from these samples gained less than 0.08% water when exposed to laboratory air for 10 min. In contrast, samples predried in an oven at 115 °C or stored in a desiccator at room temperature rapidly regained weight (corresponding to 1.5% water) when exposed to laboratory air for 10 min. At equilibrium, weight gain corresponded to weight loss after drying; gravimetric results confirmed the KF data.

To prevent bias and preserve experimental integrity in analyzing official samples, official and unofficial samples were coded in a random sequence so that the two types were indistinguishable during the official assays. Test data were regrouped into two sets and tested for statistical differences. Differences in the official and unofficial data sets were not significantly different at the 95% confidence level.

Since the results of the analysis were contrary to the expectations of the sponsor, the firm dropped its suit against the manufacturer of the foam.

Molasses Contamination Problem Unraveled

Blackstrap molasses is added to animal feed to make it more palatable and provide a carbohydrate supplement. A major mode of transportation of feed grade molasses is by barge; consequently, instances of contamination aboard ship can be anticipated. In this example, an alert marine cargo inspector observed a light-colored residue clinging to the interior walls of the ship's holds during unloading. An investigation of the ship's manifest revealed that the previous cargo was tallow. It was suspected that the molasses may have become contaminated aboard ship with tallow. Obviously, the molasses vendor, shipper, and potential customer were all interested in solving this analytical problem. Molasses samples were submitted to GSRI for analysis along with smears of the residue taken from the vessel. It was requested that both screening and confirmatory tests be run to identify, quantify, and predict the source of the suspected tallow contaminant.

An analytical team was assembled to brainstorm the problem and develop a general approach. The residue smears were used to represent the primary tallow standard. Uncontaminated molasses was spiked with standard tallow for use as a reference; likewise, fresh fats and oils were purchased and used as reference materials. It was planned to run both gravimetric and infrared screening tests with confirmation using gas chromatography. A search was performed to compile a library of ester compositions of industrial fat and oil products. A suitable ester composition algorithm was developed.

The screening test scheme was as follows. Samples were extracted with pentane, and the extracts evaporated to dryness. The weight of the residue upon evaporation was taken as an indication of tallow contamination. The dry residue was redissolved in chloroform and analyzed by IR spectroscopy. The carboxyl peak at 1750 cm^{-1} was also taken as a measure of apparent tallow concentration.

For the GC confirmation, molasses was extracted with pentane, and fatty acid esters in the extract were saponified with base (1). Methyl esters were prepared by refluxing with boron trifluoride esterification reagent. Esters were analyzed on a polyethylene glycol succinate column by use of a flame ionization detector. The GC column separates the fat and oil esters according to the sum of carbons and double bonds in the acid and alcohol moieties. Methyl ester extracts of the uncontaminated molasses (I), uncontaminated molasses plus tallow spikes (II), contaminated molasses (III), and standard tallow (IV) were prepared and assayed by GC. Order of testing was randomized to avoid systematic errors. Reference methyl ester mixtures simulating most fats and oils were injected into the GC.

Relative peak areas and retention times of the fat and oil reference (F&OR) mixture made up of lard, beef, tallow, and palm oil closely matched those of (II), (III), and (IV); similar peaks were absent or of low relative intensity for (I). These data strongly indicated that residual tallow adhering to the metal surfaces of the ship's holds had contaminated the molasses. Table II compares the methyl ester composition of the tallow of unknown origin with F&OR mixtures and library data. A simple but effective algorithm was developed to identify the origin of the tallow based on the fraction of unmatched methyl ester composition of a library and unknown composition.

An algorithm was constructed from the methyl ester composition of the seven esters shown in Table II. Slots which were empty, i.e., where the ester was not reported, were filled with zeros to enhance computational efficiency. Methyl ester compositions were compared on the basis of a dissimilarity index (DI) computed as follows. The sum of the ester composition of each fat and oil listed in the table equaled 100%. Fats and oils containing additional esters were excluded in compiling the library. Summed differences (ΣD) between the methyl ester composition of the tallow contaminant and each library entry were computed. The summed composition of each library entry and the unknown of any two listings (C_t) equaled 200%. Therefore, DI equaled $\Sigma D/C_t$. Note that DI is bounded below by 0 (a perfect match), and above by 1 (a complete mismatch).

The smallest DI was obtained with the first F&OR mixture shown in the table; the composition of the tallow contaminant more closely approached this reference mixture than the other five available standards. For comparison, the next closest reference composition is shown in the table. The "sensitivity" of the algorithm to predicting the origin of the unknown tallow is striking. The algorithm predicts that the tallow is of animal origin, probably North American beef ($DI = 0.10$).

Quantitative GC analysis of the tallow contaminant in the molasses, based on the methyl stearate peak areas, indicated contaminant levels from 0.4 to 6.6 ppm. The confirmed shipboard contamination of the molasses had a profound effect in the settlement of this dispute over product liability.

Acknowledgment

The analytical contributions of coworkers C. Eyer, I. Cabasso, K. L. Crochet, H. Hidalgo, C. G. Lee, L. Truxillo, R. Wawro, and D. Brady are gratefully acknowledged.

References

(1) "Official Methods of Analysis (1970) 11th ed.," p 454, AOAC, Washington, D.C., 1970.

Miscellaneous

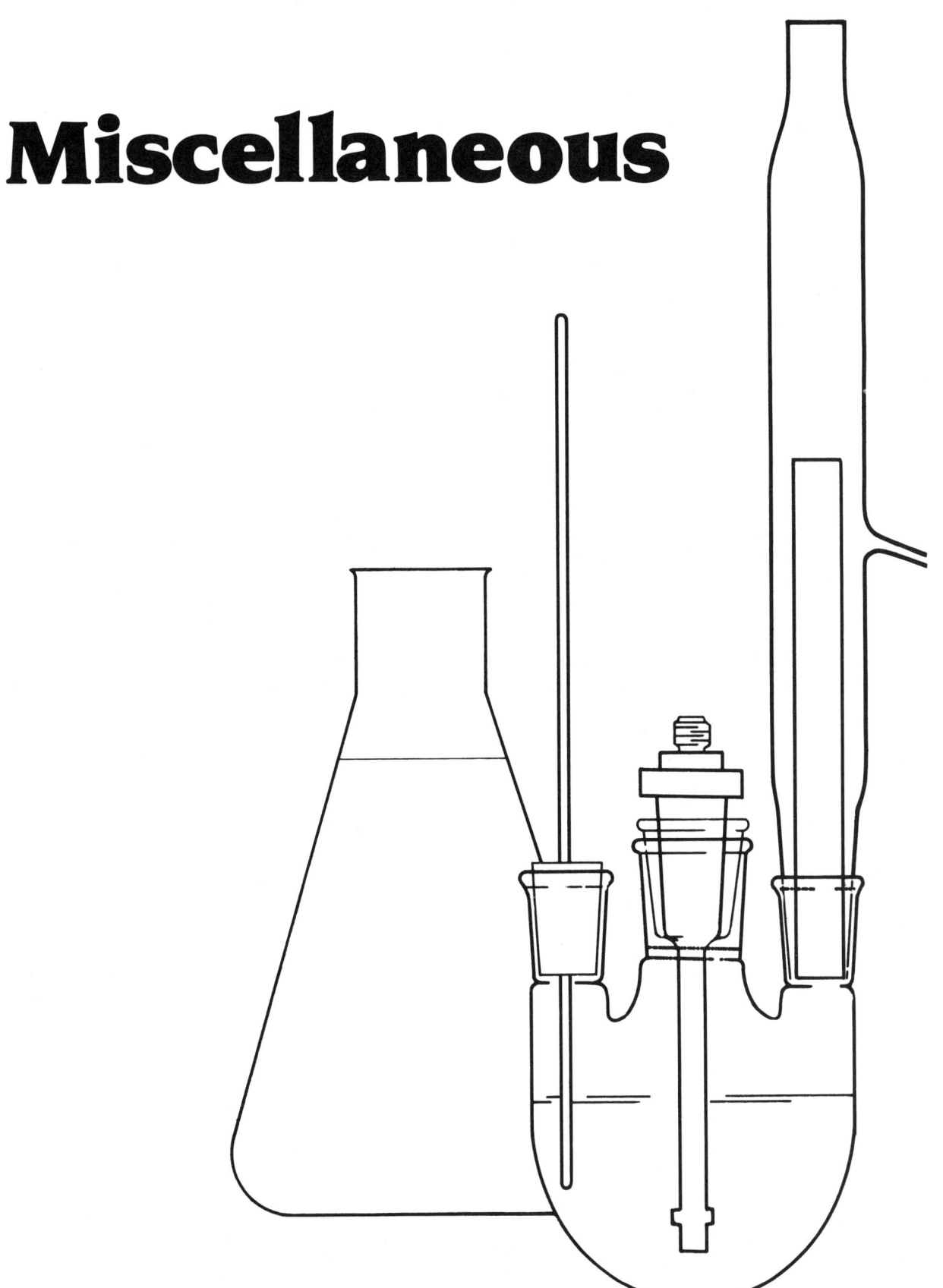

In this set of four articles are some case histories which are of general interest and could be used to enliven any course on analytical problem solving. They offer superb examples of the analytical approach in real-world problem solving.

Caution in accepting published analytical data is suggested in "**Reliability in the Analysis of Rocks and Minerals.**" The author cites examples where reputable scientists published results that can only be described as widely deviant from those of others. He emphasizes that complete descriptions of methods are vital for effective communication, and suggests terminology to improve such communication. Finally, he mentions the cross-disciplinary nature of rock and mineral analysis where geologists may do analytical work as a sideline while chemists may lack sufficient knowledge of geological materials. "There is no such thing as a 'bad' method, only bad analysts who fail to allow for its limitations."

An in-depth description of the use of an infrared interferometer spectrometer (IRIS) to make molecular analyses in remote space is presented in an Instrumentation feature article on the Voyager spacecraft equipment. The advantages of interferometric methods and the difficulties of operating in the hostile environment of space are highlighted in this discussion of the Voyager mission to Jupiter and Saturn. The Voyager may continue on a rendezvous with Uranus in 1986, and perhaps even Neptune in 1989.

Microscopy is featured in its various forms in "**Authenticity of Medieval Document.**" This study was carried out in an attempt to confirm the authenticity of the Vinland map, a map proporting to describe the New World and supposedly drawn before Columbus' voyages. Non-destructive studies and the analysis of minute samples suggest that the map is a clever attempt at forgery for reasons unknown.

In "**The Bust of Nefertiti**," the authors describe an investigation into the identity of the blue dye used in the 3,000 year old bust of Nefertiti. They were able to identify the dye as crystalline Egyptian blue, based on multitechnique analysis (X-ray, DTA, TG, microscopy) of very small samples taken from the priceless relic. As a postlude, the dye was synthesized from raw materials available to the early Egyptians. This final step adds to our knowledge of Egyptian technology.

Reliability in the Analysis of Rocks and Minerals

Sydney Abbey

Geological Survey of Canada, Ottawa, Ontario, Canada K1A 0E8

Originally published in ANALYTICAL CHEMISTRY, 1981, Vol. 53, No. 4.

We are all familiar with the many spectacular advances in analytical methodology in recent years. We have seen striking improvements in selectivity, sensitivity, precision, accuracy, and speed. The analysis of rocks for geological studies has been no exception in these developments, but the matter of overall reliability is another question.

What is so special about rocks?, you may ask. One might as well ask: What is a rock? From a chemical point of view, a rock may be regarded as a naturally occurring, essentially inorganic material, a heterogeneous mixture of solid phases, called minerals, with highly varied chemical, physical and mechanical properties. A rock may contain any or all of the elements in the periodic table. Its major and minor chemical constituents normally include Si, Ti, Al, Fe(III), Fe(II), Mn, Mg, Ca, Na, K, H, C and P, generally expressed as oxides, and possibly F and S as well. Other constituents are considered "trace elements."

How does a rock differ from an ore? After all, an ore may be considered as merely a rock containing economically significant concentrations of a useful constituent. However, the real difference lies in the purpose of the analysis. An ore is generally analyzed for its economically recoverable constituents and possibly also for others that may have effects on processing. Perhaps most important of all is the fact that ores are normally analyzed by those who have a financial stake in the quality of the results. Rocks, as mentioned earlier, must be analyzed for just about everything. There are very few laboratories, if any, that are competent to determine all the constituents of a rock that may be of interest to an earth scientist.

The classical giants of rock analysis, such as W. F. Hillebrand and H. S. Washington, were able to report extraordinarily reproducible results, from which it was assumed, perhaps questionably, that their results were very accurate as well. There was a general belief that careful work by an experienced analyst could produce results comparable to those of Hillebrand or Washington.

Then the roof fell in.

In 1951, there appeared the first detailed compilation of data on the two so-called standard rocks, G-1 and W-1 (1). All concerned were shocked by the highly discordant values reported for each constituent by the 30-odd analysts involved. For example, results reported for silica in G-1 ranged from 71.05 to 72.74%—enough to set both Hillebrand and Washington spinning in their graves. The agonized debates that followed are a matter of record. Many and varied were the explanations, excuses, and downright alibis put forward.

Inhomogeneity in the samples was blamed by many as a cause of the disparate results, but no one really tried to show just how badly the mineral constituents of the samples would have to be segregated to explain such large discrepancies.

Continued study and discussion regarding G-1 and W-1 revealed a number of shortcomings in the procedures followed in their preparation, distribution, and analysis. With the benefit of that experience, the U.S. Geological Survey embarked on the production of six new samples—G-2, GSP-1, AGV-1, BCR-1, PCC-1 and DTS-1. At about the same time, several geological groups in various other countries initiated similar programs. Many laboratories throughout the world contributed analytical data for many constituents of all of those programs, using a wide variety of methods.

With the benefit of experience with G-1 and W-1, one might have expected that the later programs would have produced more consistent results. Unfortunately, that did not happen. As shown in Table I, not only was there no improvement, but some distressing additional facts were revealed. There was, in effect, no significant improvement in the overall precision of silica determination over a 43-year interval (1931–1974) and, worse still, the disparities seemed to increase with the number of results involved!

Shortly after publication of the first compilation of data on the second set of reference rocks from the U.S. Geological Survey, we attempted to arrive at "best values" for many of the constituents of those samples (2). The purpose of the exercise was to provide "standards" for atomic absorption spectrometry. Although some progress was made toward that goal, the resulting studies served to reveal much more about the reliability of rock analysis—or, perhaps, the unreliability.

It became clear early in the game that application of rigorous statistical interpretation of the raw data would be of highly questionable value because the data were so grossly imbalanced in terms of the number of results obtained per constituent, the degree of replication per reported result, the number of constituents reported per sample—and in just about any other parameter one can imagine. To

Table I. Standard Deviation of Silica Results

Year reported	Sample type	No. of results	*Std. dev.	Ref.
1931	Glass	5	0.28**	10
1951	Granite	34	0.37	1
1963	Tonalite	14	0.26	3
1970	Feldspar	9	0.10	11
1972	Granite	30	0.18	12
1972	Syenite	36	1.06	13
1974	Granodiorite	35	0.46	14

* %, absolute, ** 0.09 after eliminating one result

Table II. Some Results for Arsenic (ppm)

Sample 1	Sample 2	Sample 3	Analyst
82	94	106	C
16.2	16.4	0.67	F
12	14	2	G
22	24	12	H
14	13	4	J
20	22	1.4	K
17.9	19.8	0.48	L
17.8	19.6	0.6	M
14	4.7	0.03	N
18	21		P

coin a phrase, we were faced with "heterogeneous data."

To illustrate these points, let me tell you a few horror stories. All of them involve comparatively respected names, working in renowned institutions.

Case A. Prof. A, a world-famous geochemist, decided, for some reason, to determine potassium in a group of proposed reference samples of rocks by means of a radiometric counting technique. His results averaged about 20% lower than the consensus of other reported results, which were based on a variety of analytical techniques. Was there no flame photometer available for Prof. A to cross-check his results?

Case B. Prof. B, another world-famous geochemist, offered to provide trace-element data on another group of reference samples. His results were far removed from all others reported by other contributors, for about half of the elements reported. As for "method used," he merely described it as "our regular spectrographic method, as used by our analyst, Mrs. X" (more about that below).

Case C. This one is really sad. Table II shows results for arsenic in three proposed reference rocks. Please note the truly sad case of Analyst C. I find it particularly distressing because I had visited C's laboratory a few years earlier and had come away with a favorable impression of his competence. Incidentally, subsequent studies revealed that the true arsenic content of sample 3 is probably close to those reported by F, L, and M. Several other reported results are therefore questionable, but none as questionable as C's. C reported similarly absurdly high results for B, Ge, Li, Nb, Pb, Sb, Sr, Y, and Zr.

It must be mentioned that after C reported his results, the coordinator of the program thanked him heartily for providing more data than anyone else (some of his data—on elements other than those shown—being fairly good). However, C was provided with median values of results reported by others for the same elements, from which he could easily see that something was wrong. He made no attempt to rectify the situation.

Case D. In a sense, this one is even sadder than Case C. The laboratory involved is in a major governmental research center in a highly developed country. Let us look at its results for cerium (in Table III). The two sets of results from D will be explained in a moment. Please note that D's results are lower than nearly all others on Samples 1 and 2, but close to most of the others on Sample 3, whose cerium content is comparatively low. Our old friend Prof. B checks in here with a high result on Sample 2 and an apparent "pass" on Samples 1 and 3. Perhaps his bid on Sample 2 should have been registered as "double," but we are not playing bridge. D reported similarly low results for Dy, Eu, La, and Tm.

As happened in the preceding case, the coordinator of the program on Samples 1, 2, and 3 provided D with the median values of results from the other participating laboratories. This happened after D had reported the first set of results. Some months later, a second report came from D. It stated that the apparatus used had been dismantled, cleaned, and recalibrated, and the analysis repeated. Since no procedural change was mentioned, it must be assumed that exactly the same procedure was followed. Not surprisingly, the second set of D's results was much the same as the first.

The method in this case involved preconcentration by ion exchange and final determination by a highly precise technique. Reference to the original literature on the method and on earlier applications clearly indicated that the method was fundamentally sound. The only problem, apparently, was that it had never before been applied to samples with such high rare-earth contents! It then appeared to be a simple case of incomplete recoveries in the ion-exchange step. That view is supported by the comparatively good results for Sample 3, and also for certain other elements whose concentrations in all three samples were much the same as in all samples tested in earlier work. So we have here a striking example of the disastrous effect of stubbornly clinging to a so-called "good method" with no thought to the possible need for changes in special applications.

I could go on for hours with stories of this nature, but one more must be told.

Case E. Dr. E has established a good reputation in one of the new analytical techniques that is particularly useful in determining trace elements. In a recent manuscript, which I had the questionable honor of reviewing for a certain journal, he reported his results for certain trace elements on a group of proposed reference samples of rocks, for which very few other values have appeared in the literature. Having been involved earlier with those samples, I was pleased to see that his results appeared to be fairly good. Unfortunately, he accompanied his data with what he called "range of other published values." I was somewhat puzzled by some of those "ranges." Reference to the only published compilation on those samples revealed that it included anywhere from three to 10 results for each of the elements reported by E. From those published results, I had assigned tentative values for each constituent for our own use—actually merely the median of those available. I was surprised to find that, for a number of elements, E's "range of published values" did not encompass my median. Further reference to the earlier compilation revealed that Dr. E, not being satisfied with merely good results, tried to make them look even better by selecting those of the other published values that nicely bracketed his own results. The laws of libel preclude my applying a label to such goings-on, so let us just say that it was a case of over-enthusiasm to the point of overstepping the bounds of objectivity.

You may have noticed that in the tables, no mention was made of the method used to produce each result. That was done deliberately, because much nonsense has been written about the relative merits of, say, method X compared to method Y. In many compilations of analytical data on reference samples of rocks and similar materials, one finds serious statistical data, such as mean, standard deviation, and coefficient of variation, for each of the "methods" used for a particular determination. Unfortunately, the "methods" used are described merely as AA, XRF, NAA,

Table III. Some Results for Cerium (ppm)

Sample 1	Sample 2	Sample 3	Analyst
300	2550		AM
	3500		B
160	2400		AC
237	2309	25	S
130	1140	26	D
214	2640	<63	V
150	960	24.6	D
215	2250	25	AJ
237	2274	24.5	S
110	2400	12	N

Color, etc. Some have even made the comparison between "physical methods" and "chemical methods." Now, if I tell you that a certain determination was done "by AA," what you are hearing is something less than a half truth. You do not know how the samples were decomposed; what separations, if any, were made; what additives were used; how the instrument was calibrated; whether flame or furnace atomization was involved, etc. Any of those variables could contribute more to the reliability of the result than the fact that atomic absorption was used for the final measurement.

There was even a case where two comparatively *low* values for zirconium in a certain sample were listed in a recent compilation of data. What was striking was the fact that the two results were identical, the "method" in one case being "spectrophotometric," the other "ion exchange." Two separate references were quoted, but it took little investigation to reveal that the two were really the same result, one reported as a private communication to the compiler, the other gleaned from a published paper. The method involved separation via ion exchange and measurement via a color reaction. To add icing to the cake, it is worth mentioning that the technique of sample decomposition actually used in that case left some doubt whether all of the zirconium had been brought into solution! I might add that the result had come from one of the laboratories involved in the horror stories mentioned earlier.

So, before we can realistically consider the concept of "reliability in analytical chemistry," perhaps we should do something about our terminology. Personally, I like to think of analytical processes in terms of unit operations, perhaps a throwback to my early training as a chemical engineer. Rock analysis and many other analyses can involve up to four unit operations:

(I) sample attack,
(II) separation,
(III) measurement, and
(IV) data reduction.

Operation (I) could include such things as acid treatment, fusion, and pelletization; (II) could involve precipitation, solvent extraction, ion exchange, etc.; (III) could mean gravimetry, titrimetry, emission spectrometry, etc.; (IV) would range from the simple use of a gravimetric factor to the highly complex computerized manipulation of the intensity measurements in an XRF spectrometer.

I would suggest that we reserve the term "technique" for characterizing a unit operation; thus "acid leaching" is a technique of sample attack; "fractional distillation" is one of separation; "spectrometry" one of measurement; and "application of interelement corrections" one of data reduction.

The word "method" could be applied to a combination of techniques, e.g., the sample was decomposed by sodium carbonate fusion (I); fluorine was recovered by water leaching (II) and determined by selective-ion electrode measurement (III), using pure sodium fluoride for calibration (IV).

"Procedure" could then be restricted to the specific details of the unit operations that make up a particular method. A procedure would not be described in a single sentence, just as a method would not be described in a word or two.

One may well object that the foregoing proposals are somewhat convoluted, that circumstances can arise in which it is necessary to provide a concise but meaningful title to a newly published method, or even to an official "standard" method. In that case, emphasis should be placed on the characteristics that distinguish the particular procedure from others. A method recently introduced in our own laboratories provides an example. The sample is heated with a flux mixture in a resistance furnace in a stream of nitrogen; the water, carbon dioxide, and sulfur dioxide evolved are determined by infrared absorption in three separate fixed-wavelength units. A title for the method could be "Infrared Determination of Water, Carbon, and Sulfur" but that would not be satisfactory, because it ignores the novel features of the method. A more meaningful title might be "Simultaneous Determination of Water, Carbon, and Sulfur in Rocks by Volatilization and Fixed-Wavelength Infrared Spectrometry." This would emphasize the important characteristics; no one has apparently combined those three determinations before. In fact, the new method permits a single operator to determine the three constituents in nearly twice as many samples per day as three operators were able to do by the methods formerly used.

I have now painted a rather gloomy picture of the state of the art of rock analysis. To be fair, one must point out that many geological laboratories do turn out a good deal of satisfactory data. The problem is how to sort the wheat from the chaff. Not all bad data are as patently obvious as those presented in my tables. Continuing programs on the preparation and evaluation of reference samples have helped to reveal the inadequacies of some data and to identify their sources. A major problem in such programs is how to deduce good consensus values from a mass of discordant data. Many statistically based procedures have been proposed (3–7), but evidence has been put forward to question their validity (8, 9). On the basis of a few rather empirical tests, it appears that the scheme used at the Geological Survey of Canada (8) gives what we think are the best results. More important, however, is the fact that the scheme is based on rating the performance of contributing laboratories and deducing the best values from those reported by a select group. Thus we have not only what we think are the best values for the concentrations of the constituents of reference samples, but also an indication of which laboratories can be depended upon to produce more reliable results.

There remain the nagging questions of why there are so many bad results and what can be done to improve the situation. A possible explanation of the former lies in the fact that the analysis of geological materials is a no-man's-land, suspended in mid-air between geology and analytical chemistry. Too often, analysis is done by geologists as a sideline, mainly because no analytical specialist is available. Conversely, it may be done by a chemist or a physicist with insufficient knowledge of geological materials. I have seen examples of both.

You will note that I have placed the emphasis on the people, rather than the method or instrument. There are far too many occasions where poorly informed decision makers are led to believe that the mere purchase of a new zillion-dollar what's-it-ometer will solve all their analytical problems. I have seen many examples of that too, and, no doubt, so have many of you.

My philosophy is simple: The reliability of a result depends more on *who* produced it than on *how* it was done. There is no such thing as a "bad" method—only bad analysts who fail to allow for its limitations.

To answer the second part of the nagging question—what can be done to improve the situation?—we can only say we are doing our best in an uphill battle. Among the positive steps has been the establishment of a new journal, *Geostandards Newsletter*, which is endeavoring to educate its readers on the importance of reference materials and how their intelligent use can help to improve the reliability of results. Unfortunately, there are cases where reference samples have been used less than "intelligently." When one sees how much questionable data are provided in programs on proposed reference samples, one cannot but despair at the inevitable amount of useless, if not downright misleading, information that is finding its way into so many data banks.

Finally, I must say that what has been described has concerned only one particular field of application of analytical chemistry where considerable vigilance is required in assessing the reliability of analytical data. I have concentrated on that field because it is what I know best. My only hope is that some of this material may be of use to those involved in other areas.

References

(1) H. W. Fairbairn, Ed., "U.S. Geological Survey Bulletin 980," 1951.
(2) Sydney Abbey, *Can. Spectrosc.*, 15, 9–15 (1970).
(3) W. K. L. Thomas, "Standard Geochemical Sample T-1, Msusulu Tonalite, Supplement No. 1," Government Printer, Dar-es-Salaam, Tanzania, 1963.
(4) O. H. J. Christie and K. H. Alfsen, *Geostandards Newsl.*, 1, 47–9 (1977).
(5) P. J. Ellis, I. Copelowitz, and T. W. Steele, *Geostandards Newsl.*, 1, 123–30 (1977).
(6) M. Sankar Das, *Geostandards Newsl.*, 3, 199–205 (1979).
(7) A. Colombo, *Geostandards Newsl.*, in press.
(8) Sydney Abbey, R. A. Meeds, and P. G. Bélanger, *Geostandards Newsl.*, 3, 121–33 (1979); see also Geological Survey of Canada, Paper 80-14.
(9) Sydney Abbey, manuscript prepared for submission to *Geostandards Newsl.*
(10) National Bureau of Standards, "Certificate of Analysis, Standard Sample No. 91, Opal Glass," 1931.
(11) Bureau of Analyzed Samples, "Certificate of Analysis, British Chemical Standard No. 375, Soda Feldspar," 1970.
(12) H. Grassmann, *Zeitschrift für Angewandte Geologie*, 18, 280–4 (1972).
(13) B. G. Russell, R. G. Goudvis, G. Domel, and J. Levin, National Institute for Metallurgy Report 1351, Johannesburg, 1972.
(14) A. Ando, H. Kurasawa, T. Ohmori and E. Takeda, *Geochem. J.*, 8, 175–92 (1974).

Presented at the symposium on "Reliability in Analytical Chemistry," at the 1980 Federation of Analytical Chemistry and Spectroscopy Societies (FACSS) Meeting.

Voyager Infrared Spectrometer

Stuart A. Borman

American Chemical Society, Washington, DC 20036

Originally published in ANALYTICAL CHEMISTRY, 1981, Vol. 53, No. 13.

Each of the two *Voyager* spacecraft currently exploring the outer reaches of our solar system carries sophisticated infrared and ultraviolet spectrometers. In this month's INSTRUMENTATION, the *Voyager* Fourier transform infrared (FTIR) interferometer spectrometer will be described.

The spectrometer was developed in a joint effort spearheaded by the National Aeronautics and Space Administration (NASA) Goddard Space Flight Center, with the participation of research and development groups from the University of Michigan and Texas Instruments, Inc. The instrument, named IRIS for infrared interferometer spectrometer, flew on the *Nimbus 3* and *Nimbus 4* meteorological satellites in earth orbit and on the *Mariner 9* spacecraft that orbited Mars, in addition to the *Voyager* spacecraft.

One individual who has participated in the development of IRIS from the very beginning is Rudolt A. Hanel of the Goddard Space Flight Center's Laboratory for Extraterrestrial Physics. Hanel is presently in charge of the infrared spectrometer investigation on the two *Voyager* spacecraft.

Interferometry

An FTIR spectrometer like IRIS acquires a signal called an interferogram in the time domain, instead of direct acquisition in the frequency domain, as with grating instruments, for example. An interferogram is a plot of the autocorrelation function of electromagnetic radiation as a function of time. If monochromatic light were involved, for example, the time domain signal would be a sine wave. A much more complicated interferogram results when polychromatic light is analyzed.

Unfortunately, there is no transducer with a sufficiently rapid response time to follow the oscillations of electromagnetic radiation in the infrared region. What is needed is a black box able to "slow down" the electromagnetic radiation so it can be detected, without distorting the information carried in the original signal.

The Michelson interferometer operates on electromagnetic radiation to "slow it down" in just this way—a process known as modulation. As mentioned above, if a single frequency is incident on an interferometer, the time domain signal generated, the interferogram, will consist of one pure sine wave whose frequency is proportional to but much lower than the frequency of the incident radiation. But most radiation sources in the real world (or real universe) contain a mixture of many frequencies. These different frequencies added together form a complex interferogram that often looks something like a decaying transient disturbance. One might think that such a pattern could not possibly contain any useful information. But this jumble of frequencies can be deconvoluted by the mathematical process of Fourier transformation, converting the information contained in the interferogram from the time domain to the frequency domain. A power vs. time signal is thus converted to a spectrum of radiance vs. frequency.

An excellent description of Fourier transform spectrometry and the Michelson interferometer can be found in Reference 1. Other books containing more detailed information include References 2, 3, and 4.

IRIS

The IRIS instrument on the *Voyager* spacecraft does not operate like a conventional infrared laboratory spectrometer, with a source, a sample, and a detector, in that order. With IRIS, the sample is also the source, because the instrument measures thermal (infrared) *emission*. Since infrared emission from molecules at low temperature is very weak, it is evident why a Michelson interferometer, with its high signal-to-noise ratio, was chosen for the *Voyager* mission. This higher S/N ratio, the "Fellgett advantage" Michelson interferometers possess over prism or grating instruments, arises from the fact that the detector in the Michelson interferometer sees all spectral intervals simultaneously instead of sequentially.

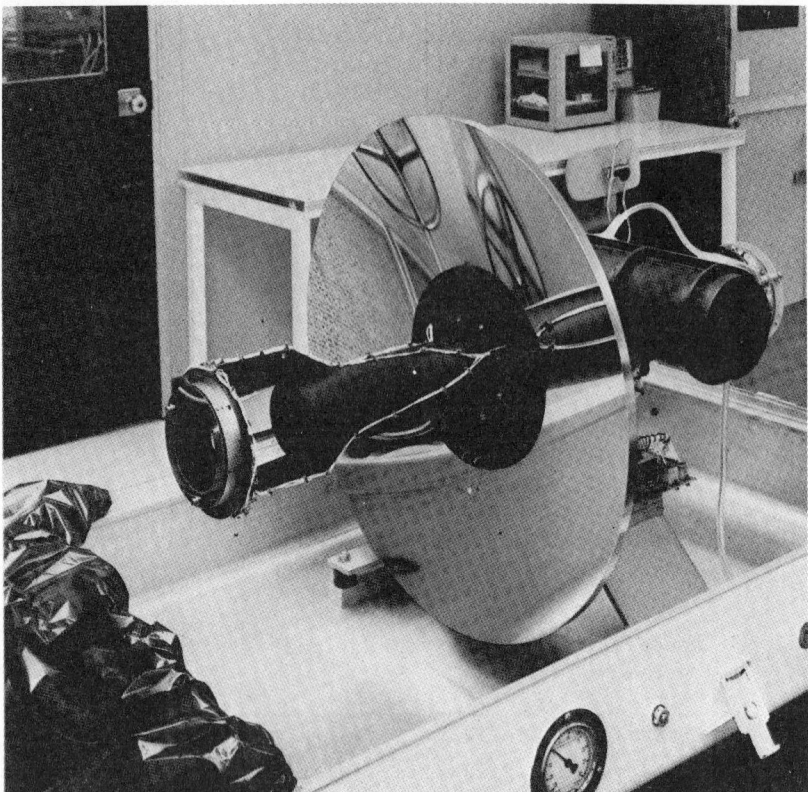

Figure 1. IRIS spectrometer. Interferometer is to right of primary mirror

Figure 2. IRIS spectrometer. Interferometer at center, radiometer at left

The instrument and its optical layout are shown in Figures 1, 2, and 3. The most prominent feature in Figure 1 is the spectrometer's telescope. Radiation collected by the 50-cm-diameter primary and the secondary mirror is directed through a field stop (Figure 3) to a spherical dichroic mirror (5).

The interferometer itself can be seen behind the primary mirror in Figure 1. It is also clearly visible in Figure 2. The device behind the dichroic mirror (on the left in Figure 2) is a radiometer for measurements in the visible and near-infrared regions. Radiation reflected by the spherical dichroic mirror is directed to a beam splitter (Figure 3), and thence to the fixed and stationary mirrors of the interferometer. The interferometer modulates the incident radiation by splitting it into two beams. One beam is directed to a fixed mirror and the other to a movable mirror called the Michelson mirror. The two beams are then recombined at the beam splitter, and an interference pattern generated by the interaction of the two beams is focused on a thermopile detector. For clarity, in Figure 3 both the main and reference interferometers are rotated 90° about an axis between the main interferometer beam splitter and the dichroic mirror; also, the reference interferometer is not to scale.

For the interferometer to operate properly, the instrument's temperature must be held constant, in this case at 200 ± 0.5 K. Therefore, the optical components of the *Voyager* IRIS (except the active surface of the primary) are wrapped in multilayer thermal blankets for the flight. A thermal radiator mounted on the interferometer cools the instrument by radiating to deep space. Three sets of heaters provide fine thermal control for the instrument. In addition to thermal protection, considerable effort was invested in radiation-hardening the instrument, to ensure proper operation in the high-energy particle environment of Jupiter's magnetosphere.

The reference interferometer (Figure 3) contains a neon source, from which the 5852 Å line is isolated by a narrow-band interference filter. The moving mirror of the reference interferometer is operated by the same linear motor that operates the moving mirror of the main interferometer. The interferogram generated by the reference interferometer is a pure sine wave. The reference interferometer serves two functions: It provides feedback control of the Michelson mirror motion, and, by sensing the zero crossings of the neon sine wave, it initiates the sample and hold commands needed for the acquisition of discrete data points in the interferogram. Each interferogram is acquired in 45.6 s with a rate of 80 words s^{-1}. Each interferogram thus contains 3648 digital words. Hundreds of individual spectra are averaged to produce an emission spectrum of high S/N ratio for scenes under study.

The output of the detector is quantized and temporarily stored in the spacecraft for transmission to earth, where the data are processed in a digital computer. The interferogram is Fourier-transformed to yield an amplitude and a phase spectrum. The amplitude in each channel is proportional to the net spectral radiance between the instrument and the scene within the field of view. This raw amplitude vs. frequency spectrum must then be corrected for phase and calibrated against an absolute radiance source. The phase of the signal in each spectral interval of the raw amplitude spectrum depends on whether the scene is warmer or colder than the instrument. The phase of a colder source is 180° from the phase of a warmer source. After the amplitude spectrum has been phase-corrected, it is calibrated. Two points are used in the calibration: deep space, assumed to be a perfect heat sink, is one point, and the accurately known temperature of the instrument (200 K) is the other. With these two calibration points, relative amplitudes can be converted to absolute radiance units.

Ice Clouds on Mars

The infrared emission spectrum of the atmosphere of a planet is a difficult thing to interpret, since spectral radiance reaching the interferometer is a complicated composite of emission and absorption. For instance, emitted radiation by CO_2 in one atmospheric layer may be reabsorbed by CO_2 molecules higher up in the atmosphere. As Hanel put it, "The primary goal of remote sensing in the infrared is to extract the chemical composition and physical parameters... from the measurement of spectral radiances.... In the early days of evolution of this technique, the problem was often compared to the task of unscrambling an omelet and reconstructing the eggs" (6).

One spectrum Hanel and his associates were able to unscramble is shown in Figures 4a and 4b (7). Figure 4a shows *Mariner 9* IRIS measurements of the lower Arcadia region of Mars under clear weather conditions and of the Tharsis Ridge region under conditions of partial cloudiness. Although it had been suspected that clouds of the type that appeared in the Tharsis Ridge region that day were composed of water ice, no direct spectral evidence had been found.

The spectrum measured over lower Arcadia showed an approximately constant brightness temperature, except for the CO_2 absorption band centered at 667 cm^{-1} and the rotational water vapor absorption lines below 400 cm^{-1}. In contrast, the Tharsis Ridge spectrum exhibited a strikingly broad absorption feature extending from 550 to 950 cm^{-1}, with a second broad absorption feature between 225 and 350 cm^{-1}. Superimposed on the latter was a sharp spike near 227 cm^{-1}. The theoretical water ice cloud spectrum (Figure 4b) had similar features. These spectroscopic data strongly indicated that the Tharsis Ridge clouds were composed of water ice.

Voyager

Voyager 1 IRIS spectra of the planet Jupiter showed clear evidence of a number of compounds in the Jovian atmosphere, including hydrogen, methane, acetylene, ethane, ammonia, phosphine, water, CH_3D, and GeH_4. Pearl et al. (8) painstakingly identified the presence of SO_2 on Jupiter's

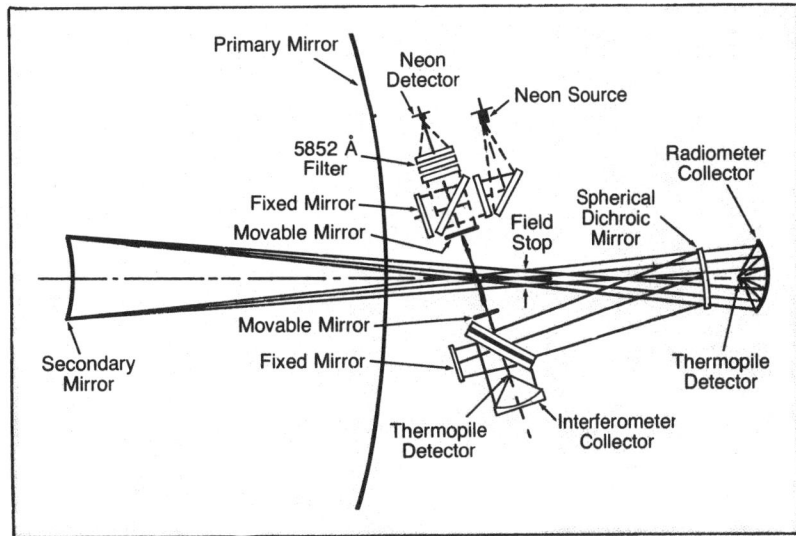

Figure 3. Optical layout of the *Voyager* IRIS. Reprinted from Reference 5 with permission of the Optical Society of America

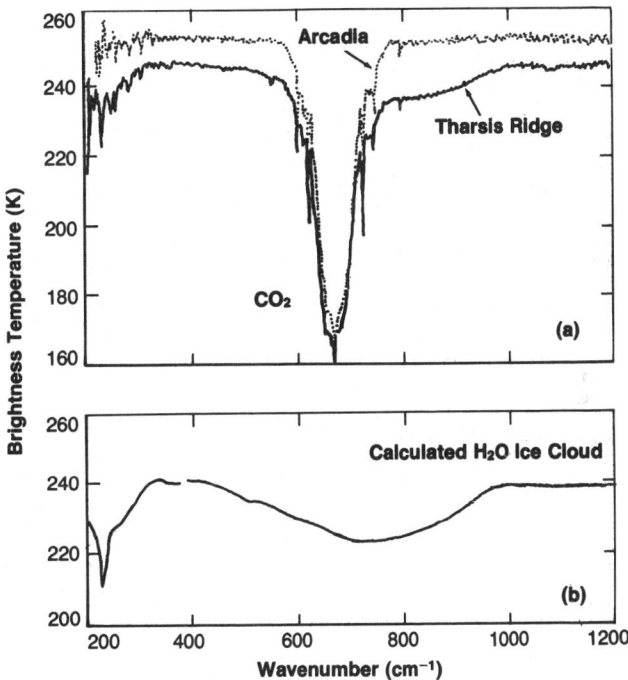

Figure 4. (a) *Mariner 9* IRIS spectra from Mars. Arcadia spectrum is offset for clarity. (b) Theoretical water ice cloud spectrum. Brightness temperature (y axis) is a function of spectral radiance. Reprinted from Reference 7, copyright 1973 by the American Association for the Advancement of Science

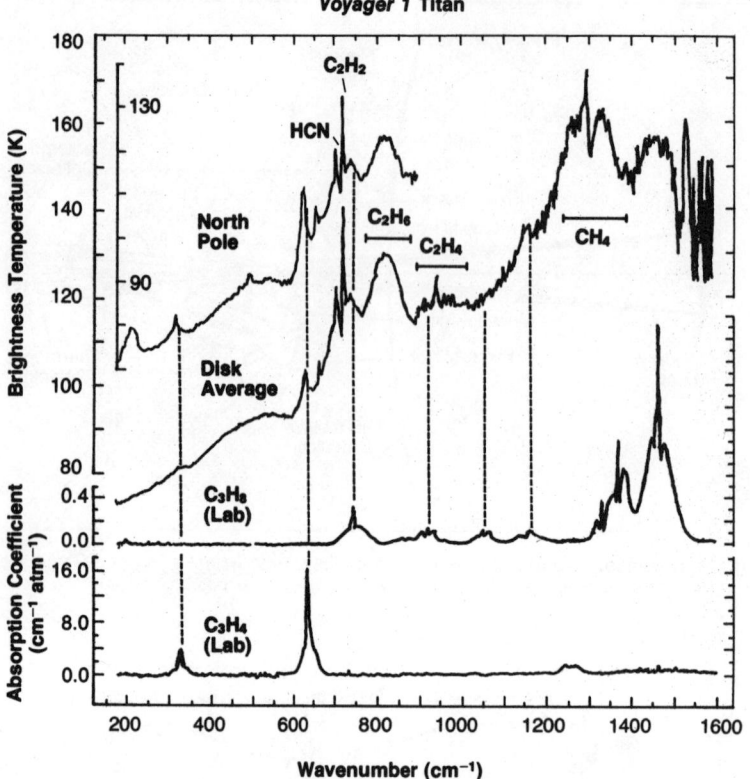

Figure 5. Comparison of observed Titan spectrum with laboratory spectra of propane and methyl acetylene. Reprinted from Reference 9 by permission of Macmillan Journals Ltd.

volcanic moon, Io, by comparing tiny blips on spectra from Io's atmosphere to peaks in a synthesized spectrum of SO_2.

During the passage of *Voyager 1* through the Saturn system, IRIS acquired data on the planet, its rings, Titan, and other satellites. Infrared spectra of Saturn indicated the presence of hydrogen, methane, ammonia, phosphine, acetylene, ethane, and possibly methyl acetylene (C_3H_4) and propane (C_3H_8). On Titan, positive identifications were made for methane, acetylene, ethylene, ethane, and HCN, but again only tentative identifications could at first be made for methyl acetylene and propane.

Positive identification of these last two components had to wait for a more detailed analysis of Titan spectra, as shown in Figure 5 (9). At the top of Figure 5 are spectra acquired from the center of Titan's disk and from its north pole. At the bottom of the figure are spectra of propane and methyl acetylene taken in the laboratory. The average disk spectrum shows weak but definite spectral features of the propane fundamentals at 748, 922, 1054, and 1158 cm^{-1}. The disk spectrum and, to a greater extent, the polar spectrum show the fundamentals of methyl acetylene at 328 cm^{-1} and at 633.2 cm^{-1}. These data provided convincing evidence for the existence of these gases in the atmosphere of Titan.

Voyager 2 made its closest approach to Saturn on Aug. 25, 1981. Due to a rare planetary alignment occurring only once every 175 years, *Voyager 2* should be able to continue on to a rendezvous with Uranus in January 1986, and perhaps even Neptune in August 1989. Eventually the *Voyager* spacecraft and their remarkable spectrometers will leave the solar system. But instruments like IRIS will no doubt fly again when new opportunities for the spectrometric exploration of space present themselves.

Thanks to Rudolf Hanel for reviewing this material and to Ron Joho of Hewlett-Packard for suggesting the topic.

References

(1) Skoog, Douglas A.; West, Donald M. "Principles of Instrumental Analysis," 2nd ed.; Saunders College/Holt, Rinehart and Winston: Philadelphia, 1980; 241–54.
(2) Bell, Robert John. "Introductory Fourier Transform Spectroscopy"; Academic Press: New York, 1972.
(3) Griffiths, Peter R. "Chemical Infrared Fourier Transform Spectroscopy"; John Wiley & Sons: New York, 1975.
(4) Griffiths, Peter R., Ed. "Transform Techniques in Chemistry"; Plenum Press: New York, 1978.
(5) Hanel, R. A., et al. *Appl. Opt.* **1980,** *19*, 1391–1400.
(6) Hanel, R. A., "Chemical Evolution of the Giant Planets"; Academic Press: New York, 1976; Chapter 13.
(7) Curran, Robert J., et al. *Science* **1973,** *182*, 381–3.
(8) Pearl, J., et al. *Nature* **1979,** *280*, 755–8.
(9) Maguire, W. C., et al. *Nature* **1981,** *292*, 683–6.

Authenticity of Medieval Document Tested by Small Particle Analysis

Walter C. McCrone

McCrone Associates, Inc., Chicago, IL 60616

Originally published in ANALYTICAL CHEMISTRY, 1976, Vol. 48, No. 8.

Perhaps the greatest contribution ever to microscopical analysis was the contribution of Sir Arthur Conan Doyle in writing the Sherlock Holmes stories. Holmes was able to solve complex and involved crimes, principally through the use of the microscope, and these accounts were so convincing and reasonable that ever since the general public has expected police laboratories to routinely match Holmes' accomplishments. With increasing success, microanalysts have been competing with Sherlock Holmes. We know now that Sir Arthur was right and that single small particles cannot only be identified but will yield a great deal of information about people, places, and activities. It is now possible microscopically to identify any small particle; many of them are considerably smaller than Sherlock Holmes could even see. We now have instruments such as the microprobes and electron microscopes that Holmes could only dream about. Fortunately, we find that the polarizing microscope, even one such as Holmes might have used, will still solve most particle identification problems. The physical microanalytical instruments save considerable time and, strangely enough, require less background and training for use in identifying particles than the light microscope.

At McCrone Associates we define microscopy as the use of any tool or technique that enables us to identify microscopic objects. An identification is based on the recognition of specific "identifying characteristics" for each particle. This includes not only morphological and optical properties of the particles but also elemental determinations by microchemical means or by electron microprobe. It even includes the use of electron or micro x-ray diffraction. We have found that there is a great need for particle identification techniques, not only in criminalistics but also for the identification of environmental pollutants or contaminants in almost every product produced by man, e.g., asbestos fibers in drinking water, exotic crystals in human biopsy samples, pigment identification in tiny samples from valuable paintings, penicillin cross-con-

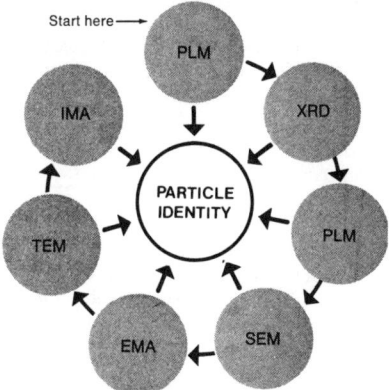

Schematic representation of the microanalytical procedure for small particle identification. One begins with the polarized light microscope, turns next to micro x-ray diffraction, returns if necessary to measurements with the PLM, and finally turns to successive use of more sophisticated microprobes and electron microscopes

224 The Analytical Approach

Figure 1. The Vinland Map

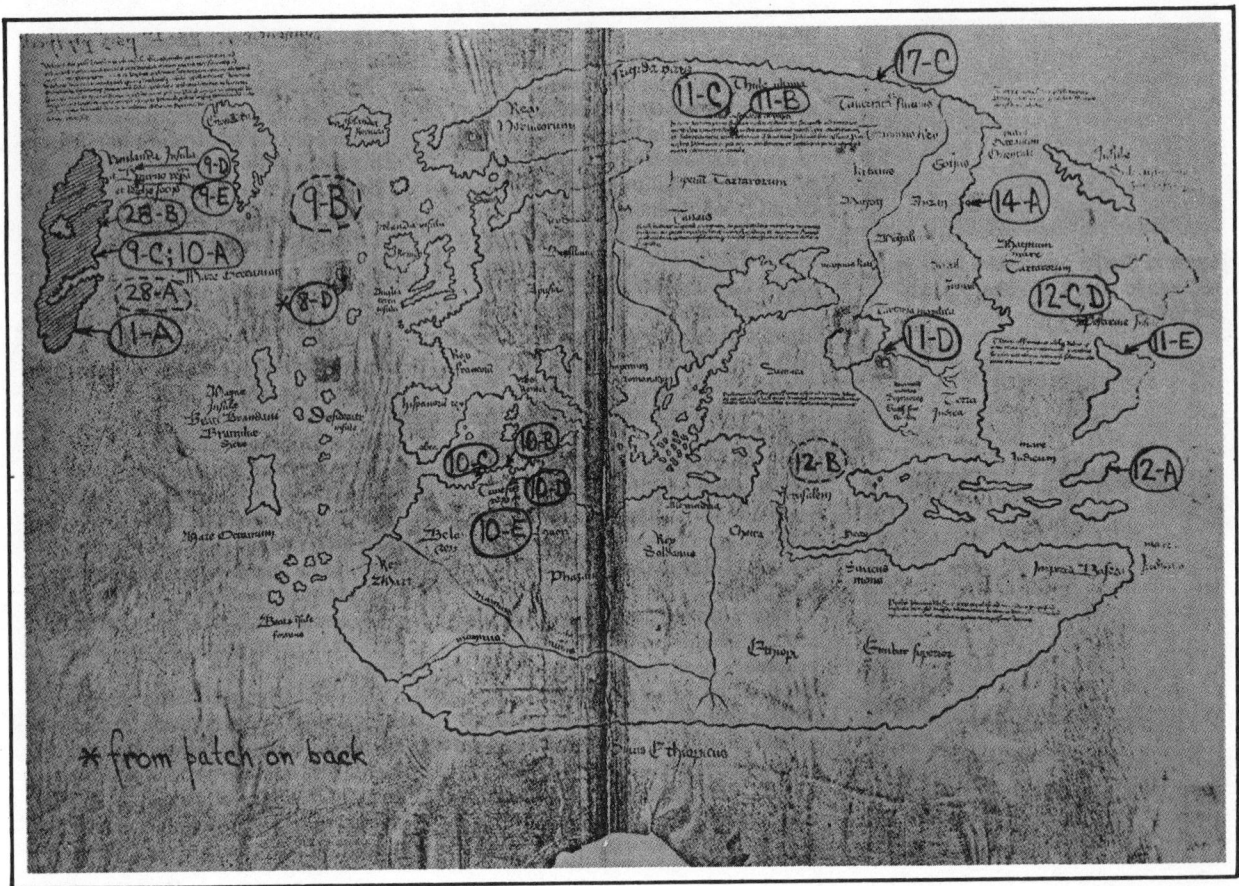

tamination in other drug products, occluded particles in metals or polymers, foreign particles in injectable drugs, particles in industrial city smog, micrometeorites or moon dust. It makes little difference to the microscopist where the small particle comes from if he is asked only to identify it by composition and source.

The Vinland Map

The problem of identifying a given particle always starts with the light microscope and progresses to more and more sophisticated techniques until success is achieved. One of the best examples of the application of these techniques involves the analytical approach to dating of the Vinland Map (Figure 1). This map, now owned by Yale University Library and purportedly drawn about 1440, shows a large island west of Greenland identified as Vinland. Since this date predated Columbus' visit to the New World, the Vinland Map was accorded considerable attention and not a little fanfare. Although it is generally accepted that Leif Ericson visited the New World nearly 500 years before Columbus, there had hitherto been no direct map evidence of this accomplishment.

The Vinland Map first surfaced in 1957 when it was purchased in Europe by a dealer in rare books and manuscripts in New Haven, Conn. The importance of the map was enhanced when it was discovered that wormholes in the map itself could be registered exactly with wormholes in two other acknowledged medieval documents, the Tartar Relation and the Speculum Historiale. This indicated that the parchment could have been of early date. It appeared that the map had at one time been bound between these two documents. The first physical analytical work done on the map was carried out for Yale at the British Museum by David Baynes-Cope. He, however, was not allowed to remove any material from the map and had to be content with the use of microscopical examination and examination with infrared and ultraviolet light. He found that although the fluorescence under ultraviolet was normal for the Tartar Relation and the Speculum Historiale, the Vinland Map ink did not quench the vellum fluorescence in the normal way for an iron gall ink. He concluded that there was a strong case for a detailed scientific analysis of the map to confirm or deny its authenticity. Partly as a result of this, the Yale University Library hired McCrone Associates to make this detailed examination. This was several years later after we reported that the instrumentation, e.g., the ion microprobe, was finally available for such an analysis.

Nondestructive Analysis

The problem with analysis of any valuable document or painting is that the method must be nondestructive or, at least, show no damage visible to the naked eye. In the course of our work we have developed many methods suitable for handling and analyzing very small samples. The delicate job of removing minute samples from the map as well as from the two related documents was done in a particle-free clean room with a tungsten needle. The needle had a fine point, 1 μm at the tip, with a small ball of rubber cement near the tip to pick up the ink particles dislodged with the needle. To avoid damaging the vellum the needle was held below 20 degrees to the map surface and drawn backward rather than pushed toward the point. The map was observed throughout this procedure with the aid of a stereobinocular microscope. The samples were set down on clean microscope slides, and the rubber cement softened and stuck to the slide by use of a tiny drop of amyl acetate from a special micropipet. Micromanipulators have been

Miscellaneous 225

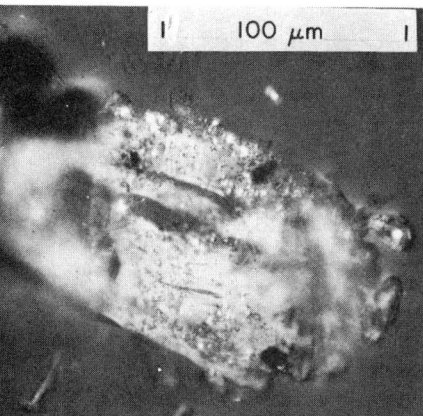

Figure 2. Single parchment fiber from the Vinland Map with adhering ink-crossed polars

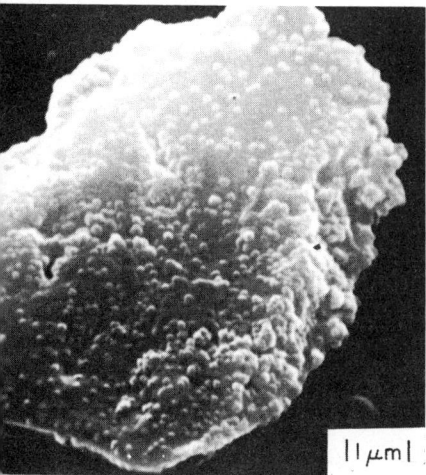

Figure 3. Scanning electron micrograph of ink particle from the Vinland Map

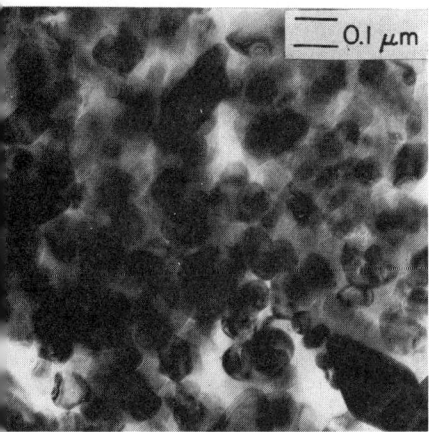

Figure 4. Transmission electron micrograph of anatase pigment particles in the Vinland Map sample 8A

found to be an encumbrance for this type of work, and these samples were taken simply with a steady hand. In all, 54 ng samples were taken; 29 from the Vinland Map, 18 from the Speculum Historiale, and 7 from the Tartar Relation. The sampling positions encircled on the map are shown in Figure 1. The total weight of all of the samples was much less than a microgram, and if they had all been put together, they would scarcely have been visible to the naked eye. Thus, the first requirement was fulfilled; there was no visible damage to the documents.

The initial optical microscopical examination showed that the lines had originally been drawn with black ink, but most of it had flaked off. The map outlines remained as a yellow stain which we initially supposed was from an organic vehicle present in the original ink. The appearance suggested the vehicle had soaked into the parchment and discolored with time, a behavior wholly consistent with 500 year-old documents.

Study of single bits of vellum on which ink could still be found (Figure 2) showed the presence of tiny, high birefringent, high refractive index particles in the yellow-stained portion of the inked line. By examination of some of the background samples (for example, 9B in Figure 1), it was determined that these particles were not present in the vellum itself. There seemed to be some association between the particles and the yellow portion of the ink line. A more careful examination of the map was then made, and it was determined that the ink line was actually composed of two lines. One was a yellow ink which simulated soaking of the vehicle into the parchment with discoloration in time. The second ink line was black and had been carefully drawn over the original yellow line. This required a skilled artistic effort although we did find at least one portion of the line where the registration was not perfect. It was perhaps even less perfect at the time of execution, but at least 80% of the black ink had been flaked off by rubbing of the map.

Electron Microscopy

It seemed then that the map had been drawn in such a way that it would appear to be old, and forgery or careful hand facsimile was therefore strongly suspected. To reenforce this conclusion we made further analyses, particularly of the pigments in the yellow line.

Examination with the SEM (scanning electron microscope) showed tiny dots or outlines of pigment particles in a bit of the yellow ink itself (Figure 3).

The SEM is fitted with an energy-dispersive x-ray analyzer which showed for this same particle a high peak for titanium. This was certainly a puzzling constituent, and we then proceeded to make every effort to identify the titanium component of the ink.

After several attempts, we were able to obtain an x-ray diffraction pattern of a yellow ink fragment only 7 μm in maximum dimension. This particle, sample 9C from the east coastline of Vinland, showed calcite and anatase (TiO_2); thus, the high refractive index, high birefringent particles seemed likely to be anatase. It became very important to confirm this conclusion since anatase had only been prepared, at least in precipitated form, about 1920.

The TEM (transmission electron microscope) was next used for further characterization of the anatase pigment. The anatase particles had an average diameter of about 0.1 μm and possessed a rounded shape characteristic of precipitated anatase pigment (Figure 4). The shape, in fact, was totally inconsistent with a ground mineral and agreed very well in shape and size distribution with commercial anatase pigments. It was also possible to confirm the identification of anatase by means of selected area electron diffraction.

To make certain that the map had not been retouched in modern times with an anatase-containing pigment, it was decided to analyze all of the ink samples with the electron microprobe. As a result, titanium was found in more than 20 samples from widely different areas of the map. Also, the vellum itself contained no titanium, nor did the all-black ink particles. Thus, the anatase is apparently only present in the brownish-yellow ink and is present in all samples thereof. Furthermore, the yellow line is under the black line where the latter is present, and we were forced to conclude that the yellowish line was intentionally added to simulate the staining to be expected from the vehicle of an old ink discoloring over the ages. The black was applied over the yellow line, dried, and flaked off to embellish the aged appearance.

Ion Microprobe Identification

Finally, we attempted to identify the vehicle in the yellow ink by means of the IMA (ion microprobe). The IMA is an extremely sensitive ion-sputtering mass spectrometer. The accelerating voltage for the ion beam is, however, considerably higher than the accelerating voltage in most mass spectrometers; hence, fragmentation of organic molecules is considerably more extensive. It is therefore difficult to interpret the mass spectra of organic materials with the ion microprobe. Nevertheless, we were able to make direct comparisons with a number of possible vehicles and found that the ink vehicle was completely different from all known inks used during medieval times. We tested India ink, tea, bearberry, sepia, tannin, celandine, gall and iron gallotannates. In comparing this spectrum with modern vehicles, we found that it resembled most closely a suspension of pigment particles in an alkyd resin. Although we could not be certain of this conclusion, it was consistent with the modern origin of the ink pigment since alkyd resins were first synthesized in the 1920's. There seemed little doubt that the Vinland Map was a very clever forgery, but that it incorporated several major mistakes. The presence of the yellow underline was an obvious effort to deceive, and the presence of anatase as a precipitated pigment was impossible before about 1920.

The Vinland Map is still available for study in Yale University's Beinecke Rare Book and Manuscript Library. Presented to Yale in 1965 by an anonymous donor, the map is now perhaps more famous than ever; it is certainly better known. The conclusion of a modern origin has been almost universally accepted although a few scholars feel that, just possibly, a 15th century mapmaker could have synthesized anatase pigment. Another noted scholar suggests the map is a copy of *the real map* which may or may not still exist. The only question still remaining and one which our analytical tools will probably not answer is who would have wanted to perpetrate such a document. This may never be known.

Acknowledgment

The author gratefully acknowledges the patience, perseverence, and skill of the McCrone Associates personnel who participated in the Vinland Map study: Lucy McCrone as project coordinator, Anna Teesov (sampling, particle manipulation, and PLM), Ralph Hinch (XRD), John A. Brown (SEM and EDXRA), John Gavrilovic (EMA, EDXRA, and λDXRA), Gene Grieger (TEM and SAED), and Michael Bayard (IMA).

The Bust of Nerfertiti

H. G. Wiedemann

Mettler Instrumente AG, CH-8606 Greifensee, Zurich, Switzerland

G. Bayer

Institute of Crystallography and Petrography, Swiss Federal Institute of Technology, 8092 Zurich, Switzerland

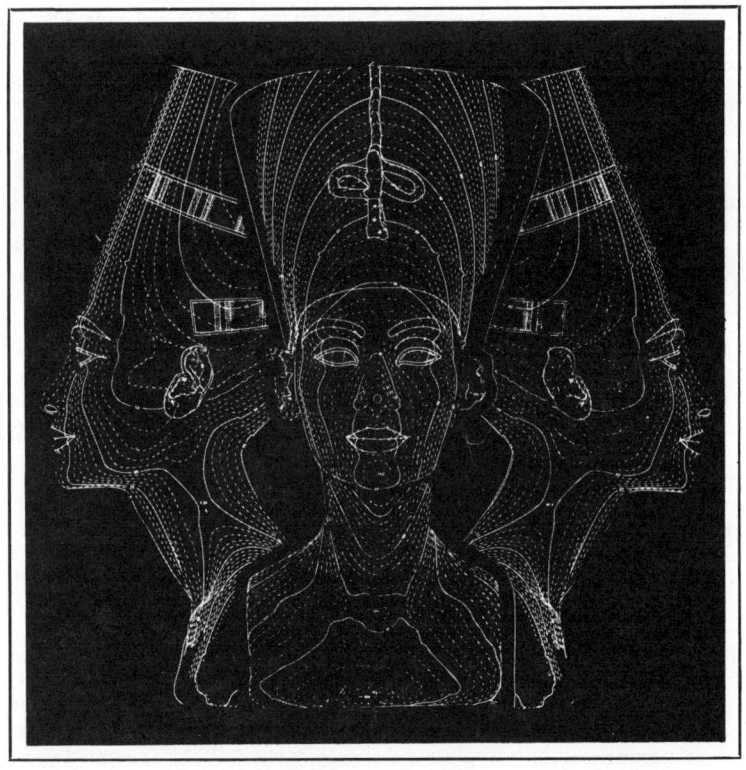

Courtesy of Ägyptisches Museum, West Berlin

Originally published in ANALYTICAL CHEMISTRY, 1982, Vol. 54, No. 4.

The bust of the famous Egyptian queen Nefertiti was made during the reign of her husband Amenophis IV-Echnaton (1364–1347 B.C.) in the XVIIIth Dynasty. There are many fascinating stories, more or less historical, about the life of Nefertiti and about the origin of the bust. This unique art piece was found in December 1912 in Tell El-Amarna (Middle Egypt) during excavations carried out by the German Oriental Society led by Ludwig Borchardt. In 1920 the bust was donated to the New Museum on the museum island (today East Berlin) by James Simon, who had financed the archeological expedition. It was hidden at the end of World War II, found by American troups in a salt mine in today's German Democratic Republic, and transported to the "Art Collecting Point" in the former German–U.S. zone in Wiesbaden. From there the bust of Nefertiti was brought back to the Egyptian Museum in West Berlin. Since 1967 it has been one of the main attractions of this exhibition (Berl. Mus. No. 21300).

The bust is about 0.5 m high and made of a soft limestone supplemented with plaster. It is richly painted and decorated, and overall its state of preservation is excellent (see cover). For decoration of the bluish crown of the collar, gold stripes and inserts made of colored frits and stones were used abundantly.

Chemical analysis of the dyes and pigments used on the bust was first carried out by Friedrich Rathgen and published by Ludwig Borchardt in 1923. To confirm or refute Rathgen's findings, we wanted to examine these pigments using some of the more modern analytical techniques that have developed since his experiments. We were especially interested in the composition of the blue pigment because of some previous studies we had made of ancient Egyptian blue pigments of various localities and dynasties. With special permission from the Egyptian Museum, it was possible to get small samples from the original object for our investigations.

Ancient Egyptian Pigments

In the ancient Egyptian murals, frescoes, papyrus illustrations, and painted objects, color played a domi-

228 The Analytical Approach

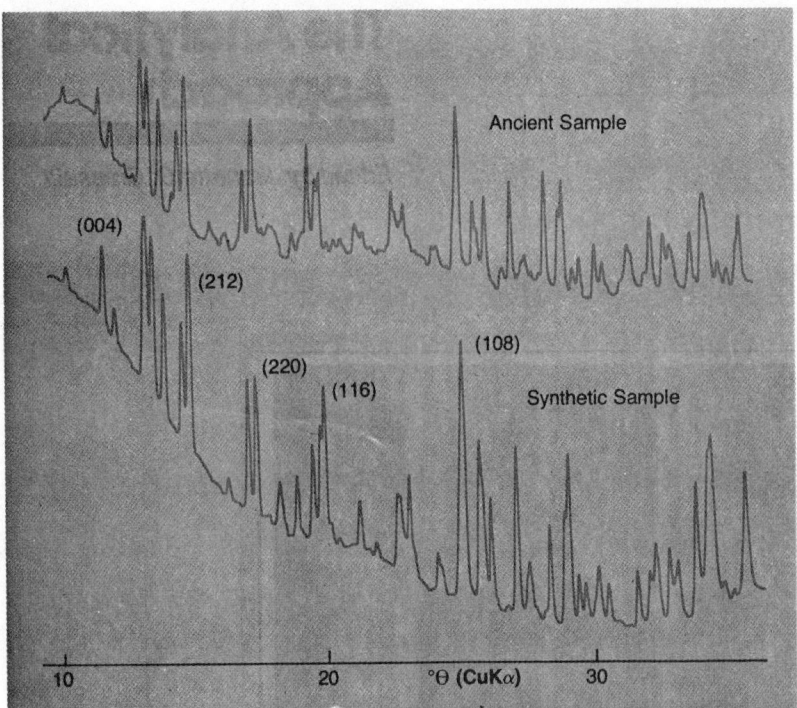

Figure 1. X-ray patterns of the original blue pigment (bust of Nefertiti, crown) and of synthetic Egyptian blue. Typical reflections are given with their (hk*l*) indices

nant role. Where did the Egyptians get their color pigments? For the most part they were produced from natural minerals, e.g., azurite (blue), malachite (green), and calcite (white). However, there was frequently a great demand for certain colors not available in large amounts in nature; therefore, many attempts were made to produce them artificially.

Blue was an especially important Egyptian color because it was regarded as the color of the gods (*1*). Blue occurred in nature as lapis lazuli, a dark-colored silicate mineral that played an important role as a gemstone. However, this mineral could not be used extensively as a pigment because it was rare in Egypt and therefore regarded as very precious. Azurite, a basic copper carbonate, was also a natural blue pigment. However, it had disadvantages; its blue color was not very brilliant, and its chemical resistance was rather poor.

Taking into account the very high level of ancient Egyptian techniques and craft skills, it is not surprising that the all-important blue color was produced synthetically. Glasses, glazes, glass frits or crystalline pigments (*2*) were produced in the forms of compact or powdered materials.

What Is Egyptian Blue?

One of the most important artificial blue pigments, Egyptian blue was already in use during the IVth Dynasty in Egypt, i.e., about 2600 B.C. It was produced exclusively in Egypt in consistent quality probably for more than 2000 years. Later it was used also in Greece and Rome (*3*). As the Roman Empire was coming to an end late in the fourth century, the process for manufacturing Egyptian blue appears to have been lost, and it was rediscovered only in the 19th century.

Scientifically, the term Egyptian blue applies to a sheet silicate with the formula $CaCu[Si_4O_{10}]$. The name was given to it because of its brilliant blue color and because it was produced synthetically only in Egypt. The raw materials necessary for its production were probably calcite ($CaCO_3$), malachite ($Cu_2[(OH)_2CO_3]$) or azurite ($Cu_3[(OH/CO_3)_2]$), and silica sand (SiO_2).

The Problem: Nefertiti's Blue Pigment

Rathgen in 1923 (*4*) identified the blue pigment on the bust of Nefertiti as a powdered glass frit colored by cupric oxide. In our opinion, however, it looked more like crystalline Egyptian blue than the noncrystalline glassy frit. It should be mentioned though that other authors have incorrectly applied the term "frit" to describe the crystalline Egyptian blue (*5*).

The analytical approach to clear up the confusion concerning the nature of Nefertiti's blue pigment and to confirm Rathgen's findings on the other pigments took two main directions: 1) X-ray, differential thermal analysis (DTA), thermogravimetry (TG), and microscopy data were collected on the original samples from the bust and compared with data from known pigments. In addition, a number of other ancient Egyptian blue samples from different dynasties were investigated by X-ray and thermal analysis. 2) Different methods of synthesis were used to make Egyptian blue from various raw materials, including original Egyptian raw materials. The main goal of this study was to find out the possible conditions under which this pigment might have been produced in ancient times.

Analysis of the Pigments

Experimental investigations on ancient materials are always linked with difficulties. Whenever possible, nondestructive methods are preferred. In certain cases destructive methods can be tolerated if they are very sensitive and require only very small amounts of material. This is the case with the X-ray methods that were mainly used in our investigations and also with special thermoanalytical instrumentation (Mettler TA 3000) that requires sample amounts below 10 mg.

The first X-ray pattern ever taken of the blue pigment from the bust of Nefertiti proved that this was indeed a very pure and well-crystallized Egyptian blue. Figure 1 shows the X-ray pattern of one of the original bust samples compared to that of pure, synthetic Egyptian blue. The agreement of these photometric tracings is perfect. Several samples from different locations on the crown all gave this same result. A blue glassy frit as assumed by Rathgen would not show any X-ray peaks.

In our study we also included some of the other pigments from the bust: red from the necklace, green from the headband, and black from the eyelash. Figure 2 shows the X-ray pattern and the energy dispersion (ED) spectrum of the red pigment. The X-ray pattern shows strong reflections of calcite and rather weak ones of quartz and probably some poorly crystallized iron oxides. These findings are confirmed by the ED spectrum, which in addition to Ca, Mg, Si, Fe, and Al, also shows the presence of S. Since a microtome section was used in this study, it is possible that S and Al come from the plaster and clay padding on which these pigments were painted. The red color is definitely not caused by Cu^{1+} but comes from the red ochre which was regularly used in ancient Egypt as a slip, as a paint, or in medicines (*6*). Compared to the clay or plaster padding (100 μm thick), the ochre layer was very thin (~5 μm).

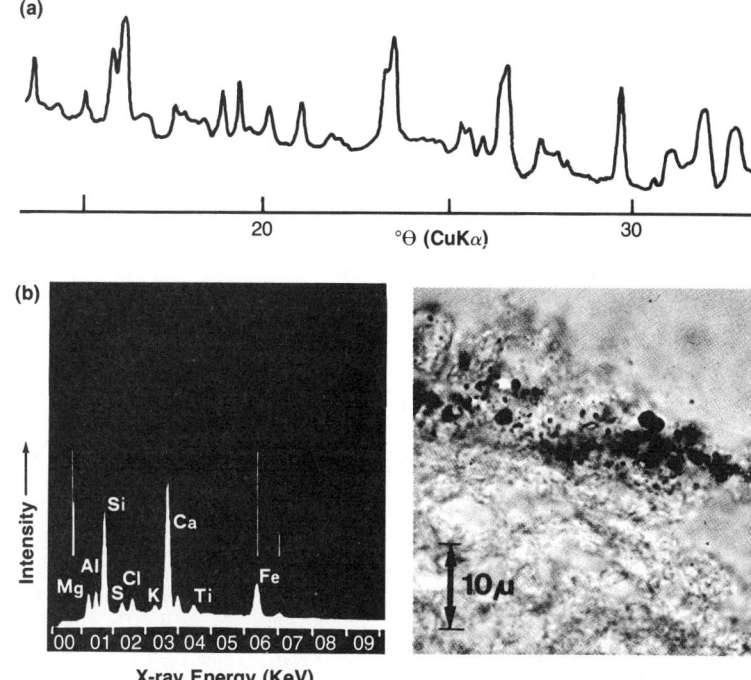

Figure 2. (a) X-ray pattern and (b) ED spectrum of the red pigment. The micrograph of the thin section shows the ochre layer on the top of the padding

The investigation of the green pigment, which is bordered by yellow (identified as orpiment, As_2S_3), was more difficult. The X-ray pattern showed the reflections of quartz, kaolinite, and many additional, weak peaks that could not be assigned unambiguously (Figure 3). They probably correspond to a green-colored copper compound, since the presence of Cu could be definitely confirmed by the ED spectrum. This compound was not malachite or greenish-colored Egyptian blue that could be formed with correct synthesis conditions, e.g., stoichiometry, temperature, and flux. The X-ray pattern shows some similarity, however, to that of verdigris, the green basic acetate of copper. It is known that this intense green-colored compound was used in ancient Egypt, frequently in diluted form, as a pigment and in cosmetics and medicines (6). It was relatively simple to prepare by soaking metallic copper in dregs of wine and scraping off the green surface layer periodically. The black pigment used for the eyelash was obviously carbon, which was almost always used in some form.

The binding medium is wax—probably beeswax—as can be seen from a comparison with fresh beeswax in Figure 4. This figure also shows a differential scanning calorimetry (DSC) curve of this original wax binder. The melting range of from 40 to 66 °C agrees well with original values given by Rathgen (4). The maximum of the melting peak was found at 64.4 °C for the original wax binder and at 63.4 °C for beeswax. The corresponding heats of fusion were 127.4 J/g and 207 J/g, respectively. Furthermore, investigations on this sample have been carried out with field desorption (FD) mass spectrometry and also with a quadrupole mass spectrometer for obtaining electron impact mass spectra. Both methods lead to similar results. The FD mass spectra of the original sample (wax binder from the bust) and of beeswax are shown in Figure 5. The mass numbers of the original sample correspond to the components typical for beeswax, e.g., higher paraffins, cerotin acid, and especially esters of C16–C30 acids and C24–C36 alcohols.

Synthesis of Egyptian Blue

For the thermosynthesis of Egyptian blue, finely powdered calcite, malachite (or azurite), and quartz were used. These raw materials were partly from Egyptian localities, for example, limestone from Tura and sand from the desert. However, the Fe content of the sand was very important. Concentrations higher than 0.5% Fe_2O_3 cause a change of the blue color of the synthesized pigment toward green.

The course of the reaction during heating could be followed by means of X-ray and thermal analysis (7). In previous experiments, mixtures with different ratios of the components were heated with and without a flux additive at different temperatures in oxidizing or reducing atmospheres. In all cases the most intense and brilliant blue color was found for the composition corresponding to stoichiometric Egyptian blue (1 CaO–1 CuO–4 SiO_2) with the addition of a flux (such as soda, borax, or sodium sulfate).

The optimum firing temperatures were in the range 870–950 °C. The atmosphere during heating must be oxidizing; otherwise a reduction of Cu^{2+} to Cu^{1+} and Cu^0 takes place, which results in a reddish-brown reaction product. But decomposition of Egyptian blue occurs even in oxidizing atmospheres at temperatures in excess of 1000 °C (7).

A variety of Egyptian blue pigments were synthesized from the stoichiometric, molar mixtures: 1 limestone, ½ malachite (or ⅓ azurite), 4 sand. Four different fluxes (4 wt %) were added—borax, papyrus ash, salt, and sodium sulfate. These mixtures were then heated 4 °C/min up to 950 °C and held for 1 h. The pigments were purified by boiling with 20% HCl. The resulting colors were quite different, ranging from bright blue to almost green, depending both on the flux and on the copper mineral. Generally, malachite gave somewhat clearer blues than azurite; the most effective fluxes were borax and soda ash. In spite of the different colors, all of these pigments gave an X-ray pattern corresponding to pure Egyptian blue.

The limestone–malachite–sand–borax mixture was also investigated with respect to the rate of formation of Egyptian blue. For this purpose, 60-mg batches were heat treated at 880 °C for 1, 4, and 15 h. Afterward the unreacted CuO and CaO and Ca

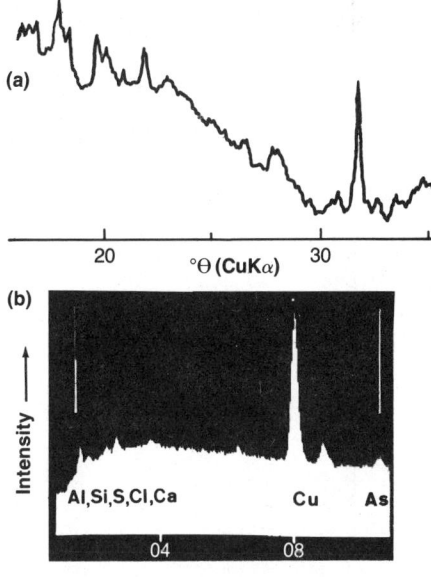

Figure 3. (a) X-ray pattern and (b) ED spectrum of the green pigment used for the headband on the bust

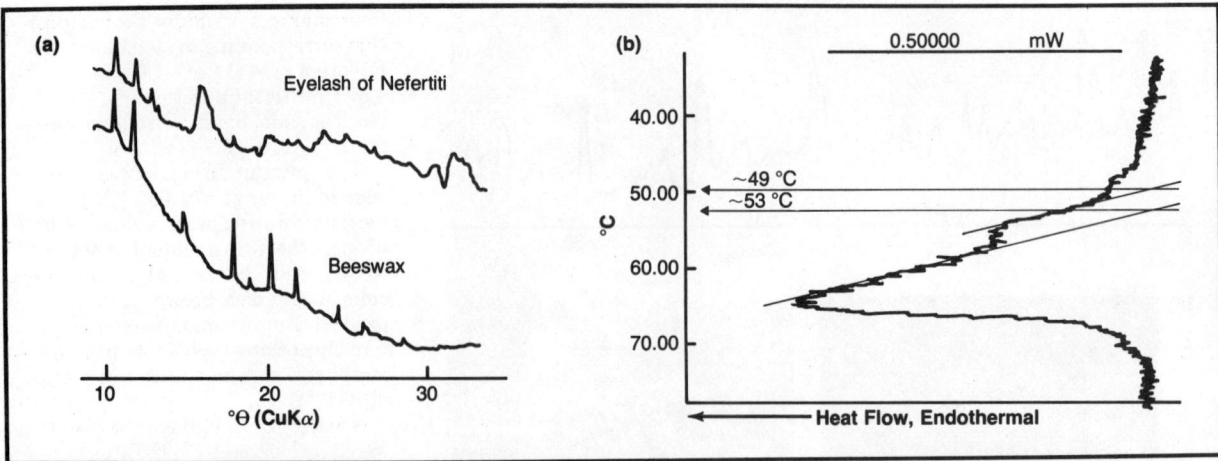

Figure 4. (a) X-ray pattern and (b) DSC curve of the black-colored beeswax used for eyelash on the bust of Nefertiti

silicates were leached out with boiling, concentrated HCl.

The bluish residue, which consists of Egyptian blue and unreacted SiO_2, was melted with sodium carbonate (1:4), dissolved in HCl, and titrated iodometrically after addition of an acetate buffer at pH 4–5. The results showed the following yields of the $CaCu(Si_4O_{10})$ reaction product: 32%, 36%, and 61% after 1, 4, and 16 h heating at 880 °C, respectively. This means that the reaction starts rather rapidly at the interfaces of the grains and proceeds gradually with time. For complete reaction, longer heating times or temperatures higher than 880 °C are necessary.

As a representative example, Figure 6 shows the thermogravimetric analysis of the formation of Egyptian blue from this same mixture of limestone-malachite-sand-borax. The TG curve proves that the decomposition of malachite to CuO occurs at about 300–400 °C followed by the decomposition of the limestone to CaO in the range 550–740 °C. The corresponding maxima of the reaction rate can be seen in the differential weight curve (DTG) at 380 °C and at 725 °C, respectively. These oxides then react with the SiO_2 at a higher temperature to form the desired compound $CaCu(Si_4O_{10})$. The values of the activation energies for these decompositions did not change significantly for the different flux additives.

In addition to Egyptian blue, other alkaline earth copper silicates, namely, $SrCu(Si_4O_{10})$ and $BaCu(Si_4O_{10})$, could be synthesized under similar conditions. They have the same brilliant color as $CaCu(Si_4O_{10})$ and identical crystal structure. They were not known or used, however, as pigments in ancient Egypt.

All of these copper silicates show a pronounced tendency to grow in the form of square, platelike crystals due to their sheet-silicate structure (Figure 7). This formation is favored by longer heating times and higher additions of flux (5–10 wt % soda, borax or sodium sulfate), and is an indication of intensive and complete reaction. Also, some of the ancient Egyptian blue pigments, especially in the form of the originally prepared balls, showed the presence of such platelike crystals (7). The Egyptian blue pigment used on the bust of Nefertiti did not contain such platelike crystals; they had probably been destroyed by fine grinding.

Conclusions

The conclusions based on our analyses of the pigments are summarized in Table I. They are consistent with Rathgen's findings except for the nature of the blue pigment, which we identified positively as Egyptian blue, not as a noncrystalline blue glassy frit.

In our studies on the synthesis of Egyptian blue, we learned that the following main points have to be observed to get the most intensive and brilliant color:

- The ratio of the components must be as close as possible to the stoichiometric composition 1 CaO–1 CuO–4 SiO_2.
- The completeness of the solid-state reaction and the crystallinity of the pigment must be catalyzed by the addition of small amounts of fluxes or mineralizers, such as salt, borax, or sodium sulfate.

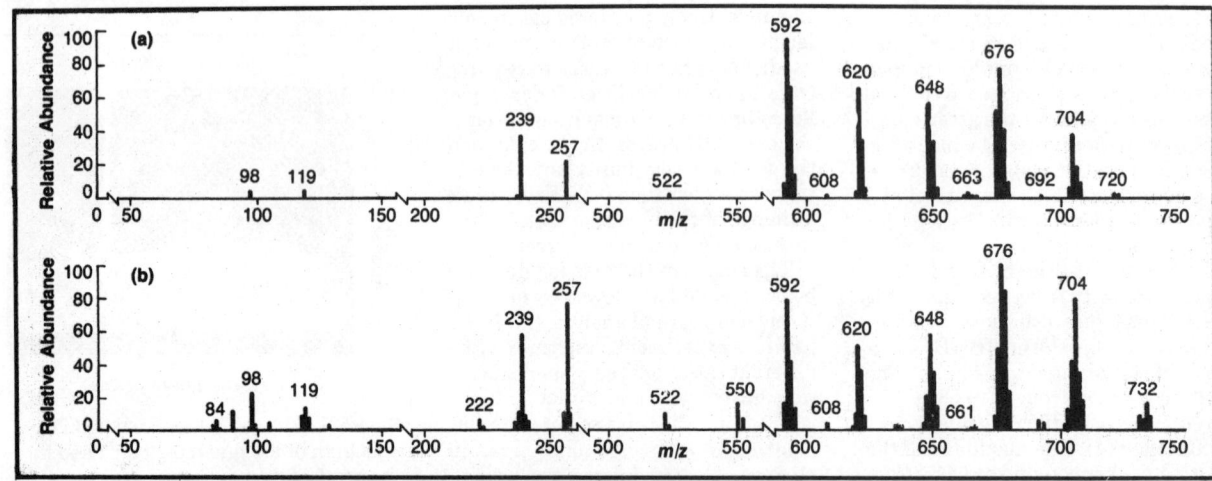

Figure 5. (a) FD mass spectra of binding medium from the bust and (b) fresh beeswax

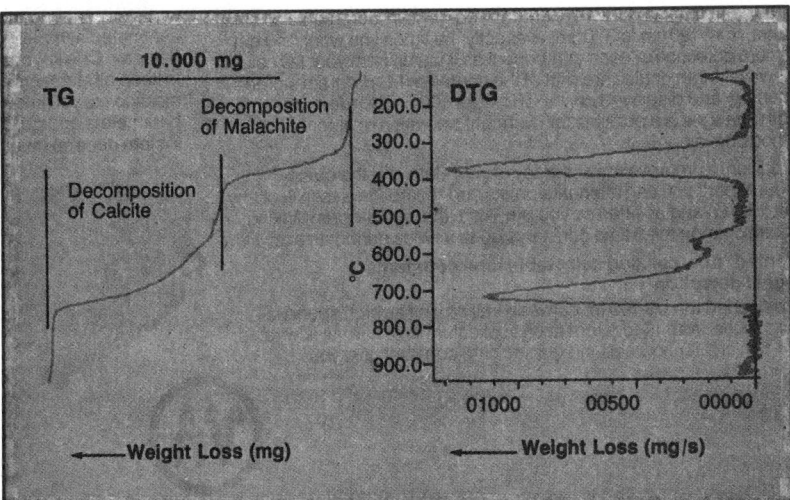

Figure 6. Thermogravimetric curve (TG) and its first derivative (DTG) for the reactions of a mixture of malachite, calcite, sand, and borax (flux) with formation of Egyptian blue (sample weight, 117.9 mg; heating rate, 6 °C/min)

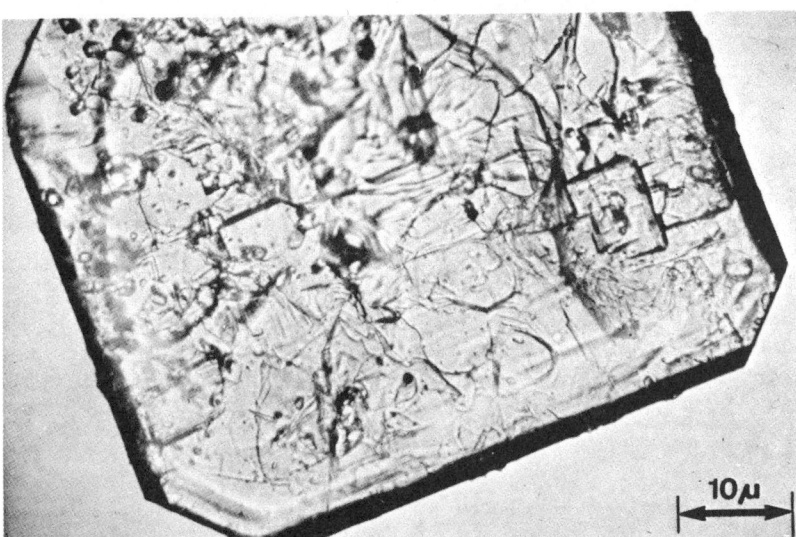

Figure 7. Microphotograph of synthetic Egyptian blue crystals (160 ×)

Table I. Results of the Analysis of the Pigments

Color	Bust location	Composition
Red	Necklace	Iron oxide (Red ochre)
Green	Headband	Possibly the green basic acetate of Cu (verdigris)
Yellow	Headband	Orpiment (As_2S_3)
Black	Eyelash	Carbon
Blue	Crown	$CaCu[Si_4O_{10}]$ (Egyptian blue)

Binding medium for carbon: beeswax

- The temperature of synthesis must be lower than ~1000 °C; otherwise, the calcium copper silicate pigment decomposes due to the degradation $Cu^{2+} \rightarrow Cu^{1+}$ and Cu^0, which results in an irreversible loss of blue color.
- The furnace atmosphere must be oxidizing because in a reducing atmosphere the degradation $Cu^{2+} \rightarrow Cu^{1+}$ takes place at 800 °C and a reddish-brown reaction product containing Cu_2O is formed.

All these results indicate that the Egyptians must have had ways of weighing the required portions of the raw materials and of maintaining relatively good control of temperature and furnace atmosphere.

They also knew about the importance of adding fluxes during the synthesis. Without these alkali additives, the blue pigments obtained are rather pale because of the very small crystals obtained. Ideally, for brilliant blue pigments, the crystals should be in the order of about 5–50 μm in size. This coarser crystallinity is achieved only by addition of fluxes. Because its intensity depends strongly on the size of the crystals, Egyptian blue is regarded today as a rather poor pigment compared to cobalt aluminate blue or zircon–vanadium blue.

While cognizant of the technological achievements of our time, we must certainly admire the skills and know-how of the ancient Egyptians, who were able to produce excellent-quality Egyptian blue, a very complex pigment, for over 2000 years and use it to adorn the bust of their queen, Nefertiti.

Acknowledgment

We thank J. Settgast, director of the Egyptian Museum, West Berlin, and J. S. Karig, managing director of the museum, for allowing us to sample the bust of Nefertiti. H.-R. Schulten, Physkialisch-Chemisches Institut der Universitaet Bonn, carried out the mass spectrometric investigations. We also thank D. Mueller, Finnigan MAT AG, Basel, for supplying GC/MS spectra, and R. Giovanoli, Laboratorium fuer Elektronenmikroskopie, Universitaet Bern, for running the ED spectra. The titrimetric analyses were carried out by K. Mooibroek, and the thermoanalytical measurements were done by D. Woodtli, both of Mettler Instrumente AG, Greifensee/Zurich.

References

(1) A. Lucas, and J. R. Harris, "Ancient Materials and Industries," 4th ed., Edward Arnold Ltd., London, 1962.
(2) W. T. Chase, "Egyptian Blue as a Pigment and Ceramic Material," by R. H. Brill, MIT Press, Cambridge, Mass., 1971, pp 80–90.
(3) W. Noll, "Anorganische Pigmente in Vorgeschichte und Antike," *Fortschritte der Mineralogie*, **57**, 203–63 (1971).
(4) L. Borchardt, "Ausgrabungen der Deutschen Orient-Gesellschaft in Tell-El-Amarna, III. Portraets der Koenigin Nofret-ete," J. C. Hinrichs'sche Buchhandlung, Leipzig, 1923.
(5) R. Giovanoli, and J. Riederer, "Egyptian Blue," Studies in Conservation, in preparation.
(6) J. Riederer, "Kunstwerke Chemisch Betrachtet," Springer Verlag, Berlin–Heidelberg–New York, 1981.
(7) G. Bayer, and H. G. Wiedemann, "Aegyptisch Blau, ein Synthetisches Farb-Pigment des Altertums, Wissenschaftlich Betrachtet," *Sandoz Bull.*, **40**, 19–40 (1976).

Index

Index

Accelerant, gas chromatograms 179f
Acetaminophen, phenylpropanolamine hydrochloride, separation 63
Acid components in orange juice 55
Acid strengths for nuclear fuel cycle elements 136f
Adenine 25, 26f
Adhesive failures 75–76
Admissibility in court, scientific testing 198–200
Admixture problem, analysis scheme 80f
Adulteration of honey 69–73
Adulteration of orange juice 55
Aerospace materials
 analysis scheme 53
 problem solving 51–54
Air pollutants analysis 131–133
Air sampling in arson analysis 180
Air-entrained concrete 80–81
Airborne isocyanate monitoring 39
Alkali additives 231
Alkoxyphenols 90
Alkyd resin in ink 226
Alkyd resin residuals 38
Allantoin analysis 62–63
Allergy remedy
 analysis 63
 components separation 63
Alpha spectrometry in plutonium detection 114
Amino acids in orange juice 56
Amobarital 189
Amphetamines 187f
Analysis scheme
 admixture problem 80f
 air pollutants 132
 ash filter 20f
 blood preservative impurity 28f
 carcinogenesis 151
 coatings 37f
 concrete 79t
 crude oil 42t
 drilling rig coating 208
 electron beam resist 66
 flavor component 8
 honey adulteration 72
 honey/corn syrup mixtures 69, 70f
 illicit drugs 188
 migration 142
 organic products 65
 packaging material 143
 reboiler solids 43t
 small particles 223
 solids mixture 16f
 trace organics in water 159

Analysis scheme—Continued
 vitamin impurity 23f
 wire coating 68
Anatase pigment, micrograph 225f
Angiosarcoma 127
Anhydride, terephthalic acid 21
Aniline–acetone reaction products, chromatogram 94f
Aniline–acetone resin 97t
Anion exchange conditions 32
Anthranilic acid 183
Antigen, detection and measurement 172–173
Antimony in bullets 193–194
Appliance problem solving 45–47
Arc emission spectroscopy, leaf analysis 45
Arsenic in rocks 216t
Arson investigation 179–182
Ash filter analysis scheme 20f
Ash, Mount St. Helens 105, 107, 109–110, 117–119
Ashing of orange juice 55
Atomic absorption spectrometry
 concrete analysis 80
 gasoline analysis 58
 rocks analysis 215–216
Auger spectrometry, coatings analysis 75

Barbiturate, blood, determination 188–189
Binder, beeswax, x-ray pattern 230f
Binder, cemented carbide 67
Binders, differential scanning calorimetry 229, 230f
Bioassay of C.I. Acid Red 14 153–154
Bioassay tests 152
Biological monitoring 147–148
Bis(4-morpholinecarbodithioato)nickle(II) 53, 54t
Bis(trimethylsilyl)adenine 26, 27f
Blood barbiturate determination 188–189
Blood preservative impurity identification 25–28
Bomb detection in arson analysis 80
Bone analysis 182
Boron
 in ground water 45
 in plant leaves 46t
 quantitative analysis 5
 SIMS determination 5
Borosilicate glass, isotopic analysis 3t

Bullet analyses, JFK assassination 191–195

Cable jacketing failure 77
Cadiz oil spill 111f
Calibration
 ion chromatography 33
 personal sampling pump 148
Capillary column gas chromatography, water analysis 160–161
Carbide binder, cemented 67
Carbon-12/carbon 13 analysis, honey 71–72
Carbon dioxide fixation in plants 71
Carbon filter for drinking water 159–160
Carbon oxidation states in corrosion scale 16, 18f
Carbon tetrachloride solubles, spectrum 177f
Carcinogenesis testing 151–155
Cemented carbide binder 67
Cereal off-flavor analysis 11–13
Cerium in rocks 216t
Characterization of
 anatase pigment 225
 drilling rig coating 209
 electron beam resist 66, 67t
Charcoal tube 149f
Chemical ionization mass spectrometry, drug analysis 203
Chine White, identification 201–205
5-Chloro-2-(2,4-dichloropenoxy)-phenol
 reaction products, 125f
 synthesis 124f
Chlorendic acid 155
 stability in feed 154
Chloride in boiler steam 33f
Chloride impurity clean-up 34
Chlorination of drinking water 158
Chlorine measurement, sulfur interference 42
Chlorine in naphtha 41
Chloroform extraction of nickel-plating solids 47
Choline salicylate analysis 61–62
Chromatopyrogram
 neoprene sheath 86f
 nitrile sheath 85f
 silicone rubber 84f, 85f
Chromatopyrography for polymer characterization 83–86
Chromosome staining 171
Citrus constituent determination 56

235

Coating materials problem solving 75–77
Coatings industry problem solving 37–39
Coatings process improvement 38–39
Cohesive failures 75–76
Color Index Red 14 153
Collection devices, personnel 148, 149f
Colorimetric titration of coolent 52
Computer-simulated oil weathering curve 102, 103f
Concentrator column efficiency 32–33
Concrete analysis scheme 79t
Concrete problem solving 79–81
Condensing agent 184
Construction materials failure 79–81
Contamination, instrument 5
Coolant precipitate analysis 53
Copper silicates 230
Corn syrup, high fructose 55, 69–73
Corroborative problem solving 207–211
Corrosion
   characterization 15
   inhibitor analysis 43t
   preventative coating 208–209
   on resistors 46
Corrosion products, stainless steel 54
Corrosion scale 18f
   nuclear reactor 15–18
   steam turbine 31–35
Cough–cold remedy analysis 63
Cristobalite, spectrum 117f, 118f
Crude oil
   analysis scheme 42t
   refining 42
Curare detection in human tissues 197–200

Detection and measurement of antigen 172–173
Detector tubes 148
2,7-Dichlorodibenzodioxin 123
Differential scanning calorimetry, binders 229, 230f
Diffusion coefficients for nuclear fuel cycle elements 136f
Digestive system, nitrosamine formation 163
Dihydroactinidiolide 8f, 9
Dimethyl terephthalate purity specifications 19
2,4-Dinitrophenylhydrazone derivatization 30
Disease agent, Legionnaires' disease 161
Disperse Blue 1 153
Display device liquid crystal analysis 89–90
4,4′-Dithiodimorpholine 53, 54f
2,2′-Dithiolbisbenzothiazole 52
Dome C glacier sampling 114
Drilling rig coating, analysis scheme 208
Drinking water, hotel, studies 157–161
Drug analyses with HPLC 61–63
Drug homicide investigation 188–189
Drug manufacturing, illegal 183–185
Drugs of abuse detected in hair 169
Dye analysis, bust of Nefertiti 227–231
Dynamite residue analysis 173

Egyptian blue 228–231
Egyptian pigments, ancient 227–228
Electrical failure 76–77, 85–86
Electron beam resists 65–67
Electron impact mass spectroscopy, electrooptic materials 88
Electron micrograph of corrosion scale 15, 16f
Electron microscopy, small particle analysis 225
Electron spectroscopy for chemical analysis
   coatings 75–77
   corrosion scale 16, 17f
Electronics raw materials analyses 65–68
Electrophoresis, enzyme separation 172
Elemental analysis
   air pollutants 131–132
   red dye 153
Emission spectrography, bullet analyses 191–192
Emission spectroscopy
   leaf analysis 45
   solids analysis 47
Energy dispersion spectroscopy, pigment analysis 228
Enzyme separation, electrophoresis 172
Enzyme-linked immunosorbent assay for antigens 172–173
Ethyl carbamate 152
Exposure levels to vinyl chloride 127
Exposure levels, human, crystalline silica 117
Exposure, industrial, determination and measurement 147f
Extract D & C Red 153
Extraction, chloroform, of nickel-plating solids 47

Failure analysis in coatings 75–77
Failure, construction materials 79–81
Failure, electrical 85–86
Fats and oils, methyl esters 209t
Fatty acids, separation 38
Fatty acids, linseed 38
FDA approval scheme for resin 142
Fibrosis, progressive massive 105–106
Field desorption mass spectroscopy 94–97
Field desorption mass spectrum, beeswax binder 230f
Fingerprinting polymer volatiles 46
Flavor component analysis, tobacco 7–9
Food packaging material, commercialization 141–143
Fourier transform IR interferometer, **Voyager** 219–222
Fourier transform NMR spectrum
   blood preservation impurity 27f
   methyl carbamate 153t
Fuel oil analysis 180
Fuel, spent, isotopic analysis 138t

Gamma-ray spectrometry, bullet analysis 193–194
Gas–liquid chromatography, honey analysis 72, 73f
Gas pyrolysis chromatography, polymer analysis 84
Gasoline dispensing hose leach test 58t
Gasoline, unleaded, phosphorus 57–59
Gas chromoatogram
   accelerant 179f
   drinking water samples 159f, 160f
   β-naphthylamine 154f
   rat urine 154f
   resin volatiles 13f
   trimethylorthoformate 29f
Gas chromatography
   air analysis 132
   ash analysis 21
   blood analysis 188
   carcinogen identification 152
   citrus analysis 56
   drug analysis 187
   impurity analysis 25–26, 29–30
   industrial hygiene 148
   molasses analysis 210
   odor analysis 11–12
   vinyl chloride analysis 127, 130
   water analysis 158–159
Gas chromatography/mass spectrometry
   blood analysis 188
   drug analysis 201–202
   gunpowder analysis 171
   impurity analysis 26, 29–30
   nitrosamine analysis 163
   wastewater analysis 121
   water analysis 159–160
Gas liquid chromatography, blood analysis 188
Gas sampling bags leaching 129f
Gel permeation chromatography
   See also Liquid chromatography
   coatings analysis 37–39, 77
   polymer analysis 66
Geochronology, 210Pb 114
Glaciers, plutonium 113–115
Glass, borosilicate, isotopic analysis 3t
Glassine liner, component limits 12f
Global atmosphere, human impact 113
Green pigment, x-ray pattern 229f
Ground water containing boron 45
Gunpowder analysis, gas chromatography/mass spectrometry 171

Hair, drugs of abuse detected in 169
Headspace analysis 13, 46, 47f
Headspace sampling devices 181f
Headspace vapor analysis 180
Health effects of Mt. St. Helens ash fall 106
Herbicides analysis 132
Heroin detected in hair 169
Heroin, synthetic substitute 201
High performance liquid chromatography
   adenine 25

High performance liquid
 chromatography—Continued
  air analysis   132-133
  curare analysis   199
  drug analyses   61-63
  wastewater analysis   121
Honey analysis   69-73
Hotel drinking water studies   157-161
Human impact on global atmosphere
  113
Human leukocyte antigens   172-173
Human tissues, curare detection
  197-200
Humidity and resistivity   77f
Hydrolysis study of ash filter   21

Ice clouds on Mars   221
Identification of amphetamine
  187-188
Identification of China White   201-205
Illegal drug manufacturing   183-185
Illicit drug analysis scheme   188
Impurity clean-up, sodium and chloride
  34
Impurity determination scheme   23
Impurity isolation, blood preservative
  25
Impurity levels in water/steam cycle
  34f
Inconel corrosion   15-18
Inductively coupled plasma emission
  spectroscopy   175
Industrial hygiene   39, 145-149
Ink analysis   225
Ink particle, scanning electron micrograph
  225f
Inorganic ocean pollutants   112
Instrument contamination   5
Instrumental neutron activation analyses,
  bullet analysis   193-194
Interferometry   219-222
Interlaboratory ash analysis   106-107,
  110-111
Interlaboratory rocks analysis   216
Ion chromatography   32f
  corrosion scale analysis   31-35
Ion emission, thermal, mechanism   5
Ion etch treatment of corrosion scale
  17
Ion exchange chromatography, glacier
  analysis   114
Ion implantation   3
Ion microprobe, small particle analysis
  225-226
Ion-pairing HPLC   61
Ion plotting   26
Ion sputtering   3, 17
IR spectrometer, Voyager   219-222
IR spectrometry
  ash analysis   106
  coating analysis   76, 77
  gasoline analysis   58
  resin   143
IR spectrophotometry, ash analysis
  118
IR spectroscopy
  ash analysis   20
  binder analysis   67
  concrete analysis   80
  coolant analysis   52-53

IR spectroscopy—Continued
  impurity analysis   27, 30
  insoluble polymer analysis   66, 67t
  molasses analysis   210
  polymer analysis   95
  solids analysis   47
IR spectrum
  carbon tetrachloride solubles   177f
  China White hydrochloride salt   202f
  oil spill   102f
  9-phenyladenine   27f
  reboiler tube solids   43f
  Saturn   222
  tar by-product   96
IRIS spectrometer   220f
IRIS spectrum
  Jupiter   221-222
  Mars   221f
Isocyanate monitoring, airborne   39
Isocyanate monomers, control   39
Isoelectric focusing   172
Isolation flow chart, flavor component
  7
Isomaltose in honey   73f
Isotope ratio mass spectrometry   71
Isotopic analysis
  borosilicate glass   3t
  SIMS   3-4
  spent fuel   138t

JFK assassination bullet analyses
  191-195
Jupiter, IRIS spectrum   221-222

Kerosine analysis   180

Laser Raman spectrometry, ash
  analysis   118
Leaching of phosphorus by unleaded
  gasoline   57-59
Lead alkyls determination   182
Lead in concrete   79-80
Lead emission analysis   132
Lead geochronology   114
Leaf necrosis   47-48
Legionnaires' disease, etiology
  157-162
Leukocyte antigens, human   172-173
Light microscopy, particle analysis
  224
Liner, glassine, component limits   12f
Linseed fatty acids   38
Liquid chromatogram
  aniline-acetone reaction products
  94f
  liquid crystal mixture   89
  polystyrene   95
  tar by-products   96
Liquid chromatography
  See also Gel permeation
    chromatography
  coatings analysis   76, 77

Liquid chromatography/mass
  spectrometry
  liquid crystal analysis   87-91
  polymer analysis   93-97
Liquid crystal mixtures   87-91
Lithium determination, SIMS   5

Maltose in honey   73f
Marine analysis, standardization
  112
Mars
  ice clouds   221
  IRIS spectrum   221f
Mass chromatogram, base peaks in liquid
  crystals   88f
Mass spectrometry
  allantoin analysis   62
  curare analysis   199-200
  impurity analysis   26, 30
  isotope ratio   71
  radioactive sample analysis   136-137
  secondary ion   3-5
Mass spectroscopy   94
  fragmentation patterns   87f, 88
Mass spectrum
  bis(trimethylsilyl)adenine   27f
  blood preservative impurity   26f
  dihydroactinidiolide   9f
  liquid crystal component   90f, 91f
  methyl carbamate   152f
  1,4,4-trimethylcyclohexan-2-one acetic
    acid enol lactone   9f
  trimethylsilyl derivative   26f
  wastewater contaminant   122
Measurement and detection of antigen
  172-173
Mecloqualone   184
Medieval documents, small particle
  analysis   223-226
Metal content of corrosion scale
  15, 16, 18f
Metal finishing solid waste disposal
  45
Methamphetamine   184, 187, 202
Methaqualone preparation   184
3-Methoxyacrolein   30
Methyl carbamate assay   152-153
Methyl esters of fats and oils   209t
Methyl (N-methoxymethyl)carbamate
  153
4-(2″-Methylbutoxy)-4′-cyanobiphenyl
  90
3-Methylfentanyl   204
Migration analysis scheme   142
Migration levels for packaging materials
  142
Minerals analysis   215-218
Molasses contamination   210-211
Monitoring airborne vinyl chloride
  127-130, 133
Monitoring program, personnel
  146-149
Morphine detected in hair   169
Morpholine metabolite, salivary
  163-165
Mount St. Helens ash   105, 107,
  109-110, 117-119

Naphtha containing chlorine 41
Naphthylamine 155
  gas chromatogram 154f
Necrosis, leaf 47–48
Neoprene cable jacketing 77
Neoprene sheath, chromatopyrogram 86f
Neutralization of wastewater 121
Nickel-plating problem solving 46–47
Nitrile sheath, chromatopyrogram 85f
Nitrosamine formation 163
N-Nitrosomorpholine 163
N-Nitrosooxazine 163
NMR spectroscopy
  coatings analysis 208–209
  drug analysis 202–203
  impurity analysis 27, 30
  polymer analysis 96
NMR spectrum
  China White 202f–203f
  trimethylolpropanetriacrylate 96
Nondestructive analysis 8, 17–20, 19–20, 83–86, 132, 224–225, 228–229
  See also IR spectroscopy, UV spectroscopy, NMR spectroscopy
Nonvolatiles in blood 188f
Nuclear fuel cycle elements 136f
Nuclear reactor corrosion scale 15–18
Nuclear reactor, TRIGA Mark I 192f, 193
Nuclear safeguards, international 135

Ocean pollution 111–112
Odor/flavor transfer 12
Odor source detection in cereal 11–13
Odor, piney, component identification 13t
Oil slick sampler 102
Oil spill identification 101–103
Oils and fats, methyl esters 209t
Orange juice analysis 55–56
Organic analysis of air 132–133
Organic compounds in industrial wastewater 121–125
Organic ocean pollutants, synthetic 112
Organic products analysis scheme 65
Organics analysis 176–178
OSHA regulations for vinyl chloride 127–130
Over-the-counter drug analyses, HPLC 61–63

Packaging material analysis scheme 143
Packaging material, food, commercialization 141–143
Paint adhesive failure 76
Parchment fiber 225f
Pin connector failure 76–77
PCP detected in hair 169
Pentachlorophenol 155

Personnel monitoring program 146–149
Personnel sampling unit, vinyl chloride 128
pH of boiler steam 33f
Phase solubility analysis 23
  thiamine mononitrate 24f
Phenyladenine 27
Phenylpropanolamine hydrochloride acetaminophen, HPLC separation 63
Phosphoglucomutase 172
Phosphorus in unleaded gasoline 57–59
Photoelectron spectrum, corrosion scale surface 17f
Photomicrograph of air-entrained concrete 80f
Picose determination in honey mixtures 70
Pigment analysis, bust of Nefertiti 227–231
Pigment, anatase, micrograph 225f
Pigments, ancient Egyptian 227–228
Pilot plant problem solving 19–21
Piney odor component identification 13t
Plant leaves, boron levels 46t
Plastic wire coating 67–68
Pleochroic dyes 88, 90–91
Plutonium
  equilibration 135
  in glaciers 113–115
  pollutant 113
  standards, atom ratios 138t
Polarographic analysis or resin 143
Polymer characterization by chromatopyrography 83–86
Polymer source identification 46
Polymer volatiles, fingerprinting 46
Polymers, insoluble, analysis 65–67
Polymorphs, crystalline silica 106, 117–119
Polysaccharide fractions in honey 71
Polystyrene, liquid chromatogram 95
Polyurethane oligomers 95
Portable air sampler 147
Powder diffraction ash analysis 106–107
Problem solving
  aerospace materials 51–54
  appliance 45–47
  coating materials 75–77
  coatings industry 37–39
  concrete 79–81
  construction materials 79–81
  corroborative 207–211
  gasoline 57
  nickel-plating 46–47
  pilot plant 19–21
  raw material 29–30
  referee 207–211
  refinery corrosion 41–43
  vitamin production 23–24
Progressive massive fibrosis 105–106
Proline in honey 70–71
Propionanilide 203–204
Purification, terephthalic acid 19f
Purine impurities 25
Pyrolysis gas chromatography, polymer analysis 84

Quadropole mass spectrometry 184–185
Qualitative analysis, flavor component 8
Quartz, Raman spectrum 117f, 118f

Radiation, low-level, analysis 3–5
Radioactive fallout 113–114
Radioactive solutions, sampling and analysis 135–138
Radioimmunoassay
  curare 199
  hair 169
  sexual identity from bloodstains 171–172
Radionucleic ocean pollutants 112
Raman spectrometry, laser, ash analysis 118
Raman spectrum
  cristobalite 117f, 118f
  quartz 117f, 118f
  volcanic ash 118f
Raw material problem solving 29–30
Reboiler solids analysis scheme 43t
Reboiler tube blockage 43
Reconstructed ion chromatogram, methyl carbamate 152
Red pigment, x-ray pattern 229f
Referee problem solving 207–211
Refinery corrosion problem solving 41–43
Refractive index detection in liquid chromatography 38
Regulation, food packaging 142–143
Research needs, ocean pollution 112
Resin
  alkyd, residuals 38
  aniline–acetone 97t
  coating, intermediate 38–39
Resistor corrosion 46
Resists, electron beam 65–67
Rocks analysis 215–218
Ross ice shelf glacier sampling 114
Rubber and plastics polymers 93–97
Rat urine, gas chromatogram 154f

Salicylate, choline, analysis 61–62
Salivary morpholine metabolite 163–165
Sample preparation, bullet 193
Sampler, oil slick 102
Sampling in Antarctica 114
Sampling devices
  direct 149t
  headspace 181f
Sampling medieval documents 224–225
Sampling system, hotel water 158
Saturn, IR spectrum 222
Scanning electron micrograph
  ink particle 225f
  resin bead 137f

Scanning electron microscopy
  coatings analysis   75
  corrosion analysis   46
Scientific testing admissibility in court   198–200
Secobarbital   189
Secondary ion mass spectrometry   3–5
Separation
  allantoin   62
  allergy remedy components   63
  liquid crystal mixture   88–90
  phenylpropanolamine hydrochloride acetaminophen   63
  salicylate   61
Silica, crystalline, in volcanic ash   105–107, 109–110, 117–119
Silica in concrete   79
Silica in rocks   215t
Silicone rubber chromatopyrogram   84f, 85f
Silicosis   105, 117
Siloxanes   66
Silver determination in bullets   194–195
SIMS (Secondary ion mass spectrometry)   3–5
Sludge in nickel-plating baths   47
Small particle analysis, medieval documents   223–226
Small particle analysis scheme   223
Sodium in boiler steam   33f
Sodium impurity clean-up   34
Sodium mercaptobenzothiazole   52
Solid waste disposal, metal finishing   45
Solids mixture analysis scheme   16f
Solubility analysis, phase   23
Solvent extraction, arson analysis   180–181
South Greenland glacier sampling   114
Spalled concrete   81
Spinels in corrosion scale   17
Stainless steel corrosion products   54
Standardization of marine analysis   112
Starch-gel electrophoresis, bloodstain analysis   172
Steam generator corrosion   15–18
Steam turbine corrosion scale   31–35
Structure determination from mass spectra   123–125
Sublimation purification   19f
Sugar analysis of orange juice   55
Sugar in concrete   80
Sulfate in boiler steam   33f
Sulfur corrosion   43
Sulfur emission analysis   132
Sulfur interference with chlorine measurement   42
Synthesis of Egyptian blue   229–230
Synthetic organic ocean pollutants   112

Talvitie ash pretreatment   106–107, 109–110, 118
Tar by-products, liquid chromatogram   96
Terephthalic acid anhydride   21
Terephthalic acid purification   19f
Terpenes and piney off-flavor   13
Testosterone/progesterone ratios, blood   172
Thermal energy analyzer, nitrosamine analysis   164–165
Thermogravimetric analysis of binders   67
Thermogravimetric cure for Egyptian blue   231f
Thiamine   23
Thiamine mononitrate   23
  phase solubility analysis   24f
Thiamine thiazolone   24
Thin-layer chromatography
  drug analysis   187
  honey analysis   72, 73f
  impurity analysis   23–24
Thiothiamine   23
Three Mile Island sample analysis   5
Titanium in ink   225
Titration of coolant, colorimetric   52
Tobacco flavor component analysis   7–9
Toxicity of thiamine thiazolone   24
Trace metals analysis   175–178
Trace organic in water, analysis scheme   159
Transmission electron micrograph, anatase pigment   225f
Transmission of Legionnaires' disease   161
TRIGA Mark I nuclear reactor   192f, 193
1,4,4-Trimethylcyclohexan-2-one acetic acid enol lactone   8f, 9
Trimethylorthoformate impurity analysis   29–30
Trimethylsilyl derivatives of adenine   26
Tubocurarine chloride   198f

Uranium equilibration   135
Uranium salts analysis   175–178
Uranium standards, atom ratios   138t
Urethane foam, water   209–210
UV detection in liquid chromatography   38
UV spectrophotometry
  blood analysis   188
  drug analysis   187

Vacuum distillation, coatings analysis   208–209
Vinland map dating   224
Vinyl chloride, OSHA regulations   127–130
Vitamin B1—See Thiamine
Vitamin impurity analysis scheme   23f
Vitamin production problem solving   23–24
Volatiles in blood   188f
Volcanic ash, Raman spectrum   118f
Volcanic ash, silica crystalline   105–107, 109–110
Voyager IR spectrometer   219–222

Warren commission   191
Wastewater, industrial, organic compounds   121–125
Water dip   33
Water in urethane foam   209–210
Water, ultrahigh purity   33
Weathered oil analysis   101–103
Wire coating, plastic   67–68
Wire wrap connector   77f
Workplace stresses   145

X-Ray diffraction, ash analysis   106, 118
x-Ray diffraction, coolant analysis   52–53
x-Ray energy spectrometry, corrosion analysis   46
x-Ray fluorescence spectroscopy
  air analysis   131–132
  concrete analysis   80
  gasoline analysis   58
x-Ray induced x-ray fluorescence spectroscopy, corrosion scale   15
x-Ray pattern
  beeswax   230f
  failure point in resistor   47f
  green pigment   229f
  nichrome wire   47f
  red pigment   229f
x-Ray spectrography, ash analysis   20

Yellow cake analysis   175–178
Yellowing of coatings   38

Book design and layout by Kathleen Schaner and Martha Sewall
Cover design by Kathleen Schaner
Production by Robin Giroux and Anne G. Bigler
Elements typeset by Service Composition Co., Baltimore, MD
Printed and bound by Maple Press Company, York, PA